VIOLENCE AND HARM IN THE ANIMAL INDUSTRIAL COMPLEX

This book grapples with multispecies violent exploitations embedded in corridors of power within the animal-industrial complex (A-IC). The A-IC is a useful framework for understanding how exploitative human-animal relations are central to capitalist relations and profit accumulation. 'A-IC-related-violence' – killing animals for economic gain – has a ripple effect which results in profound consequences for humans as well.

This collection of international scholarship explores topics as varied as how A-IC-related-violence is reproduced and sustained through rapidly changing discursive strategies, ideological architecture, and particular cultural forms that elide and legitimize animal cruelty. Several chapters expose collusion between governments, corporations, and academia as central to maintaining dominance of A-IC-related-violence. Other scholars explore the trouble with making the conditions of "meat" production visible – of de-fetishizing meat commodities. The scholarship critically explores dynamic components of an apparatus that enables A-IC-related-violence and harm but is situated within the capitalist order and charts A-IC-related-violence as the key profit-generating practice in select domains of the A-IC.

The book unmasks inherent cruelties in a proliferation of social forms that ultimately reflect a socioeconomic system that centralizes capitalist life characterized by endless growth, competitiveness, and profligate consumption. This is essential reading for those engaged in critical criminology, green criminology, violence studies, peace and conflict studies, critical animal studies, or animal rights-oriented scholars.

Gwen Hunnicutt is Professor of Sociology at the University of North Carolina Greensboro. She studies gender violence – its varieties, causes, consequences, interspecies entanglements, and politicizations. She is the author of *Gender Violence in Ecofeminist Perspective.*

Richard Twine is Reader in Sociology and Co-Director of the Centre for Human-Animal Studies (CfHAS) at Edge Hill University, UK. He is the author of *The Climate Crisis and Other Animals* (2024), *Animals as Biotechnology – Ethics, Sustainability and Critical Animal Studies* (2010) and he co-edited (with Nik Taylor) *The Rise of Critical Animal Studies – From the Margins to the Centre* (2014). He has also published several articles on ecofeminism, vegan transition, the food system, and the animal-industrial complex.

Kenneth Mentor is Senior Lecturer in the Department of Sociology and Criminology at the University of North Carolina Wilmington. His published research includes peer-reviewed papers in the disciplines of criminology, organizational behavior, public administration, law and society, and online learning.

"Violence, of course, is not only found on the streets, in pubs and taverns, or in domestic/household settings. It is frequently directed at different types of nonhuman animals (NHAs). However, critical criminologists have not been fleet at foot in addressing this major world-wide problem. Thus, this groundbreaking anthology helps fill a major research gap. It is destined to become a classic that should be mandatory reading for all progressive criminologists concerned about the protection and preservation of NHAs."

– **Dr. Walter S. DeKeseredy**, *Anna Deane Carlson Endowed Chair of Social Sciences, Director of the Research Center on Violence and Professor of Sociology at West Virginia University*

"The volume strikes the right balance between theoretical and applied content and showcases the impressive expertise of scholars across multiple disciplines and countries. This volume will be of interest to many, including (critical) animal studies scholars, ecofeminists, and political economists (to name but a few), and more generally, anyone interested in developing a future where human actions no longer threaten the species and environments around us."

– **Professor Amy Fitzgerald**, *University of Windsor*

VIOLENCE AND HARM IN THE ANIMAL INDUSTRIAL COMPLEX

Human-Animal Entanglements

Edited by Gwen Hunnicutt, Richard Twine and Kenneth Mentor

Routledge
Taylor & Francis Group
LONDON AND NEW YORK

Designed cover image: Animal Terror by Paul Klee, The Louise and Walter Arensberg Collection, 1950

First published 2025
by Routledge
4 Park Square, Milton Park, Abingdon, Oxon OX14 4RN

and by Routledge
605 Third Avenue, New York, NY 10158

Routledge is an imprint of the Taylor & Francis Group, an informa business

British Library Cataloguing-in-Publication Data
A catalogue record for this book is available from the British Library

ISBN: 978-1-032-57977-1 (hbk)
ISBN: 978-1-032-57978-8 (pbk)
ISBN: 978-1-003-44190-8 (ebk)

DOI: 10.4324/9781003441908

Typeset in Times New Roman
by KnowledgeWorks Global Ltd.

To all nonhuman animals in captivity

CONTENTS

CONTRIBUTORS

Daniel Allen is an animal geographer at Keele University, UK. His research interests include human-animal relations in policy and practice. He is founder of the evidence-based Pet Theft Reform campaign, and patron of the Stolen and Missing Pets Alliance (SAMPA).

Núria Almiron is co-director of the UPF-Centre for Animal Ethics and a full professor in the Department of Communication at Pompeu Fabra University (UPF). Her main areas of research include and combine the ethics and political economy of communication – particularly interest groups and persuasive communication – from the critical animal studies and critical public relations perspectives. She is the director of the research project COMPASS (Lobbying and Compassion: Interest Groups, Discourse and Nonhuman Animals in Spain) (2021–2025). Her latest edited book is *Animal Suffering and Public Relations. The Ethics of Persuasion of the Animal Industrial Complex* (2024), co-authored with the COMPASS research team.

Olatz Aranceta-Reboredo is a PhD researcher in the Department of Communication at Universitat Pompeu Fabra (UPF) and a CritiCC communication research group member at the same university. Their research areas include critical animal studies, interest groups, and the representation of nonhuman animals in media. They graduated in English Studies from the University of the Basque Country and hold an MA in Media, Power and Difference (UPF). Olatz is the project manager of COMPASS (PID2020-118926RB.I00) and is currently conducting PhD research on the discourse of lobbies of the Spanish Animal-Based Entertainment Complex under the supervision of Prof. Núria Almiron.

Stacy Banwell is Professor in Criminology at the University of Greenwich. Her research interests include conflict-related sexual and reproductive violence; gender, climate change, and atrocity crimes; and multispecies justice. These themes are explored in her recent monographs *Gender and the Violence(s) of War and Armed Conflict: More Dangerous to be a Woman?* and *The War against Nonhuman Animals: A Non-speciesist Understanding of Reproductive Violence*. Her forthcoming article, *Rape, Sexual Violence and Forced Pregnancy* (*Animal Liberation*), explores consent and forced pregnancy in relation to the International Criminal Court Statute: *Other Serious Violations of the Laws and Customs Applicable in Armed Conflicts not of an International Character*.

Anat Ben Yonatan is a social-sciences informed Human-Animal Relations researcher. She is co-coordinator of The Coller-Menmon Program for Animal Rights and Welfare in the Buchmann Faculty of Law, and a teaching fellow in Human-Animal Studies at The Multidisciplinary Program in the Humanities at The Lester and Sally Entin Faculty of Humanities, Tel Aviv University. Anat is the co-chair of The Community for Human-Animal Studies in Israel (HASI), and co-founder and member of the steering committee of both HASI and the Israeli Human-Animal Connections Forum.

Michael D. Briscoe is Assistant Professor of Sociology at Colorado State University Pueblo. His work focuses on the intersections of human, animal, and environmental well-being. Most recently, he has published articles in *The Sociological Quarterly*, *Society & Animals*, and *Environmental Sociology*.

Cindy Brooks Dollar is an Associate Professor in the Department of Sociology at the University of North Carolina-Greensboro. Her research investigates inequalities, harms, and social control. Professor Dollar's research has been published in numerous journals, including *Critical Criminology, Humanity & Society*, *Journal of Quantitative Criminology, Deviant Behavior*, and *Women Studies International Forum*. She has also contributed to edited books on crime causation, alternative governance, and societal trauma. The release of her first book, *I Never Wanted to Be a Stereotype*, was met with significant academic praise.

Jessica Carey-Webb is an Assistant Professor in the Department of Spanish & Portuguese at the University of New Mexico. Her research specializes in the historical development, environmental representation, and sustainable future of Latin America and the Amazon region. Her book, *Eyes on Amazonia: Transnational Perspectives on the Rubber Boom Frontier* (2024), examines how race and gender inform structures of empire in the Amazon during a period of heightened extractivism (the first rubber boom, 1875–1913). Her current research addresses environmental impact in an Amazonian context. Previously, she was a Mellon/ACLS public fellow for the Natural Resources Defense Council working on environmental policy in Chile, Mexico, and Brazil.

Júlia Castellano is a PhD candidate and researcher in the Department of Communication at Universitat Pompeu Fabra (UPF), a member of the CritiCC research group and the UPF-Centre for Animal Ethics. Her research focuses on Vegan Studies, particularly on strategic communication for vegan brands and political consumerism, under the supervision of Prof Núria Almiron. Júlia holds a psychology degree from the University of Barcelona (UB), a postgraduate in Socio-environmental Analysis and Intervention (UB/UAB), and a MA in Media, Power, and Difference (UPF).

Estela M. Díaz is Professor in the Department of Management at the Universidad Pontificia Comillas. Estela holds a PhD in Economics and Business Administration (Universidad Pontificia Comillas), a Masters in Sustainability and CSR (UNED and UJI), a Masters in Research in Economics and Business Administration (Universidad Pontificia Comillas), and a degree in law (University of Granada). Critical animal studies, ethical and transformative consumption, gender, sustainable transitions, and education are her main research interests. She has published in high-impact journals such as *Human Ecology Review*, *Psychology & Marketing*, *Journal of Macromarketing*, *Heliyon*, *Anthrozoös*, and *Society & Animals, and Humanities and Social Science Communications.*

Laura Fernández is a Juan de la Cierva postdoctoral fellow at University of Barcelona, where she works with the Centre of Research in Information, Communication and Culture (CRICC) within the research lines of gender, inclusion, and diversity. Laura's research interests include critical animal studies, strategic visual communication, social movements, fat studies, and feminist media studies. She is the author of more than ten academic publications and a book. She is a team member of the research project COMPASS, a board member of the UPF-Centre for Animal Ethics and a member of the European Association for Critical Animal Studies (EACAS).

Melanie Flynn is a Senior Lecturer in Criminology and Associate Dean for Research, Scholarship and Knowledge Exchange in the Faculty of Business, Law and Criminology, Liverpool Hope University, UK. She has been working in the field of green criminology and nonhuman animal harm since 2008. In 2017, she proposed, with Matthew Hall, the concept of nonhuman animal victimology and much of her recent work focuses on this approach. She is currently working on several projects relating to dog abduction, nonhuman animals as victims of crime, and the development of vegan criminology.

Kirsten L. Grieve is a Registered Veterinary Technician in Toronto. In addition to her veterinary nursing degree, she holds an undergraduate degree in psychology and completed post-graduate work in biological chemistry. She has been a member in good standing of the Ontario Association of Veterinary Technicians for almost 20 years. Kirsten has special interests in parasitology, geriatrics, and palliative/

end-of-life care. She has worked in a variety of research, veterinary medicine and animal care settings, including a medical school, a small-animal hospital and feline specialty veterinary practices. She currently works for a veterinary hospice and in-home euthanasia service and volunteers in the surgical theatre of Toronto Street Cats' Trap, Neuter, Release Program at the Toronto Humane Society.

Jason Hannan is Professor in the Department of Rhetoric, Writing, and Communications at the University of Winnipeg. His books include *Trolling Ourselves to Death: Democracy in the Age of Social Media* (2023) and *Meatsplaining: The Animal Agriculture Industry and the Rhetoric of Denial* (2020).

Josh Hendrix is a research scientist at RTI International in the Justice Practice Area. He has more than 10 years of experience of conducting impactful and actionable mixed-methods research, specializing in violence prevention research and program implementation and with a specific focus on juvenile justice involving youth, victimization, and school violence and safety. He currently leads or supports multiple federally funded projects examining crime trends and violence prevention efforts in schools and communities and his recent research has been published in the *Journal of Threat Assessment and Management*, *Preventing School Failure*, and the *Journal of Forensic Sciences*.

Gwen Hunnicutt is Professor of Sociology at the University of North Carolina Greensboro. She studies gender violence – its varieties, causes, consequences, interspecies entanglements, and politicizations. Her research is located at the intersections of interdisciplinary violence studies, gender, peace and security studies, and ecofeminism. Her 2020 book, *Gender Violence in Ecofeminist Perspective*, explores how exploitation of the earth and its nonhuman inhabitants informs a host of other harmful social practices, including gender-based violence. Chiefly interested in nonviolence, her current project is titled, *Non-Violent Futures: "Ecologies of Care" Practices in More Than Human Spaces.*

Robinson Matthew is Professor of Justice Studies at Appalachian State University. He is author of 27 books in the fields of Criminology and Criminal Justice as well as more than 100 other publications. He is Past President of both the North Carolina Criminal Justice Association and the Southern Criminal Justice Association. He currently serves as Treasurer of the Justice Studies Association.

Corina Medley is Assistant Professor in the Criminology Program in the Department of Sociology, Anthropology, and Social Work at Kansas State University. She is an interdisciplinary writer interested in criminology, theory, culture, sexuality, and animals/animality. Her work is published in *Crime, Media, Culture* and a range of edited collections.

Kenneth Mentor is Senior Lecturer in the Department of Sociology and Criminology at the University of North Carolina Wilmington. With a Master's Degree in Psychology, a Juris Doctor, and PhD in Social Science, he has enjoyed a multidisciplinary teaching and research career. While his interest in closed access corporate publishing is limited, his list of published research includes peer-reviewed papers in the disciplines of criminology, organizational behavior, public administration, law and society, and online learning.

Stephen L. Muzzatti is Associate Professor of Sociology at Toronto Metropolitan University. His research lies at the intersection of cultural and ultra-realist criminologies, focusing on crime, social harm, consumerism, and human interaction with animals. He is particularly interested in the intertwining social dynamics of neoliberalism, brutality, and victimization. He has published over 50 journal articles and book chapters on an array of topics, including terrorism, violence against animals, state and corporate crime, crimes of globalization, narcissism, working-class spaces and identities, motorcycle culture, and advertising and the monetization of violence. He is a member of the Animals in Society Research Cluster and the Veterinary Humanities Network.

Angus Nurse is Professor of Law and Environmental Justice and Director of the Centre for Access to Justice and Inclusion at Anglia Ruskin University. He is a member of the Wild Animal Welfare Committee (WAWC). His work focuses on green criminology, animal harm, corporate environmental crime, and environmental rights. His current work includes a review of wildlife law and corporate breaches of human rights and redress for the environmental harms that Global North corporations commit in the Global South. His books include *Animal Harm*, *Wildlife Criminology* with Tanya Wyatt, and *Cleaning Up Greenwash*.

Patricia Puente-Guerrero is Assistant Professor of Criminology in the Department of Criminal Law and Criminology of the National University of Distance Education (UNED, Spain). She earned her PhD in Law and Social Sciences in 2021 for her thesis titled "Invisible and Invisibilized. The Special Vulnerability of People Experiencing Homelessness to Violence," which received the Extraordinary Doctoral Award. She holds a Bachelor's Degree in Criminology, a Bachelor's Degree in Psychology, and a Master's Degree in Criminology and Criminal Justice System. Among her research interests, violence against vulnerable groups and green criminology stand out. She is a member of the Working Group on Green Criminology of the Spanish Society of Criminological Research (SEIC).

Chiara Stefanoni is guest scholar and lecturer at the Institute of Philosophy and Art History (IPK) of Leuphana University of Lüneburg. With a philosophical education, she obtained her doctoral degree in Transcultural Studies in Humanities from the University of Bergamo, focusing on the topic of animal oppression within capitalist societies. Drawing from continental practical philosophy, particularly in the Marxist tradition, and incorporating feminist perspectives, her work in Critical Animal Studies delves into sociopolitical aspects of human-animal relations. She has been a visiting researcher at the Centre for Animal Ethics at Pompeu Fabra University in Barcelona and is a member of the editorial board of anti-speciesist Italian journal *Liberazioni*.

Richard Twine is Reader in Sociology and Co-Director of the Centre for Human-Animal Studies (CfHAS) at Edge Hill University, UK. His latest book is *The Climate Crisis and Other Animals* (2024). He is also co-editor (with Nik Taylor) of *The Rise of Critical Animal Studies – From the Margins to the Centre* (2014) and author of *Animals as Biotechnology – Ethics, Sustainability and Critical Animal Studies* (2010). He has also published several articles on ecofeminism, vegan transition, the food system, and the animal-industrial complex. His website is http://www.richardtwine.com

John Walliss is a senior lecturer in criminology at the University of Derby. His research interests include green criminology, crime history, and the final statements of condemned prisoners in the contemporary United States. He is the author of *The Bloody Code in England and Wales, 1760–1830* (2018) as well as a number of articles on crime and punishment during the long eighteenth century. His current research is focused on crimes committed against companion animals, such as puppy farming, dog fighting, and pet theft.

Tanya Wyatt is a green criminologist with nearly 20 years of experience researching crimes that affect the environment. During her time in academia, she published nearly 100 books, journal articles, book chapters, and reports on green harms and harms. She holds a BA in Biology from Mills College in Oakland, California, United States and a Masters in Criminology and Criminal Justice from Eastern Michigan University in Ypsilanti, Michigan, United States. She earned her Doctorate in Criminology from the University of Kent in Canterbury, UK.

1

TOWARD MULTISPECIES JUSTICE

Unveiling Violence and Exploitation in the Animal-Industrial Complex

Gwen Hunnicutt, Richard Twine, and Kenneth Mentor

This book grapples with multispecies violent exploitations embedded in corridors of power within the Animal-Industrial Complex (A-IC). The A-IC constitutes an enormously complex constellation of relationships between powerful institutions. The dynamic components of this apparatus are all impacted by a proliferation of social forms that ultimately reflect a socio-economic system that centralizes capitalist life characterized by endless growth, competitiveness, and profligate consumption. Capitalism is an intrinsically violent enterprise marked by the human instrumental destruction and consumption of nonhuman animal bodies. In this book, we chart the violent expropriation of nonhuman animals as the key profit-generating practice in select domains of the A-IC, but we situate it within the capitalist order. The A-IC gives "conceptual weight to the multiple scales and sites in which capitalism expropriates nonhuman animal life" (Twine, 2024, p. 78). Moreover, it emphasizes exploitative human-animal relations as partly constitutive of capitalism (itself often ignored by conventional theorists of capitalism) and provides a heuristic focus for critical interdisciplinary animal research.

This collective work advances scholarship at the intersection of critical animal studies, criminology, and interdisciplinary violence studies. Each chapter collected in this volume stands on its own as an individual study of associated phenomena. However, in this editor's introduction, we aim to identify broad connections across the entire compilation that unmask inherent cruelties in the pursuit of perpetual profit. We employ classic theoretical concepts that speak to the contemporary financialization of nonhuman animals. When referring to "capitalism," we are speaking of not just an economic system, but rather an entire social order (Skotnicki, 2021), including an acknowledgment of the Capitalocene – how capitalism has shaped our current, unsustainable ecological conditions (Moore, 2015).[1]

DOI: 10.4324/9781003441908-1

A Brief Genealogy of the Animal-Industrial Complex Concept

Recognized now as a key conceptual term (Calarco, 2021) in the broad field(s) of human-animal studies, the concept was first coined by anthropologist Barbara Noske (1989). Noske's focus was to underline that exploitative human-animal relations were underpinned by and embedded within capitalist relations and profit accumulation. Noske employed the concept to situate modern animal production within monopolistic financial interests rather than those of farmers, consumers, or workers (1989). Thereafter, the concept gradually was used within ecofeminist (Adams, 1997; Haraway, 1992) and critical animal scholarship but did not undergo closer conceptual development until later (e.g., Nibert, 2011, 2013, 2017a, 2017b; Twine, 2012, 2013, 2024). Both sociologists, Nibert (2011) set the complex within its historical, political, and economic contexts, while Twine (2012) was concerned to improve its conceptual clarity. The work of Noske, Nibert, and Twine has been united by a focus upon the historical development of animal domestication and a desire to locate critical animal work within broader anti-capitalist critique. Furthermore, all three have sought to construct a wide understanding which overlaps animal production with the pharmaceutical industry, and Twine (2012, 2024) in particular, has intended the A-IC to cover all areas in which capitalism exploits nonhuman animals for profit. This further attention to the concept appears to have influenced its wider use, with it being cited only 82 times between 1989 and 2009, but 900 times between 2010 and April 2024. It is also important to acknowledge that a broad body of critical animal work has been unavoidably working on similar concerns without directly employing the concept as an overarching heuristic (e.g., Harrison, 1964; Nibert, 2002; Sanbonmatsu, 2011; Shukin, 2009; Twine, 2010; Wadiwel, 2023).

Concerned that its lack of definitional clarity lent itself merely to a rhetorical use, Twine (2012) set about underlining relations between governments, corporations, and academia as central to maintaining the dominance of the A-IC. After considering its relation to other industrial-complex ideas, the notion of "complex" was explored for its useful and appropriate triple meaning in terms of being complicated or hidden, constituting a network or system, and by highlighting the affective and psychological dimensions involved in the exploitation of nonhuman animals. Twine then arrived at a working definition of the A-IC as:

> a partly opaque and multiple set of networks and relationships between the corporate sector, governments, and public and private science. With economic, cultural, social and affective dimensions it encompasses an extensive range of practices, technologies, images, identities and markets.
>
> *(2012, p. 23)*

This broad definition was intended as a basis for further work that both helps to explicate understandings of the everyday workings of the complex (it should not be confused as only referring to industrial scale exploitation) and acts as a conceptual resource to theorize transition away from it. It has helped to embed attention given by critical animal studies to capitalism and political economy and provided a useful framework for thinking about, for example, cultural, media, public relations, legal, economic, scientific, and educational infrastructures that continue to foster anthropocentric ideologies that naturalize and co-constitute the A-IC (e.g., Almiron, 2023; Almiron et al., 2016; Arcari, 2020; Cole & Stewart, 2014; Duxbury, 2021; Fitzgerald & Taylor, 2014; Hannan, 2020; Pedersen, 2019; Rowley & Prisco, 2021; Taylor & Fitzgerald, 2018; Wadiwel, 2015). More recently, Twine (2024) has drawn upon the theoretical framework of practice theory (e.g., Shove et al., 2012) and Capitalocene theory (e.g., Moore, 2015) as further ways to frame understandings of the A-IC. The concept has also made its way into criminological research, beyond the work of Taylor and Fitzgerald cited above, including Griffin and Griffin's (2021) overview of non-speciesist approaches to harm, Beirne's work on wildlife and Covid-19 (2021), Beirne and Lynch's work on speciesist policing (2023), and Banwell's exploration of gendered reproductive violence and war crimes in relation to the war against animals (2023).

Multispecies A-IC-Related-Violence and the Exploitation Pipeline

In this chapter, we are using the term "A-IC-related-violence" to refer to corporeal violations – physical harms against the body; both large-scale, systematic killing, and smaller-scale injuries that support certain consumerist lifestyles. Yet killing animals for economic gain has a ripple effect of broader "A-IC-related-harms." Moreover, "A-IC-related-violence and harm" against animals and humans are deeply intertwined. Therefore, in defining "A-IC-related-harms," we follow Nibert (2013, p. 5) who writes:

> This widespread violence and destruction engendered by such uses of large numbers of 'domesecrated' animals encompasses both the violence experienced by the animals and the ways in which this harm has been entangled with related forms of violence against free-living animals and groups of devalued humans. These include invasion, conquest, extermination, displacement, repression, coerced and enslaved servitude, gender subordination and sexual exploitation, and hunger. Accompanying such violence have been deadly zoonotic diseases that have contributed to the destruction of entire cities, societies, and civilizations.

Violent commodification and exploitation of animals are woven into the design of our political-economic order where large-scale, systemic violence against animals is part and parcel to maintaining particular social and economic hierarchies and lifestyles (Hunnicutt, 2020).[2]

It is not a new revelation that the destruction of nonhuman animal bodies is necessary in the creation of profit. Carol Adams (2010), Jason Hribal (2010),

David Nibert (2002), Gary Francione (1995), and Barbara Noske (1989), among others, in various ways, explore how A-IC-related-violence against animals is a logical outcome of a system devoted to endless material growth. Indeed, Moore (2017) claims that capitalism itself is circumscribed by dual destructions – necrotic and productive violence. The exploitation of nonhuman animals to profitably serve human interests and desires remains socially normative; so much so that humans have virtually no accountability and the status quo remains one of silence and non-response. Additionally, this violent enterprise is sustained by habitual myths, propaganda (Hannan, 2020), and a legal apparatus that secures human domination over nature (White, 2022).

Social theorists often deny multispecies violence as an antecedent of capitalism (Moore, 2017). And when corporeal violence is given scholarly consideration, it typically attends to the suffering of *either* human or nonhuman animals. When nonhuman animals do appear in multidisciplinary violence studies, including criminology research, they are commonly framed with speciesist assumptions, such as investigating the predictive link between animal and human abuse (Cazaux, 1999). Sub-disciplinary areas, such as the sociology of violence and criminology research focused on victimology, tend to ignore other animals altogether (Cudworth, 2015). In these scenarios, nonhuman animals are not acknowledged as sentient beings, worthy of our attention, care, and protection. Conversely, discourse among animal rights scholars is largely silent on the destructive consequences for humans who are caught up in spaces marked by A-IC-related-violence and harm (Fischer, 2020). Green criminology has encouraged the discipline to move beyond a focus on individual offenders to explore state failures regarding environmental protection, corporate offending, and environmentally harmful business practices (Nurse, 2017) and arguably achieved the most success in addressing A-IC-related-violence inflicted on nonhuman animals and to humans in close proximity to animals (Beirne, 2020; Nurse, 2021; Nurse & Wyatt, 2020; Sollund, 2019; White & Renzetti, 2021; Wrenn & McKie, 2020).

Recognizing that we exist on a plane of continuity, this volume intentionally brings simultaneous and co-constituted A-IC-related-violence and harm to the foreground. While this collection concentrates a greater focus on the more frequent, extreme, and excessive destruction of nonhuman animals, we acknowledge that A-IC-related-violence and harm experienced by human and nonhuman animals are co-constituted in a vast array of interspecies entanglements (Smulewicz-Zucker, 2012). Haraway (2003) argues that because of our complex interspecies relations, human and nonhuman animals cannot be analyzed in isolation from each other. Stacy Alaimo (2010) evokes the term "trans-corporality" to assist in "thinking across bodies" since humans are, after all, bound to and with the more-than-human world. But human and nonhuman animals are continually interacting in relationships that are preconfigured by the A-IC. These interconnected lives, filled with pain and pleasure, play out within systems of domination. We exist on a plane of contingency where brutality directed toward animals within normative capitalist production is also connected – albeit differently – to human harm.

For example, A-IC-related-violence and institutionalized terror against animals ignite a process of slow violence (Nixon, 2011). Climate change is accelerated, resulting in water scarcity and deforestation, which in turn exacerbates structural violence against humans, spurring climate conflict (Kramer, 2020). Workers who are subject to the trauma of executing animals may themselves be claimed by harmful substance abuse (Hendrix & Dollar, 2018), or harming others (Fitzgerald et al., 2009; Jacques, 2015). Slaughterhouse workers, mostly male non-citizen residents, are at risk of death and injury since slaughterhouse work is among the most dangerous jobs in the world. In cases of joint domestic violence toward women and pets, threats to harm animals may be used by a batterer to create a climate of terror (Adams, 1996).

Reproducing A-IC-Related-Violence Through Discursive Strategies

Like the military-industrial complex and the prison-industrial complex, the A-IC profits off of death and captivity. The enormous industrial installations, and the conspirators that enable the methodical extermination of nonhuman animals, require intentional effort and adaptation in order to maintain itself. In *Meat scientists fight back! What the Dublin declaration tells us about the role of academia in the animal-industrial complex*, Twine documents the role of academic "meat scientists" in both legitimizing the violent practices of animal agriculture and attempting to fend off multiple criticism of the A-IC. While animal production industries have long sought license to operate by tapping into the cultural esteem and scientific authority of university researchers (Morris & Jacquet, 2024), these alliances have historically operated in more covert ways (Jacobs, 2019). The Dublin Declaration – a manifesto in defense of eating meat signed by meat scientists embedded within the animal agriculture profit apparatus – is a bold, overt, and public embrace of false assertions. Twine debunks the Dublin Declaration's main claims and exposes the instigator's enmeshment within the A-IC. Twine's chapter pulls back the curtain on the collusion between private industry and the academy – which is charged with serving a public good, not predatory capitalism.

In addition to the empirical documentation of the private sector's use of "experts" to undermine established science on the deleterious effect of animal agriculture, Twine unpacks a document riddled with falsehoods. The style of overt, provocative content, labeling their opponents "zealots" and disregard for facts, marks an adaptation among meat industries who are possibly employing strategic disinformation campaigns to preserve their status quo. Disinformation is the intentional spread of false information with the intent to deceive, creating doubt, denial, and confusion and may influence public discourse and public policy. There has been a marked increase in disinformation in the last 5–7 years (Chowdhury et al., 2023), and the Dublin Declaration suggests that some academic abettors may no longer be attempting to conceal their involvement. This "leveling up" of disinformation is alarming because it may likely escalate A-IC-related-violence; and as the

Freedom Food Alliance report asserts (Carter, 2024), the public is ill-equipped to recognize and respond to these manipulation tactics.

The A-IC is held together, in part, by an ideology that includes myths of meat consumption and rhetorics of denial (Hannan, 2020). Anti-vegan discourses are becoming increasingly common as plant-forward movements gain momentum (Gregson et al., 2022). While disinformation and propaganda have become part of capitalist dogma, it is being deployed in new ways. In *"But Bacon!" The Performative Violence of Anti-Vegan Trolling*, Hannan articulates a particularly pernicious and recent phenomenon enabled by social media – the anti-vegan troll. We are existing in a historical moment characterized by an unprecedented vitriolic climate where disinformation is boldly used to attack animal emancipation efforts. The social media anti-vegan troll engages in flagrant cyberbullying, a hate speech movement that, as Hannan points out, appears to be intertwined with inflammatory discourse of the alt-right. Whether online or the performative consumption of raw meat at anti-vegan demonstrations, the antagonisms bear a distinctly violent strain. As Skinazi (2019) wonders,

> Do jokes about killing vegans belie some widespread cultural discomfort with the killing of other animals that vegans oppose? The logic might go something like this: It's too painful to take seriously vegan claims against the daily massacre of other animals and it would be too uncouth to mock the suffering of other animals, so instead the culture mocks those who sympathize with that suffering. In this way, the jokes make the claim that the suffering itself is not subject to ridicule, but the act of sympathizing with it is ridiculous—therefore we don't really need to think about the suffering.
>
> *(p. 36)*

We exist in a socio-political moment where the effectiveness of our usual methods of "distributing knowledge" and "making social problems visible" has become diluted. Downplaying harms, greenwashing, weaponized proposed remedies, and distractions intended to undermine opposition, preserve the status quo and sway public opinion are all reflective of a larger consumer culture under capitalism. These distortions themselves follow the structural logics of capitalism. Communication – as well as morality and culture – may themselves be shaped by capitalist forms (Skotnicki, 2021). The industry backlash involves influential stakeholders controlling knowledge about how such violence is utilized in distorted ways. In *Displaying Compassion to Hide Harms: An Analysis of the Visual Communication Strategies of the Spanish Animal Industrial Complex*, Fernández, Díaz, Aranceta-Reboredo, and Almiron focus on visual imagery deployed by key collaborators in the A-IC to sanitize their domination of animals, recasting their role in visually beneficent compositions. Fernández and colleagues ultimately expose this pseudo-care imagery as a manipulative and murky communication tactic, another iteration of disinformation in our polymorphous information systems.

The Trouble With "Visibility" in Unveiling Commodity Fetishism

A key Marxian analytical device, commodity fetishism, is commonly deployed in critical animal studies (Adams, 2010). When "meat" magically materializes on our plates, the long history of its brutal production is completely separated from its consumption. The work of making the conditions of "meat" production visible, of de-fetishizing "meat" commodities, has been an urgent priority of "consciousness raising" around the gratuitous suffering inflicted by the A-IC. Yet the process of making such A-IC-related-violence visible itself presents challenges.

An obvious strategy to expose "commodity production" is to reveal scenes from the killing floor of the slaughterhouse. But showcasing the absolute domination and abject agony experienced by animals inside slaughterhouses has not had the intended effect. The spectacle of A-IC-related-violence may land on a desensitized public already habituated to the extreme. If we are indeed living amidst a crisis of individualism and indifference, interminable iterations of animal suffering meant to provoke empathy might instead result in detachment and fatigue. Despite the hope that others are compelled by such imagery on a visceral level, these visuals may paradoxically encourage greater distance and diminishment of nonhuman animals' being-value. Slovenian philosopher Slavoj Žižek (2008) points out that violence needs to be approached obliquely. It is now so hypervisible that the public may be inured to the spectacle of A-IC-related-violence, so perhaps indirect entry points need to be explored. In *The Politics of Smell and The Morality of Sight: Challenging "Slaughterhouses with Glass Walls" in Animal Advocacy,* Stefanoni draws on both Marx and Foucault to complicate activist assumptions around a self-evident relationship between making visible A-IC-related-violence and its eventual demise. Her historical analysis also refutes the belief that the sequestration of animal slaughter away from town centers was bound up in concealing the visibility of slaughter, arguing, instead, that this can be better understood in terms of emerging hygiene discourses during the 19th century. In certain respects, Stefanoni's chapter troubles the comforting humanist assumption that a rational moral response will simply prevail with greater knowledge, truth, and visibility. The actual truth may be more disturbing for how we view morality and the power of the A-IC.

Another challenge associated with the study of violence against nonhuman animals is that by focusing so narrowly on bodily harm, we run the risk of reducing the rich, complex lives of nonhuman animals to their suffering alone. In *Beef, Bible, Bullets: Suicidal Cows and the Ecological Imaginings of Brazil*, Carey-Webb at once reveals the harm experienced by cows within the context of their intricate emotional and social existence. Carey-Webb offers an analysis of the novel *De gados e homens* or *Of Cattle and Men*, written by Brazilian author Ana Paula Maia. A nuanced entry point, Carey-Webb utilizes a medium that grants passage past desensitized psychological barriers. Edgar Wilson, the protagonist in the novel, is economically dependent on a "murderous job" in a slaughterhouse. Edgar becomes a sympathetic killer though, expressing unexpected care in the bloody deed. As the novel progresses, a fragmentation of human and animal occurs where humanistic categories cease to make sense. Carey-Webb's analysis radiates with Derrida's (2008) appeal to consider the existence of multiplicity of living creatures who cannot be categorized under a singular concept of "animality" that is positioned opposite of "humanity." Ultimately, the cattle in *De gados e homens* commit mass suicide – a startling act that communicates at once the ultimate unknowability of cows, but also the devastating consequences of capitalist-consumer culture. Carey-Webb situates the novel against the backdrop of a powerful "beef culture" in Brazil where ranchers are key players in the massive cattle industry and agriculturalists wield tremendous power as members of the Brazilian legislature. Carey-Webb ultimately links the expansive economic interests of the A-IC to the "river of blood" witnessed by Edgar who himself is animalized and confined to an occupation of torture.

Humans who are paid to execute nonhuman animals are recruited from the most precarious social locations where the extraction of their labor comes with the added risk of injury, death, and trauma (Noske, 1997; Parr, 2013; Torres, 2007). In *Reexamining the Meatpacking Methamphetamine Hypothesis*, Dollar and Hendrix help to elucidate how groups of people are needed for "criminalization" as part of the material infrastructure to support a single commodity. Meth may be ingested as a way to both cope with the trauma of killing and as a way to keep up with the accelerated pace of production. Slaughterhouse workers, often undocumented, are already admonished as trespassers, failures in the ways of economic "success," and unworthy of workplace rights to safety. Already criminalized, industry-sanctioned methamphetamine use is justified. Dollar and Hendrix ultimately show, through quantitative analysis, the social order behind these exploitative relations of production. And yet revealing the ruthless instrumental use of human labor in slaughterhouses may provoke greater denial and distance. We are all frequently, albeit unknowingly or unwillingly, influenced to devalue various aspects of life because recognizing their value could threaten our own fragile sense of worth and our uncertain desire for social status (Cacho, 2012). The human value of slaughterhouse workers becomes transparent as we understand how such bodies are situated on the interspecies killing floor. In the United States, slaughterhouse workers are likely to be drawn from populations who are excluded from rights to protection, citizenship, and even personhood; individuals who exist in a perpetual state of what Lisa Cacho (2012) calls "social death." Individuals are not marked as socially dead based on individual behavior, but rather based on status and social location, primarily "racial Others" who are viewed as inherently criminal or dangerous and therefore deserving of punishment.

Cultural Hegemony of Animal Cruelty Commerce

Many different species are exploited, tortured, held captive, and killed in various industries, such as research labs and testing facilities, and entertainment industries where animals are the subject of theatrics and spectacle (Francione, 2008;

Torres, 2007). Yet the factory farming industry is distinct in its starkly distressing volume – in the billions each year – of systematic and gruesome destruction of animal bodies. The achievement of such institutionalized cruelty speaks to the power of the commodity form in depersonalizing and rationalizing the horrific origins of "food." Dennis Soron (2010) writes about the coercive power of markets where commodity fetishism is not limited to consumer goods, but where the "expansionary logic of capitalism saturates life with commodities…" to such an extent that all aspects of existence and relationship to other beings, or even ourselves, are fashioned by this capitalist social order (p. 57). This "commodity culture" is part of the infrastructure of the A-IC that naturalizes nonhuman animals as disposable things.

Zygmunt Bauman (2009) names our "consumerist rendition of freedom" as incompatible with caring and being cared for. The pursuit of a "consumerist utopia" is encouraged through technologies of rapid transmission. Selfie shots with captive or conquered nonhuman animals fuse celebrity culture, mastery of nature, subjugation of "wild" elements, and celebration of the violent subjugation of animals. In *Selfie Safaris: The Violence of Contemporary Camera Hunting & Trophy Shot Selfies*, Medley explores how violence and the animal economy are co-constitutive as exemplified by the conflation of the neoliberalization of wildlife tourism and the rise of social media exhibitionism. Medley's work ultimately points to the selfie safari as a cultural form that, when stripped of its context, elides the cruelty of such performances. Instead, these images, which communicate "pleasure" and "leisure" are manufactured consumerist displays that run parallel to a history of "camera hunting."

Author Anat Ben Yonaton reveals subtle ways that a cultural hegemony of carnism operates through a collusion of Jewish dietary laws and a capitalist commodity economy to legitimize animal slaughter, enabling its expansion by imbuing it with meaning informed by religious practices. In *Following the Cultural Traces of Normalized and Legitimized Violence by Israeli Kosher Slaughterers toward Nonhuman Animals*, Yonatan explores the slaughter of nonhuman animals in Israel through in-depth interviews with Jewish slaughterers who collectively reveal a social order of A-IC-related-violence that is carefully supported by narratives of the sacred, noble, and heroic. Yonaton uncovers Jewish slaughters who are romanticized as heroes – similar to manipulating soldiers to fight in war (Ehrenreich, 1997; Hedges, 2002). They may be regarded as esteemed experts, likened to doctors who at times must slice and euthanize for the greater good. As in the case of animal slaughter in the United States, status hierarchies perform a critical function in enabling production by transforming killing into a sacred act and thereby extinguishing anxieties and opposition.

Bullfighting is another arena where cruel commerce is normalized through the manipulation of long-standing cultural norms, rituals, and historically meaningful practices. Blood sports involving the killing of animals exist in a landscape of "biopolitics," where habits and behaviors that seek to subjugate and control both the human and nonhuman may inform other aspects and operations of culture and institutions (Kalof et al., 2004). Puente-Guerrero in *The Arena of Controversy: Bullfighting and Its Implications in Modern Spanish Society* explores the rifts between the state and increasingly divergent public opinion on bullfighting in Spain. Puente-Guerrero explores the brutality of this ritual in terms of its representations, symbols, longevity in Spain's imagination, and the performative virility of the bullfight. This chapter recalls Jody Emel's (1995) deconstruction of the killing of wolves, both of which reveal how widely practiced, ritualized killing is a complexity that can be fed by class dynamics, bureaucracy, loyalty to historical narratives, commodity production, and even superstition. Puente-Guerrero shows how ruptures over bullfighting result in the state's reassertion of power by passing dubious legislation meant to appease citizens, while effectively maintaining the violent status quo.

Legal Collusion and the Violence of "Animal Capital"

Our legal system safeguards the capture, control, and destruction of animal bodies as wealth-producing assets, which facilitate a capitalist social order that serves as the fulcrum of the A-IC. Flynn and Nurse in *Horseracing as regulated cruelty: A nonhuman animal victimology perspective* consider the speciesist limitations of the subfield victimology, but also the tendency of existing scholarship in criminology to focus narrowly on harms prohibited by criminal law. Such loyalty to legal definitions ignores state-sanctioned A-IC-related-violence. Indeed, most corporeal harm against nonhuman animals is regulated by legislative bodies inside the A-IC. Flynn and Nurse point to the management of whipping during horseracing where regulatory boards outline the parameters of acceptable and excessive whipping. Racing horses, bodies torn and broken for their entertainment value, are always subjected to violent practices and social harms because they are institutionalized. Flynn and Nurse situate the "legitimization" of race horses always subjected to possible abuse made "permissible" by authoritative political entities. Their work reminds the reader that we are addressing the violence of capitalism, but also the ways in which the routine, repetitive, and normalized features of these practices obscure their impact.

In the chapter titled, *Nonhuman animals as property: what this means when companion animals are stolen,* Wyatt and Allen explore several facets of pet abduction – from legal dimensions to impacts on relationships from a species justice perspective. In their data analysis and critique of existing scholarship on pet abduction, Wyatt and Allen find that techniques employed to interrupt these crimes are themselves decidedly anthropocentric. Legally, the whole framework is about redress for the human "owner" without much, if any, consideration given to the needs, interests, or suffering of the nonhuman. Their work evokes deeper questions about the social and personal value that nonhuman animals must accrue to receive even minimal legal consideration. Animals must be lovable to receive interdictions against being stolen. And even as they are in a community with human animals, they are regarded as commodities by the criminal justice system – objects signaling market value and monetary gain or loss for the property "owner" (Haraway, 2003).

In another chapter that challenges legal definitions surrounding A-IC-related-violence against human and nonhumans, Banwell and Walliss consider reproductive violence and coercion as both phenomena of industry and as crimes against humanity. *"They have literally given up on life"; A review of the experiences of nonhuman animals subject to reproductive violence and coercion on factory and puppy farms* considers the United Nations' definitions of crimes against humanity as limited, speciesist projects. Banwell and Walliss proceed to rehabilitate these legal discursive constructions by showing their direct applicability to nonhuman animals. They refuse to position humans above or below, instead situating sentient beings in horizontal relation. By mapping the extreme exploitation and enslavement of nonhuman animals onto the institutional language of "atrocity crimes" and "core international crimes," Banwell and Walliss reveal that these brutalities toward humans and nonhumans are indistinct.

"Food" production is the primary mechanism of A-IC-related-violence. The factory farming industrial sector systematically exploits the reproductive systems of female animals, endangering both the health of animals and humans who consume animal flesh and secretions (Adams and Donovan, 1995). Economically precarious communities have suffered some of the worst health consequences of consuming cheap, low-quality, animal flesh (Jou, 2017). In *Which Animals Did Noah Eat? An Animal-Centric Focus on Food Crime*, Robinson considers how the industrial transmogrification and dismemberment of nonhuman animal bodies for human consumption triggers an entire network of injuries and misconduct. While there is prolific documentation and analysis of the manifold atrocities associated with animal agriculture, Robinson points out that only a handful of scholars have argued for a dedicated legal framework that formally names and recognizes this cluster of mutually reinforcing harms as food crime. Food crime can manifest as fraud, labor exploitation, environmental degradation, injury, and death, just to name a few. Robinson points to the layers of food crimes associated with industrial animal agriculture, from the sordid conditions of captivity inside concentrated animal feeding operations (CAFOs) to the torment and distress leading up to the ultimate dismemberment of nonhuman animal bodies. By naming these abuses, "crime" carries a unique legal weight that can lead to concrete action and a shift in public opinion. Creating this definitional structure also invites a critical analysis of the collaborations of powerful entities inside the industrial food engine of the A-IC.

The Ideological Architecture of Profitable A-IC-Related-Violence

Being-value here is underscored in spatialized relations of production. Human bodies which exist alongside disavowed, disposable and unprotected nonhuman animals, are also subject to routine harm. Aranceta-Reboredo and Castellano in their chapter, *Inside the Spanish Zoological Park Industry: Worker Insights on Human-Animal Relationships and Shared Vulnerabilities*, explore the simultaneous occurrences of harm and care in the context of the hierarchy of captivity.

Through qualitative research on zookeepers charged with animal care, these interviews reveal that the exploitation of human workers is predicated on their proximity to nonhuman animals. Spaces of captivity give humans' control over the conditions which cause suffering for animals. And such spaces are ordered in a speciesist hierarchy where human workers who are recognized as caring for and sharing pain with nonhuman animals in captivity experience "weaponized compassion." In this case, the zookeepers' capacity for empathetic connection with animals is manipulated to exhaust human labor and jeopardize well-being. The interplay of violence and hierarchy is evident in the reports from zookeepers on killing animals. Those animals in captivity who are less adaptive, cooperative, able-bodied, or "valuable" are more likely to be euthanized.

Several chapters in this collection activate possibilities for capsizing species-based hierarchical ideologies in surprising and creative ways. In *"If I Broke Down the Wall of Flesh": Blurring the Human/Animal Distinction in the Slaughterhouse through Ivano Ferrari's Poetry*, Stefanoni offers poetry as an indirect point of access to interrupt violent practices through its "anti-speciesist potential." Through an analysis of Ivano Ferrari's poetry, Stefanoni shakes loose socially produced certainties about the human/animal divide through a "zone of indistinction" where Ferrari's stanzas "broke down the wall of flesh" to the point of indiscernibility regarding human and nonhuman corporeal forms. Mythical boundaries of human and nonhuman collapse into a continuously open space. Such collapse of the human/animal distinction precludes recovery of the human in any way that violates nonhuman life. This zone of indistinction offers a new framework of possibility for severing our compulsive identification with the human. Once destabilized, hopeful possibilities for all life forms have room to be reimagined. The zone of indistinction is made legible by the haunting imagery of slaughter. In this way, Stefanoni opens up theoretical possibilities of reconceptualizing living creatures through corporeality. Reminiscent of Stacey Aliamo's concept of "trans-corporeality," Stefononi's analysis extends the promise that such hierarchical erasures may extend outward in unforeseen ways to the environment.

Speciesist hierarchies – the ideological connective tissue of the A-IC – may depend upon inviting humans to enter realms where human/animal binaries can no longer be sustained. Stefanoni's "zoopoetique" – where poetry enables non-anthropocentric perceptions – is one such realm. In *Embodying Non-speciesism through Altered States of Consciousness*, Dollar explores embodied awareness of the consumption of animal flesh. Through in-depth interviews with participants who shared their experience with ingesting psychedelics, Dollar documents a recurring realization of a sort of "ecological consciousness." Participants reported sudden realizations of their complicity in speciesist projects where their altered states of consciousness sharpened their understanding of the ethical stakes of their consumption habits and perceived superiority over the natural world. For each of the participants, an alternative perspective on existence as a whole destabilized previously held certainties about the inferiority of nonhuman life forms.

Brooks-Dollar's study prompts the reader to consider if our widespread human collaboration to continually reify animals as objects, constantly pointing to our "differences," in fact stems from desires to secure our access to nonhuman animals, preserving our privileged position of proximity.

The violent exploitation of animals for profit produces a biopolitical sphere where animals are rendered simultaneously disposable and loveable. The maintenance of such hierarchies of worthiness is marred with confusion and crisis in the case of loving animals. The profit apparatus responds to and manipulates human desires for animal-derived companionship, eating, and pleasure (Salleh, 1997). Profound cognitive dissonance understandably results from these sets of contradictions where any given animal may be loved, killed, held in captivity, consumed, or cared for. Muzzatti and Grieve, in their chapter, *A Time to Kill: Cruelty and Compassion with Companion Animals and Urban Wildlife*, illustrate the unpredictability of human feelings and impulses surrounding lethal violence toward nonhuman animals. A discourse analysis of media accounts of killing animals uncovers an ideological map where the outcomes of either killing or compassion are closely aligned with human interests and wants – including the desire to be free of "nuisances." Muzzatti and Grieve connect these discourses of killing and compassion to the logics of late capitalism, where ranking, roles, and the hierarchical organization of certain lives is necessary. The presentation of such contradictions invites the reader to a deeper reflexivity, opening a pathway for ethical responses.

While hope remains for individuals to resolve these contradictions, several chapters in this volume offer caution when appealing to the potential for ethical clarity of individuals. In *The Lennie Small Paradox: Loving Animals to Death*, Briscoe wrestles with the disjointed and conflicting impulses and feelings we express about nonhuman lives and the arbitrariness of what it means to love nonhuman animals. Briscoe's chapter underscores how love and empathy themselves are erratic and may indeed result in violence through their own expression. The Lennie Small Paradox becomes here an allegory of how rampant anthropocentrism – an expression of capitalist culture – subjects nonhuman animals to the potentially destructive whims of humans. This allegory lends insight into the larger human-induced destruction of the nonhuman-life world in the Capitalocene.

Conclusion

Acknowledging the impossibility of excavating all systemic A-IC-related-violence and harm against human and nonhuman in its enormous territory, this collection explores select institutional spaces in the A-IC. This compilation highlights entangled oppressions, where all species are situated along a continuum of multispecies vulnerability. But this precariousness is situated within a capitalist social order, and these chapters are collectively attuned to the instrumental use of A-IC-related-violence as institutional scheme and strategy; yet the A-IC is not a totalizing "structural force" in the traditional sociological sense, even as the scale of its violence overwhelms. The ongoing embedding of critical animal knowledge production, of which this volume represents a small fragment, is itself testament to significant epistemic shifts and a contribution to the prefiguration of an alternative future. We are proud to present these voices for your critical engagement so that you may feel compelled to join and agitate in the struggle for that future.

Notes

1 See, for example, Chapter 1 in Twine (2024) for a lengthier exploration of the historicization of capitalism regarding multispecies relations.
2 For a lengthier discussion of domination as a complex system of gendered social relations, see Cudworth (2005) and Merchant (1980). For elaboration on physical violence as a coercive tool used to sustain relations of domination and exploitation, see Žižek (2008).

References

Adams, C. J. (1996). Bringing peace home: A feminist philosophical perspective on the abuse of women, children and pets. In K. Warren & D. Cady (Eds.), *Bringing peace home: Feminism, violence, and nature* (pp. 69–87). Indiana University Press.
Adams, C. J. (1997). "Mad cow" disease and the animal-industrial complex: An ecofeminist analysis. *Organization & Environment*, *10*(1), 26–51.
Adams, C. J. (2010). *The sexual politics of meat: A feminist-vegetarian critical theory* (20th Anniversary Ed.). Continuum International Publishing Group.
Adams, C. J., & Donovan, J. (1995). *Animals and women: Feminist theoretical explorations*. Duke University Press.
Alaimo, S. (2010). *Bodily natures: Science, environment, and the material self.* Indiana University Press.
Almiron, N. (Ed.). (2023). *Animal suffering and public relations: The ethics of persuasion in the animal-industrial complex*. Routledge.
Almiron, N., Cole, M., & Freeman, C. P. (Eds.). (2016). *Critical animal and media studies*. Routledge.
Arcari, P. (2020). *Making sense of 'Food' animals – A critical exploration of the persistence of 'meat'*. Palgrave MacMillan.
Banwell, S. (2023). *The war against nonhuman animals: A non-speciesist understanding of gendered reproductive violence*. Springer Nature.
Bauman, Z. (2009). *Does ethics have a chance in a world of consumers?* Harvard University Press.
Beirne, P. (2020). Non-human victims: The harmful consequences of human-animal interactions. In *Environmental crime and social conflict* (pp. 157–176). Routledge.
Beirne, P. (2021). Wildlife trade and COVID-19: Towards a criminology of anthropogenic pathogen spillover. *The British Journal of Criminology*, *61*(3), 607–626.
Beirne, P., & Lynch, M. J. (2023). On the geometry of speciesist policing: The Federal Bureau of Investigation's animal cruelty data. *International Journal for Crime, Justice and Social Democracy*, *12*(2), 137–149.
Cacho, L. M. (2012). *Social death: Racialized rightlessness and the criminalization of the unprotected*. New York University Press.
Calarco, M. R. (2021). *Animal studies: The key concepts*. Routledge.
Carter, N. (2024). Harvesting denial, distractions, & deception: Understanding animal agriculture's disinformation strategies and exploring solutions. *Freedom Food Alliance*. Retrieved April 25, 2024, from https://www.freedomfoodalliance.org/thereport

Cazaux, G. (1999). Beauty and the beast: Animal abuse from a non-speciesist criminological perspective. *Crime, Law and Social Change*, *31*(2), 105–125.

Chowdhury, A., Kabir, K. H., Abdulai, A. -R., & Alam, M. F. (2023). Systematic review of misinformation in social and online media for the development of an analytical framework for agri-food sector. *Sustainability*, *15*(4753), 4753.

Cole, M., & Stewart, K. (2014). *Our children and other animals: The cultural construction of human-animal relations in childhood*. Routledge.

Cudworth, E. (2005). *Developing ecofeminist theory: The complexity of difference*. Palgrave Macmillan.

Cudworth, E. (2015). Killing animals: Sociology, species relations and institutionalized violence. *The Sociological Review*, *63*(1), 1–18.

Derrida, J. (2008). *The animal that therefore I am*. Trans. David Wells (Ed.), Marie-Louise Mallet. Fordham University Press.

Duxbury, C. (2021). *Science, gender and the exploitation of animals in Britain since 1945*. Routledge.

Ehrenreich, B. (1997). *The ecstasy of war*. Oxford University Press.

Emel, J. (1995). Are you man enough, big and bad enough? Ecofeminism and wolf eradication in the USA. *Environment and Planning D: Society and Space*, *13*(6), 707–734. https://doi.org/10.1068/d130707

Fischer, B. (Ed.). (2020). *The Routledge handbook of animal ethics*. Routledge.

Fitzgerald, A., Kalof, L., & Dietz, T. (2009). Slaughterhouses and increased crime rates. *Organization & Environment*, *22*(2), 158–184.

Fitzgerald, A., & Taylor, N. (2014). The cultural hegemony of meat and the animal industrial complex. In N. Taylor & R. Twine (Eds.), *The rise of critical animal studies - From the margins to the center* (pp. 165–182). Routledge.

Francione, G. (1995). *Animals, property, and the law*. Temple University Press.

Francione, G. (2008). *Animals as persons*. Columbia University Press.

Gregson, R., Piazza, J., & Boyd, R. L. (2022). 'Against the cult of veganism': Unpacking the social psychology and ideology of anti-vegans. *Appetite*, *178*.

Griffin, S., & Griffin, N. (2021). Beyond meat? Taking violence against non-human animals seriously as a form of social harm. In P. Davies, P. Leighton, & T. Wyatt (Eds.), *The Palgrave handbook of social harm* (pp. 281–310). Palgrave MacMillan.

Hannan, J. (Ed.). (2020). *Meatsplaining: The animal agriculture industry and the rhetoric of denial*. Sydney University Press.

Haraway, D. (1992). Otherworldly conversations; terran topics; local terms. *Science as Culture*, *3*(1), 64–98.

Haraway, D. J. (2003). *The companion species manifesto: Dogs, people, and significant otherness*. Prickly Paradigm Press.

Harrison, R. (1964). *Animal machines: The new factory farming industry*. Vincent Stuart Publishers.

Hedges, C. (2002). *War is a force that gives us meaning*. Public Affairs.

Hendrix, J. A., & Dollar, C. B. (2018). American slaughterhouses and the need for speed: An examination of the meatpacking-methamphetamine hypothesis. *Organization & Environment*, *31*(2), 133–151.

Hribal, J. (2010). *Fear of the animal planet: The hidden history of animal resistance*. Counter Punch.

Hunnicutt, G. (2020). *Gender violence in ecofeminist perspective: Intersections of animal oppression, patriarchy and domination of the earth*. Routledge.

Jacobs, A. (2019, September 16). A Shadowy Industry Group Shapes Food Policy Around the World, *New York Times*.

Jacques, J. R. (2015). The slaughterhouse, social disorganization, and violent crime in rural communities. *Society & Animals*, *23*(6), 594–612.

Jou, C. (2017). *Supersizing urban America: How inner cities got fast food with government help*. University of Chicago Press.

Kalof, L., Fitzgerald, A., & Baralt, L. (2004). Animals, women, and weapons: Blurred sexual boundaries in the discourse of sport hunting. *Society & Animals*, *12*(3), 237–251.
Kramer, R. C. (2020). *Carbon criminals, climate crimes*. Rutgers University Press.
Merchant, C. (1980). *The death of nature: Women, ecology, and the scientific revolution* (1st ed.). Harper & Row.
Moore, J. (2015). *Capitalism in the web of life: Ecology and the accumulation of capital*. Verso Books.
Moore, J. W. (2017). The Capitalocene, Part I: On the nature and origins of our ecological crisis. *The Journal of Peasant Studies*, *44*(3), 594–630.
Morris, V., & Jacquet, J. (2024). The animal agriculture industry, US universities, and the obstruction of climate understanding and policy. *Climatic Change*, *177*(3), 41.
Nibert, D. (2002). *Animal rights human rights: Entanglements of oppression and liberation*. Rowman & Littlefield.
Nibert, D. (2011). Origins and consequences of the animal industrial complex. In Best, S. et al. (Eds.), *The global industrial complex: Systems of domination* (pp. 197–210). Lexington Books.
Nibert, D. (2013). *Animal oppression and human violence: Domesecration, capitalism and global conflict*. Columbia University Press.
Nibert, D. (Ed.). (2017a). *Animal oppression and capitalism, vol. 1*. ABC-CLIO.
Nibert, D. (Ed.). (2017b). *Animal oppression and capitalism, vol. 2*. ABC-CLIO.
Nixon, R. (2011). *Slow violence and the environmentalism of the poor*. Harvard University Press.
Noske, B. (1989). *Humans and other animals: Beyond the boundaries of anthropology*. Pluto Press.
Nurse, A. (2017). Green criminology: Shining a critical lens on environmental harm. *Palgrave Communications*, *3*, 10.
Nurse, A. (2021). *Animal abuse and unlawful killing: An examination of national and international perspectives on animal harm*. Routledge.
Nurse, A., & Wyatt, T. (2020). *Wildlife criminology*. Bristol University Press.
Parr, A. (2013). *The wrath of capital: Neoliberalism and climate change politics*. Columbia University Press.
Pedersen, H. (2019). *Schizoanalysis and animal science education*. Bloomsbury.
Rowley, J., & Prisco, C. (Eds.). (2021). *Law and veganism: International perspectives on the human right to freedom of conscience*. Rowman & Littlefield.
Salleh, A. (1997). *Ecofeminism as politics: Nature, Marx, and the postmodern*. Zed Books.
Sanbonmatsu, J. (Ed.). (2011). *Critical theory and animal liberation*. Rowman & Littlefield.
Shove, E., Pantzar, M., & Watson, M. (2012). *The dynamics of social practice: Everyday life and how it changes*. Sage.
Shukin, N. (2009). *Animal capital: Rendering life in biopolitical times*. University of Minnesota Press.
Skinazi, S. (2019). *Animal people: Freaks, elitists, fanatics, and haters in U.S. discourses about veganism*. Dissertation.
Skotnicki, T. (2021). *The sympathetic consumer moral critique in capitalist culture*. Stanford University Press.
Smulewicz-Zucker, G. R. (Ed.). (2012). *Strangers to nature: Animal lives and human ethics (Ser. Logos: Perspectives on modern society and culture)*. Lexington Books.
Sollund, R. (2019). Species justice and harm: An examination of the role of non-human animals in criminology. In *Routledge handbook of critical criminology* (2nd ed., pp. 263–275). Routledge.
Soron, D. (2010). Road kill: Commodity fetishism and structural violence. In J. Sanbonmatsu (Ed.), *Critical theory and animal liberation* (pp. 40–62). Rowman & Littlefield Publishers.

Taylor, N., & Fitzgerald, A. (2018). Understanding animal (ab)use: Green criminological contributions, missed opportunities and a way forward. *Theoretical Criminology*, *22*, 402–425.
Torres, B. (2007). *Making a killing: The political economy of animal rights*. AK Press.
Twine, R. (2010). *animals as biotechnology - ethics, sustainability and critical animal studies*. Routledge.
Twine, R. (2012). Revealing the "animal-industrial complex": A concept and method for critical animal studies? *Journal for Critical Animal Studies*, *10*(1), 12–39.
Twine, R. (2013). Animals on drugs: Understanding the role of pharmaceutical companies in the animal-industrial complex. *Journal of Bioethical Inquiry*, *10*(4), 505–514. https://doi.org/10.1007/s11673-013-9476-1
Twine, R. (2024). *The climate crisis and other animals*. Sydney University Press.
Wadiwel, D. J. (2015). *The war against animals*. Brill.
Wadiwel, D. J. (2023). *Animals and capital*. Edinburgh University Press.
White, R. (2022). *Theorising green criminology: Selected essays*. Routledge.
White, R., & Renzetti, C. (2021). Animal harm: Perspectives on why people harm and kill animals. In *International handbook of animal abuse studies* (pp. 29–48). Routledge.
Wrenn, C., & McKie, R. E. (2020). Criminology and animal studies: A sociological approach. In *The Palgrave handbook of critical animal studies* (pp. 89–104). Springer International Publishing AG.
Žižek, S. (2008). *Violence: Six sideways reflections*. Picador.

PART I

Reproducing 'A-IC-Related-Violence' Through Discursive Strategies

2

MEAT SCIENTISTS FIGHT BACK! WHAT THE DUBLIN DECLARATION TELLS US ABOUT THE ROLE OF ACADEMIA IN THE ANIMAL-INDUSTRIAL COMPLEX

Richard Twine

The multitude of practices that comprise the global animal-industrial complex (A-IC) require active work for the system of nonhuman animal exploitation to be reproduced. This has been more or less successful via such institutions as the family, sites of education and knowledge production, and in overlaps between media, marketing, and advertising. Under conditions of even mild legitimation crisis, the naturalisation of harm and violence against nonhuman animals becomes less secure. This tempts new protective practices to emerge from those with the most direct ideological interests in the A-IC. It is within such a context that the 2022–2023 meat science affirmation of the 'Dublin Declaration' (DD) may be understood.

This chapter critically explores the goals, claims, and interests of the DD, a form of push back against the problematisation of the consumption of animal-sourced foods (ASFs). Push backs arise when practitioners perceive threats to their dominance. In this case they are a visible manifestation of the relative early success of social movements for plant-based or plant-forward diets. Of particular interest is the attempt by the declaration authors to leverage scientific authority and to label opponents as 'ideological' (Leroy & Ederer, 2023). I argue that this attempt to construct a rather conventional dichotomy between science and ideology ultimately backfires due to the unique characteristics of the 'meat sciences'. Furthermore, although historically the A-IC has relied to an extent upon concealment and secrecy, attempts at push back such as the DD, can give rise to new forms of visibility around, in this case, the academic infrastructure which is integral to it. My earlier definition of the A-IC termed it as

> a partly opaque and multiple set of networks and relationships between the corporate (agricultural) sector, governments, and public and private science.

DOI: 10.4324/9781003441908-3

> With economic, cultural, social, and affective dimensions it encompasses an extensive range of practices, technologies, images, identities, and markets.
>
> *(Twine, 2012, p. 23)*

Whilst academia has always played a role in perpetuating the A-IC, the declaration affords new ways in which to understand these relations and networks involved in accomplishing the hegemony of the complex.

The DD brings into relief previously unseen, or hard to discern, connections and alliances between different actors and inadvertently ends up aiding in the analysis and understanding of the A-IC. It allows critical animal scholars to map some of the major research nodes involved in underpinning industrialised animal agriculture. Thus, this chapter examines where the concentrations of research signatories are, and which countries are most represented in the declaration. The work of those involved in the declaration can be seen as similar to those in the fossil fuel sector who have sought to obfuscate and delay transition. The chapter ends by arguing that Universities should start to defund and disincentivise research embedded in a sector so embroiled in the climate, health, and biodiversity crises.

The Dublin Declaration and the Animal-Industrial Complex

The proto-normalisation of plant-based eating and veganism more broadly has been ruffling a few feathers. This is unsurprising given both the profitability of animal commodification and the deep ideological investments that meat cultures have in cultural anthropocentrism.[1] As with other necessary transitions, notably energy, the key obstacles turn out to be political rather than practical. Emerging towards the end of 2022, and rising to prominence in 2023, the DD represented perhaps the most interesting example to date of particular social actors within the A-IC attempting to fight back against calls even for the reduction of meat and dairy consumption. Signed by over 1000 individuals worldwide[2] emerging from a conference in Dublin (October 2022) and a special issue of the journal *Animal Frontiers* (April 2023) – the declaration argues for the importance of consuming ASFs on both health and cultural grounds. In June 2023, the two key instigators of the declaration also published a short correspondence piece in *Nature Food* further making their case (Leroy & Ederer, 2023). The DD also evidenced a well-thought-out strategy of dissemination with prior (Leroy et al., 2022) and simultaneous publications written by signatories addressing the Australian context (Pethick et al., 2023) and a further article placed in the *Italian Journal of Animal Science* (Leroy et al., 2023).

The professed aim of the declaration was ‘to give voice to the many scientists around the world who research diligently, honestly and successfully in the various disciplines in order to achieve a balanced view of the future of animal agriculture’.[3] This tells us a little more precisely where the push back is coming from. The signatories are largely the global community of animal production/meat-related scientists whose work has historically and biopolitically secured profitability for the 'livestock' sector. This includes a wide array of specialisms, such as animal production geneticists, veterinary science, animal feed specialists, experts on animal reproduction, and agricultural economists. In turn this is part of a far broader set of expertise around, for example, marketing, animal housing, ventilation, and killing essential to the farming of nonhuman animals. The broadening of meat science specialisms followed and enabled the intensification and globalisation of animal agriculture during the twentieth century. This industrialisation of animal production would have been impossible without the animal/meat sciences. Within the meat sciences there is much occupational mobility between academia and the private corporate sector, and they have survived and grown through a combination of public and private funding. These are very much livelihoods embedded within the business of farmed animal commodification and we must make the important first point that the declaration represents actors acting in their own self-interest.

The declaration positions ASFs as the solution to malnutrition and as needed due to projected population growth. This despite the fact that increased human population places an even greater question mark around the profligate use of land inherent to 'livestock' production. On human health the declaration makes bold claims about the unsuitability of a diet low in ASFs for particular groups such as children, the pregnant, women of reproductive age and older adults. This despite three respected national dietetic organisations stating that a plant-based diet is suitable for all.[4] The declaration then attempts to argue for the environmental benefits of animal agriculture. Given that the sub-title of the declaration places the emphasis on the 'societal role of livestock' it then argues from the point of tradition. Of course, it is true that animal agriculture is socially and economically embedded to a greater or lesser extent across societies. It is a weak argument for the status quo however. Not only is it facile to use tradition as an argument for normativity, other environmental damaging sectors in need of transition (fossil fuels) are just as embedded within the social. The declaration is sprinkled with dubious moralisations in which 'livestock' becomes the answer for world hunger, sustainability, gender equality, and food security. Near the end it states, 'Livestock is the millennial-long-proven method to create healthy nutrition and secure livelihoods, a wisdom deeply embedded in cultural values everywhere'[3]. This is a bizarre pseudo-mystical assertion given the advances that have been seen in nutritional science in even the last few decades. Indeed, it is the increase in ASF consumption in many societies during the twentieth century that is often partly linked to the disease burden associated with the nutrition transition (Popkin, 2009).

At the end of the declaration a link is inserted, as follows 'For scientific evidence, please refer to presentation recordings from the 19/20 October 2022 International Summit on the Societal Role of Meat and the Special Issue of *Animal Frontiers*'[3].

The issue contains various papers from the meat science community and signatory list that try to make the health and environmental case for ASFs. Yet this is also problematic because the journal *Animal Frontiers* is the official journal of four major international animal/meat science organisations: the American Society of Animal Science (ASAS), the American Meat Science Association (AMSA), the World Association of Animal Production (WAAP), and the European Federation of Animal Science (EAAP). At least two of these organisations, EAAP and AMSA declare their private funding on their websites. AMSA, for example, are funded by some of the largest meat corporations in the world such as Smithfield, Cargill, and Tyson. This underlines some of the special characteristics of the animal/meat sciences field pertaining to just how embedded within industry it is. Animal scientists directly work to improve the profitability of the sector and they have an interest in the growth and productivity of the sector for the perpetuation of their own research community, livelihoods, and infrastructure. Privately funded science runs the risk of encouraging a tunnel vision wherein the utilisation of knowledge production for profit accumulation takes precedence over a broader evidence-based approach (Legg et al., 2021) to the food system.

If we situate the DD within a broader visualisation of the A-IC (Figure 2.1) we can see that it consists of relationships (see also Figure 2.2) and practices most clearly within the research and development segment. However, it overlaps with government (e.g., Teagasc which hosted and funded the declaration conference – with 40k euros[5] – is, in turn, funded by the Irish state), with corporations and associations and also with the social institutions of media via the broad dissemination strategy for the declaration. It draws upon strategies and meanings common to the broader animal production industries. Figure 2.2 provides a useful overview of the affiliations of the core group of authors of the DD illustrating their memberships of organisations and points where meat industry funding have direct input. The circularity of these relationships also underpins how the community constructs a sense of scientific authority for their claims. Moreover, it provides a flavour of the global interconnectedness of these relationships.

Are Meat Scientists Really Scientists?

The intention here is not to simplistically argue that animal/meat scientists are not scientists. This would be to construct a false picture of scientific knowledge generally as existing apart from culture. Indeed, animal scientists take the dominant anthropocentric values practiced within cultures, magnifying, and intensifying them. Arguably there is a discomforting, more extreme, instrumentalism at play within the animal sciences which may even trouble prosaic anthropocentrism. This is seen, for example, in the way that the animal sciences have been used to maximise pig litter size, to create animal genotypes that suffer inherently due to their muscle mass, within the calculative intricacies involved in the science of animal slaughter, or the reduction of an animal to the quality of their meat. This suggests that the ideology of anthropocentrism

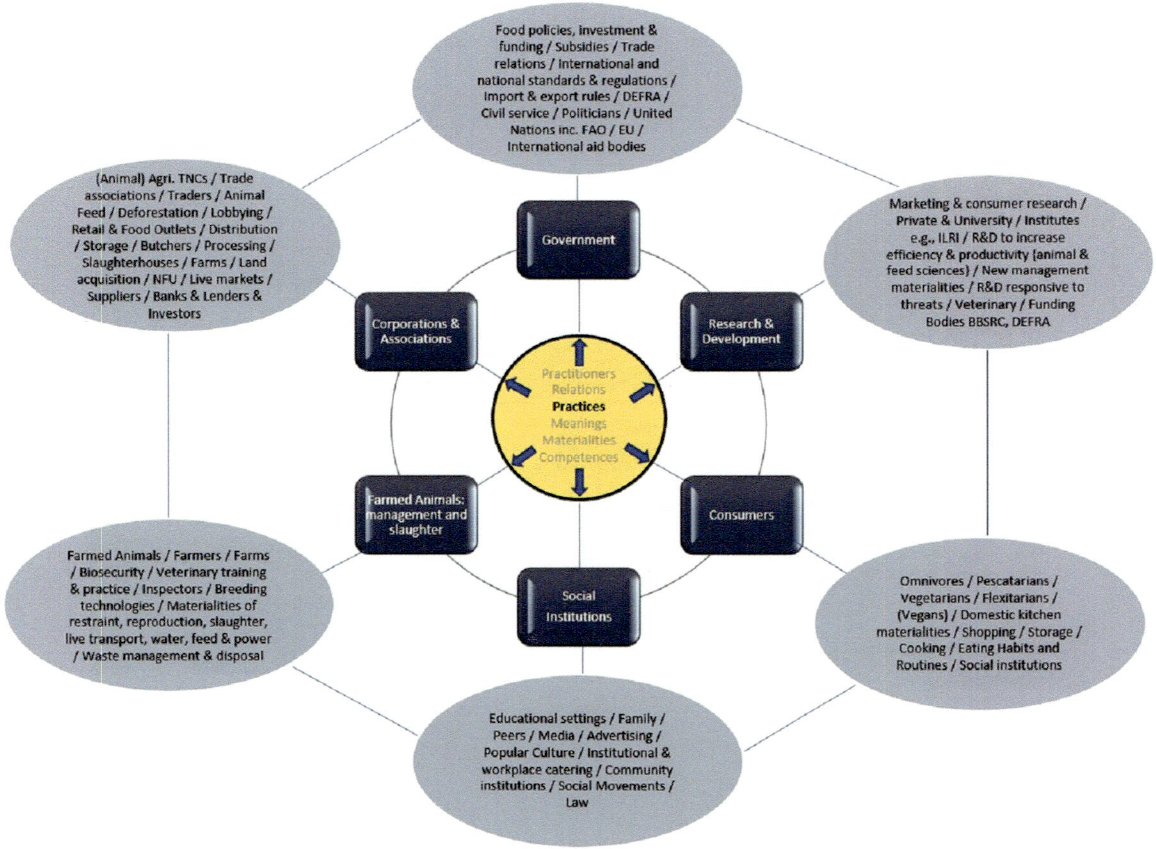

FIGURE 2.1 Picturing the animal-industrial complex through a practice theory lens (Twine, 2024).

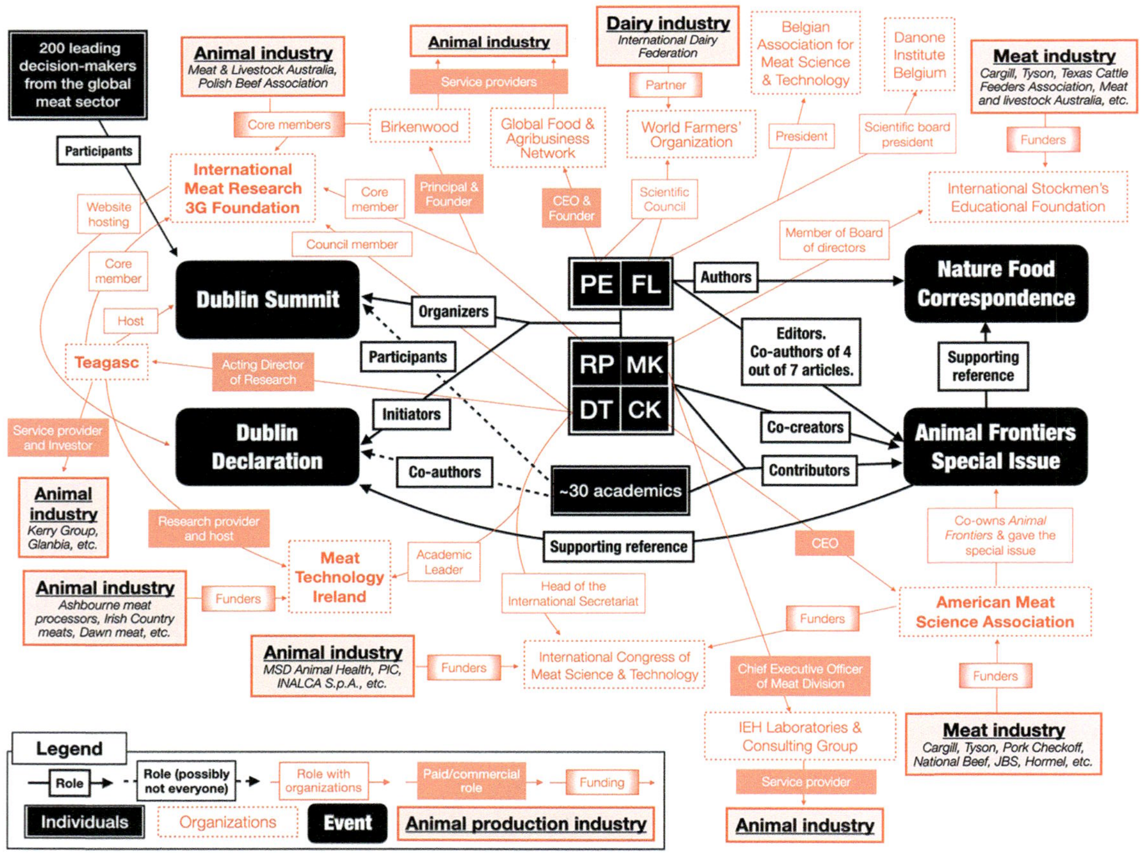

FIGURE 2.2 Visual summary of main roles and industry associations relating to the Dublin Summit, DD, the Animal Frontiers Special Issue (AFSI), and the Nature Food Correspondence (NFC) at the time of publication of the DD (based on publicly available information, mainly the authors' statement on the DD's website, but also information from journalistic investigations, the institutions' websites, and the AFSI's editorial).[6–11]

can take on an extreme form in the communities of animal science and is in constant tension with other ways of valuing nonhuman animals. Here it embodies well what the philosopher Val Plumwood (2002) referred to as sado-dispassionate reason.[12] This is an important concept because it contends that within the much discussed 'emotional neutrality' of scientific practice is the potential not only for moral disengagement but also for sadism. This commitment to 'neutrality' (which is often used to disavow the agency and abilities of nonhuman animals) furthermore, as Plumwood outlines, '*reduces the ability to resist co-option by economic forces and works systematically against a science committed to social responsibility*' (2002, p. 41). Such have been the historical affinities between this construct of science and the requirements of bio-capitalism acting together to shape, not without resistance, world ecology (Moore, 2015), most clearly in the enormous land-use transformations which animal agriculture is responsible for. Needless to say, this has meant the large-scale deforestation of the planet that shapes the biodiversity and climate crises.

Whilst it might be tempting to label animal scientists as anthropocentric ideologues, for they are certainly caught stubbornly within that worldview, they are it could be argued, an almost parodic (if it were not for the harm that they are enmeshed within) and dated version of a scientist stereotype. Affirming animal scientists as scientists does not endow them with the authority they may be hoping for. I already alluded to the shortcomings of the text of the declaration above and I want to argue that the peculiarly co-opted nature of the animal sciences degrades their own quality as scientists and highlights questions concerning the professional ethics of their disciplines.

Scientific and Ethical Tunnel Vision: Must the Animal Give Its (Sic) Life to the Human?

In the brief elaboration of the declaration published in Nature Food the authors reiterate their fears that policy may lead to a downsizing of animal agriculture, and they self-cite (from papers in their own issue of Animal Frontiers) to argue that the health and environmental problems of animal agriculture have been overestimated. The short piece also contains the curious phrase where they talk of the '*the ethical reality that the animal needs to give its life in favour of the human species*' (Leroy & Ederer, 2023, p. 438). This is an odd mystical assertion that forecloses any ethical debate and naturalises anthropocentric ideology. A critical counter-correspondence was published by the same journal a few months later arguing that Leroy and Ederer (2023) '*contained unsubstantiated generalizations and statements that go beyond the focus of the Declaration, particularly regarding the expansion of livestock production*' (Herzon et al., 2023, p. 926). Moreover, the response also questions why both the declaration and Leroy and Ederer (2023) are not in favour of reducing meat consumption in the Global North. This is an achievable policy, yet the declaration's failure to advocate for it suggests that it is not seriously interested in a 'balanced' conversation but rather in protecting the 'livestock' sector generally. I have personally documented 50 key journal articles in top science journals that have recommended reductions to animal consumption for climate mitigation and other co-benefits (such as human health and biodiversity promotion) published between 2003 and 2022 (see Twine, 2024, Chapter 4). They provide ample evidence for a consensus on downsizing animal agriculture.

The co-opted nature of the animal sciences gives rise to a compartmentalised tunnel vision which compromises both scientific credibility and quality. The DD is aloof to climate science, to nutritional science, to animal welfare science, and to the ethological exploration of animal minds. Furthermore, it ignores social science critiques of the role of the A-IC in human injustices such as environmental racism, the philosophical development of animal ethics, the harm and violence perpetrated against nonhuman animals, and the interdisciplinary emergence of posthumanism and critical animal studies which together mount powerful critiques of anthropocentrism. The narrative of the declaration is notable for its labelling of opponents as zealots. Elsewhere opponents are associated with 'ideology' rather than evidence (Leroy & Ederer, 2023).

Whilst the term ideology certainly has an innocent meaning to describe belief systems generally here it is being deployed somewhat differently to indicate some sense of bias, and zealotry to indicate some sense of fanaticism. If only (!) people would listen to '*the many scientists around the world who research diligently, honestly and successfully in the various disciplines in order to achieve a balanced view of the future of animal agriculture*'[3] to quote the declaration. Yet using the more classic sociological understanding of ideology to mean dominant views shaped by economic interests, or by gender or culture for example, it becomes inappropriate to view vegans or mere advocates of ASF reduction as ideological and far more appropriate to consider animal scientists – a privileged class after all that represents a highly dominant set of practices – as the *ideological* protectors of anthropocentrism. Indeed, given the concerted nature of the backlash to criticisms of the A-IC the accusations of ideology and fanaticism appear more like projections. Not only does anthropocentrism nourish the ontological security of especially conservative and hierarchical worldviews (Dhont & Hodson, 2014), economically the meat industry is geopolitically powerful, valued in 2021 at $897.5 billion.[13] That is a lot of profit to protect and is why I have argued before that the A-IC is a major constituent part of global capitalism itself (Twine, 2020, p. 5). However, apparently it is those animal rights and climate activists we have to fear lest they give us a safer, healthier, more caring, and more just world.

Who Signed the Declaration?

The protection of economic self-interest also appears at play when the nationality of declaration signatories is examined more closely. Figures 2.3 and 2.4 show the major countries represented by the signatories and Figures 2.5–2.7 outline the relationships between the nationalities of signatories and the GDP, meat supply per person, and beef exports in those respective countries. The animal/meat science infrastructure involved in supporting the declaration also provides for an interesting analysis which

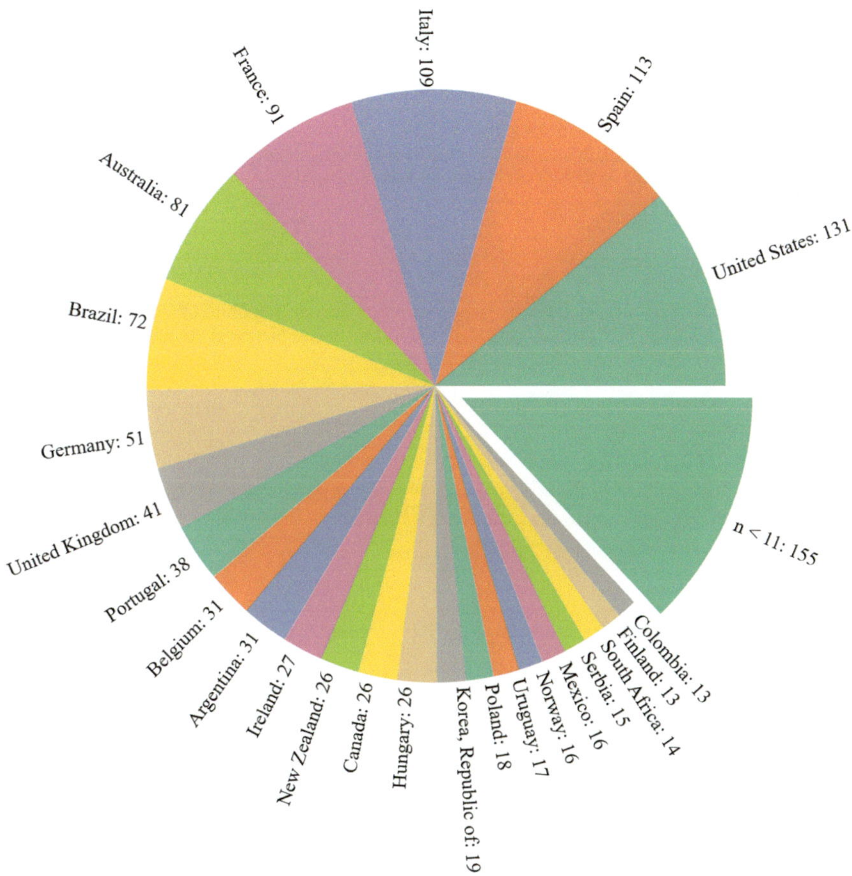

FIGURE 2.3 Country locations of signatories of the Dublin Declaration

illuminates centres of concentration.[14] Narrowing the analysis down to affiliations with ten or more signatories (Figure 2.4) indicates those Universities or research centres with likely research specialisms in animal production science or allied sub-disciplines. In total, 18 Universities or research centres provided 10 or more signatories. Australia, Spain, Italy, France, and Brazil provided 14 of these affiliations. Again, this adds to the understanding of national research infrastructure and the academic contribution to the A-IC, as well as the core contributors to the declaration. In this way initiatives like the DD promote the visibility of these networks as the animal science community feels the need to contest challenges to it. The highest countries represented in Figure 2.4 mostly resemble those in Figures 2.5–2.7, but the data provides additional information about the specific Universities and research centres that felt strongly enough to sign the declaration. Those represented add credence to the assertion that the majority of signatories were from the animal production sciences and allied disciplines.

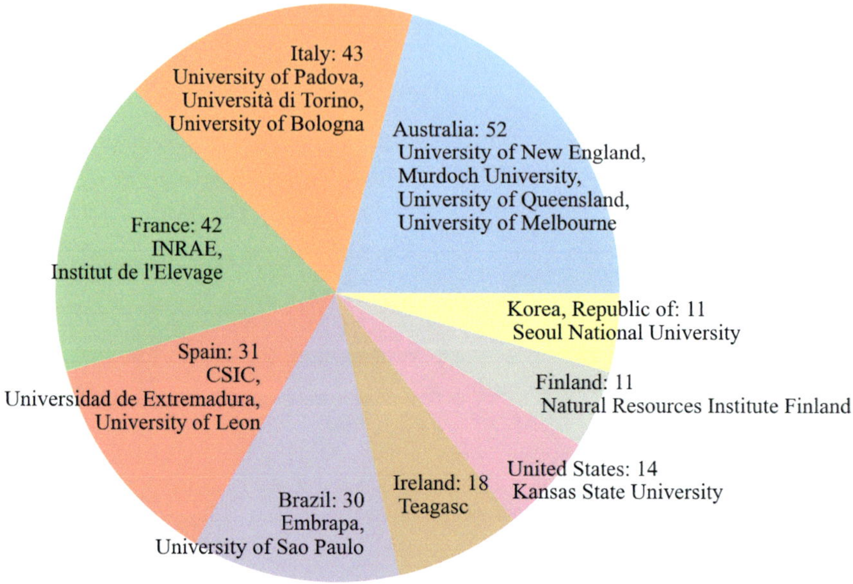

FIGURE 2.4 Affiliations, by country, with 10+ signatories

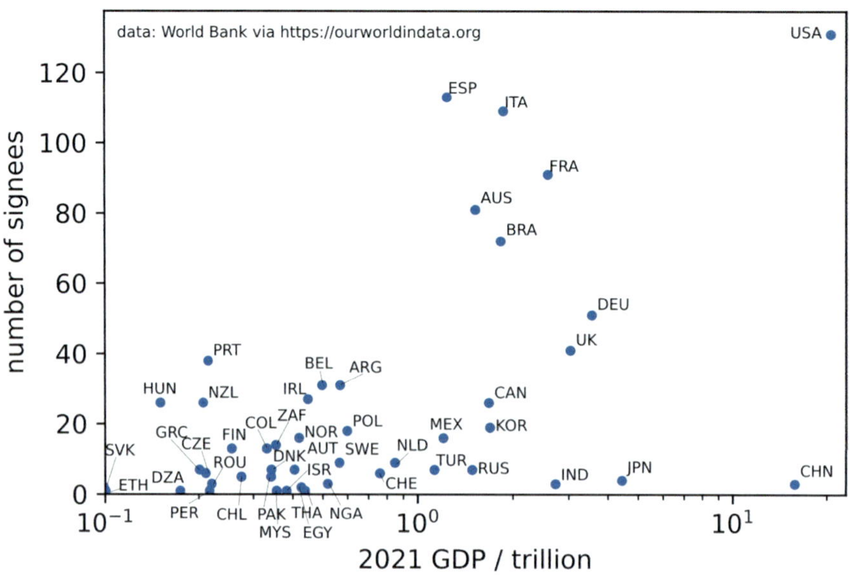

FIGURE 2.5 Relationship between number of signees and national GDP ($ trillion)

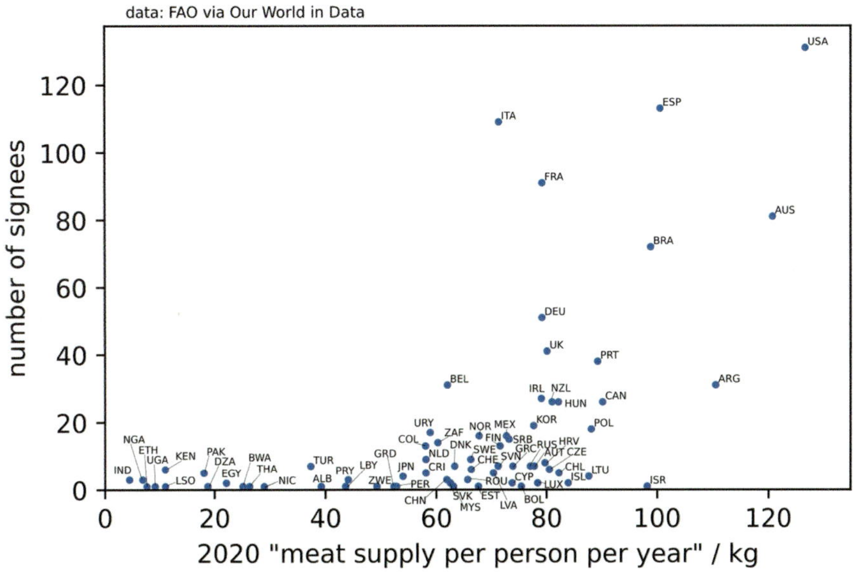

FIGURE 2.6 Relationship between number of signees and national meat supply per person per year (kg)

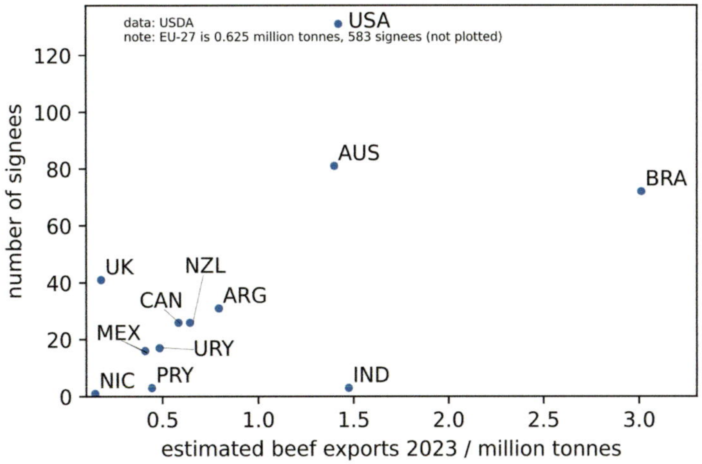

FIGURE 2.7 Relationship between number of signees and national beef exports (million tonnes)

Whilst these figures underline a failure by the organisers of the declaration to garner many signatories from the most populated countries on the planet, India and China, they throw up some interesting relationships. Figure 2.5 underlines that there are generally higher numbers of signatories from wealthier countries (the USA, Germany, the UK, Spain, Australia, France, Brazil, and Italy). Indeed, these eight countries comprise over 57% of all signatories. Figure 2.6 underlines a relationship between higher numbers of signatories in countries where meat consumption is higher (adding Portugal and Argentina). Finally, Figure 2.7 shows a relationship between number of signatories and countries with higher beef exports (the USA, Brazil, Australia, and Argentina). Whilst it could be speculated that the animal science infrastructures of China and India are less integrated into global relationships, overall, these figures underline that generally higher numbers of signatories (and by implication research infrastructure) are present in countries which are more solidly described as 'meat cultures' in terms of consumption and profit from the sector. This adds evidence to the claim that the declaration pertains to economic and cultural bias rather than scientific evidence as purported.

Given the broad claims of the declaration one might have expected a more diverse scientific representation from the signatories. However, for example, in the domain of health and nutrition it is hard to locate many nutritionists, dietitians, or related health professionals in the list of signatories. I was able to identify three such people. However, none of these can be said to represent mainstream scientific views and all have been associated with controversy. These are Bradley Johnston, Carrie Ruxton, and Trudi Deakin. Johnston, in fact an epidemiologist, was the lead author of a controversial study (Johnston et al., 2019) which argued, contra widespread scientific advice, against the limiting of the consumption of unprocessed red meat and processed meat. Johnston was also a co-author on a similar controversial study which contested public health guidelines about sugar consumption (Erickson et al., 2017). Two main concerns have been raised, firstly about Johnston's relationship with the International Life Sciences Institute (ILSI) and secondly about the methodology Johnston has used in his work. In a *New York Times* piece about Johnston ILSI is referred to as '*an industry trade group largely supported by agribusiness, food and pharmaceutical companies*', which '*has long been accused by the World Health Organization and others of trying to undermine public health recommendations to advance the interests of its corporate members*'.[15] ILSI funded the aforementioned sugar study, though not the meat one. Again, we are back in the territory of corporate interests trying to leverage scientific authority to shape dietary guidelines in a political economy where profits take precedence over public health. In the NYT piece, Frank Hu, the chair of the nutrition department at the Harvard T.H. Chan School of Public Health is critical of the methodology used by both the sugar and meat studies because it uses an approach from drug trials which are not applicable to nutrition studies and can be used to discredit a wide suite of public health recommendations.

Ruxton, a UK-based signatory and dietitian with a high media profile, has also courted controversy. Interestingly, these also relate to both sugar and meat. In 2017, it was reported that Ruxton was an advisor to chocolate company Ferrero and had advised against a sugar tax policy to Food Standards Scotland, a public body of which she was a board member.[16] In 2019, Ruxton again hit the headlines in relation to her membership of the Meat Advisory Council, a body funded by the meat industry. She was criticised after disagreeing with a WHO scientist who had called for the removal of nitrites from processed meat.[17] Trudi Deakin, meanwhile, another UK signatory, is a dietitian who goes against prevailing dietary advice to advocate for diets low in carbohydrate and high in saturated fat. It is fair to say that the paucity of nutritional expertise on the DD signatory list and the inclusion of individuals with a history of controversial practices and views fail to lend it scientific credibility.

The declaration was also criticised by Unearthed (part of Greenpeace) and reported in *The Guardian* in October 2023 drawing attention to the industry links of key signatories. The investigation also underlined the extreme views of key instigators of the declaration. For example, Peer Ederer claimed on his social media that veganism was 'an eating disorder requiring psychological treatment' and Frédéric Leroy had referred to plant-based meat alternatives as 'dietary comedy'.[18] It was also revealed that Ederer had given a talk in 2020 for the Global Warming Policy Foundation, a notorious British think tank engaged in climate denial. The Unearthed investigation also uncovered a wide range of industry and lobbying links with the Declaration and made clear how its suspect claims were being used to influence EU agricultural policy at the highest level. Public relations for the declaration were handled by PR companies with a track record of working with the meat industry. Leroy's attempts at influence also extend to the UN Food and Agricultural Organisation (FAO), a body which has a longstanding and complex relationship with the animal agricultural industries. Leroy serves on the FAO/COAG (Committee on Agriculture) Sub-Committee on Livestock and was also on the advisory committee for the FAO report 'Contribution of terrestrial animal source food to healthy diets for improved nutrition and health outcomes' (FAO 2023) which presented a strong pro-ASF message. The declaration and this broader context would seem to indicate an articulated strategy that stretches across industry and academia to fend off growing scientific and cultural voices critical of ASF consumption and to thwart transformational change towards plant-based diets. It is notable that declaration instigators Leroy and Ederer are active on social media given that the investigations into ASF lobbying groups demonstrate they have been very strategically active on social media to push their message. Much of this social media activity is specifically against plant-based diets rather than pro-ASFs. Indeed in 2023, Leroy was proud to tweet that the Global Data Social Media Analytics Platform had found that he was the second top social media 'influencer' on plant-based diets.[19]

The Broader Context of Backlash

Online ASF activism has been analysed in terms of the attempt to quash EAT-Lancet's message of meat reduction (Garcia et al., 2019) and an exposé of beef lobbying activities online (Fassler, 2023). Incidentally, Garcia et al. (2019) identified main DD instigator Frederic Leroy as playing an important and early role in the online #yes2meat campaign to discredit EAT-Lancet. Also, relevant has been the increasing attention that meat and dairy lobbyists have given to the COP process, turning out in record numbers at COP28 in December 2023.[20] To an extent the push-back of the DD is an acceleration of a process which began after the FAO published *Livestock's Long Shadow* in 2006 (Steinfeld et al., 2006). That report, which critically underlined the environmental impacts of animal agriculture, broke a longstanding relationship between the FAO and the meat industry,[21] and ever since industry players have been trying to get the FAO back on side.

One way in which this has been achieved has been through the FAO LEAP (Livestock Environmental Assessment and Performance) partnership which has welcomed partners from the private sector, including the International Meat Secretariat, the International Dairy Federation, the International Poultry Council, and the International Egg commission. The LEAP steering committee also has national country partners which consist of many of those major meat exporting nations listed above. LEAP has been an important instrument for embedding the FAO's pro-business ideology of seeking technical efficiency improvements to 'livestock' production, rather than advocating for reducing animal production.[22] It is revealing that subsequent FAO estimations for the contribution of animal agriculture to greenhouse gas emissions have followed a downward trajectory since the 18% of Livestock's Long Shadow (Steinfeld et al., 2006, see Twine, 2024). In 2013, FAO LEAP was chaired by Frank Mitloehner, a signatory of the DD. Furthermore, he is an animal production scientist who has long tried to downplay the environmental impact of the A-IC and has been questioned over his industry ties (Morris & Jacquet, 2024). The work of FAO LEAP is neatly packaged in the language of sustainability and even within the sustainable development goals of the overall United Nations. Such 'greenwashing' makes the FAO look like a willing enabler of neoliberal approaches to regulation in which the private sector is invited to regulate itself. The longstanding relationship between the FAO and the A-IC might preclude conclusions that industry and academic actors have 'captured' the FAO, but it appears that the close relationship has been successfully re-established during the past 18 years. This is also reminiscent of the way that national and fossil fuel interests have been able to successfully shape and infiltrate the IPCC and COP process.

In late 2023, the FAO was criticised by an investigation in *The Guardian* (Neslen, 2023a; 2023b). This was an important contribution as it interviewed authors of Livestock's Long Shadow (Steinfeld et al., 2006), revealing the backlash to the report. Authors showed that in the aftermath of its publication they were '*censored, sabotaged, undermined and victimised*' (Neslen, 2023a). Furthermore,

'*The big meat-producing countries – Brazil, Argentina, Uruguay and Paraguay, Australia, and the US – all complained to the FAO's higher echelons, according to Steinfeld, while protests also flooded in from "the private sector, the large-scale meat, feed and dairy producers"*'. The investigation also quotes Steinfeld, lead author of Livestock's Long Shadow, as recalling '*being told by a senior official in the (FAO) director general's office: "Even if livestock contributes 18% to climate change, the FAO shall not say that. It's not in the interest of the FAO to highlight environmental impacts"*'. Another part of the investigation found that the work of some scientists examining the methane emissions of animal agriculture was censored and that work was suppressed. Specifically,

> Serving and former FAO experts said that between 2006 and 2019, management made numerous attempts to suppress investigations into the cow/climate change connection. Top officials rewrote and diluted key passages in another report on the same topic, "buried" another paper critical of big agriculture, excluded critical officials from meetings and summits, and briefed against their work.
>
> *(Neslen, 2023b)*

These revelations highlight the highly politicised context in which scientists have tried to do their work. They show that the economic interests of the A-IC (be that industry, government, or complicit academia) have fought hard to close down criticism and analyses of the environmental impacts of animal agriculture; and that subsequently they have enrolled 'their' animal/meat scientists to try and set an agenda which downplays that impact. The DD should be understood within this broader context of the last 18 years of pushback and backlash.

This pushback has been about deflecting changes to the status quo for the purposes of economic protectionism and has been documented in the meat industry before (Clare et al., 2022). This protectionism includes not only the protection of capital, but the sectors (such as the meat/animal sciences) that have grown around the historical development of the A-IC. In a deeper sense this is also about protecting the ideology of anthropocentrism which is an important part of the classed and gendered character of meat cultures. The protection of this ideology not only acts to sustain a conservative worldview of a strict and unscientific human-animal hierarchical dualism but acts also as a proxy for the protection of a dispassionate industrial masculinity (Hultman, 2017) and a food system that funnels low quality, unhealthy food into the societal poor.

The declaration affords a new visibility to an aspect of the A-IC that invites further analysis. With visibility potentially comes vulnerability. For scientists not shaped by self-interest or siloed in-group norms, but by ecological public health and the centring of a transformation in human-animal relations as a necessary approach to the overlapping climate and biodiversity crises (Twine, 2024), there is a need to expose these profit-driven knowledge claims. Furthermore, we must question the trustworthiness of these claims and the idea that 'sustainability' can really be married to an anthropocentric ideology. Social institutions (such as schools and hospitals) should be alerted to the push back strategies outlined in this chapter. Instead, we can point to the scientific consensus that has emerged around the benefits of meat/dairy reduction for ecological public health, and to the emerging scientific evidence base for plant-based diets, not just as an alternative diet but one that can promote *better* human health outcomes, help to address the climate and biodiversity crises, and bring about a more ethical (caring and just) set of relations with other species.

The DD got one thing absolutely right. ASF consumption has become deeply embedded in our cultures. However, given the consequences of this, it is an argument for transition not the status quo. For a just transition to take place Universities should now consider a *gradual* defunding of the animal production sciences as part of their social and environmental responsibility remits. Planning should be put in place for how this community of scientists and their skill sets can be redeployed into other areas of research. Universities should lead on the plant-based dietary transition (Krattenmacher et al., 2023) rather than entering into relationships with the private sector that further lock societies into unsustainable and high greenhouse gas emitting futures. Universities should lead conversations and practices for moving towards post-anthropocentric societies.

Notes

1 I defer to Calarco's comprehensive definition of anthropocentrism as 'the view that human beings (in opposition to animals and other nonhuman beings) are of supreme importance in ethical, political, legal, and existential matters. Among the primary characteristics of anthropocentrism are: (1) a narcissistic focus on human exceptionalism; (2) a binary account of human–animal differences; (3) a strong moral hierarchy that ranks human beings over animals and other nonhuman beings; (4) a tendency to de- and sub-humanize certain populations; and (5) institutions that aim to protect and give privilege to beings deemed fully human' (2021, p. 18).
2 There were 1190 signatories at the time of writing (January 11, 2024). However, 6 of these are double entries so the correct number is 1184.
3 https://www.dublin-declaration.org/
4 The British Dietetic Association (2017); Australia's National Health and Medical Research Council (2013); American Dietetic Association (2009). All have recognised a vegan diet as healthy. Specification within the category of a 'plant-based diet' is important for optimal health and all healthy eating habits require degrees of planning. For a repository of nutrition studies related to both plant-based diets and ASF consumption, see https://docs.google.com/document/d/1viaSP-WCOrneuKE6lH42cgaWB4vz9QKQ7RqK29303FU/edit compiled by Dr. Tushar Mehta, MD.
5 https://aleph-2020.blogspot.com/2023/11/opinion-piece-activist-tactics-and.html
6 Figure 2.2 was created by Romain Espinosa, my co-author in Krattenmacher et al. 2024. The six individuals in the centre refer to Frederic Leroy, Peer Ederer, Collette Kaster, Mohammed Koohmaraie, Rod Polkinghorne, and Declan Troy. See https://www.dublin-declaration.org/authorship
7 Leroy & Ederer (2023).
8 Ederer & Leroy (2023).
9 Sustaining Partners. American Meat Science Association (2022). https://meatscience.org/membership/sustaining-partners

10 Authorship. The Dublin Declaration. (2023). https://dublin-declaration.org/authorship
11 Our Rules. International Meat Research Association 3G Foundation. (2023). https://imr3g.org/our-rules/
12 This is an elaboration of the original use of the term by Brennan (1993).
13 http://tinyurl.com/32snfcws
14 It is unclear if many important centres for animal science globally provided no signatories, but readers should especially bear in mind that China and India only had three signatories apiece.
15 Parker-Pope & O'Connor (2019), see https://www.nytimes.com/2019/10/04/well/eat/scientist-who-discredited-meat-guidelines-didnt-report-past-food-industry-ties.html. For more criticisms of ILSI, see Jacobs (2019). https://www.nytimes.com/2019/09/16/health/ilsi-food-policy-india-brazil-china.html
16 Hutcheon (2017). https://www.heraldscotland.com/news/15314027.food-watchdog-member-linked-confectionary-industry-advised-sugar-tax/
17 Hutcheon (2019). https://www.heraldscotland.com/news/17455283.food-standards-board-member-conflict-interest-row-processed-meat/
18 Carrington (2023). https://www.theguardian.com/environment/2023/oct/27/revealed-industry-figures-declaration-scientists-backing-meat-eating. See also https://unearthed.greenpeace.org/2023/10/27/dublin-declaration-meat-livestock-industry/
19 https://x.com/fleroy1974/status/1666383739306360834?s=20
20 Sherrington et al. (2023, December 9). Big meat and dairy lobbyists turn out in record numbers at Cop28. https://www.theguardian.com/environment/2023/dec/09/big-meat-dairy-lobbyists-turn-out-record-numbers-cop28
21 For example, the aforementioned World Association of Animal Production (WAAP) has had a relationship with the FAO since the 1960s.
22 This was even the favoured framing within *Livestock's Long Shadow* (Steinfeld et al. 2006). For more on this efficiency framing, see Twine (2010, 2024).

References

American Dietetic Association. (2009). Position of the American Dietetic Association: Vegetarian diets. *Journal of the American Dietetic Association, 109*(7), 1266–1282. https://doi.org/10.1016/j.jada.2009.05.027.

British Dietetic Association. (2017). British Dietetic Association confirms well-planned vegan diets can support healthy living in people of all ages. https://tinyurl.com/yhhv2zbf.

Calarco, M. R. (2021). *Animal studies: The key concepts*. Routledge.

Clare, K., Maani, N., & Milner, J. (2022). Meat, money and messaging: How the environmental and health harms of red and processed meat consumption are framed by the meat industry. *Food Policy,* 109. https://doi.org/10.1016/j.foodpol.2022.102234.

Dhont, K., & Hodson, G. (2014). Why do right-wing adherents engage in more animal exploitation and meat consumption? *Personality and Individual Differences*, *64*, 12–17. https://doi.org/10.1016/j.paid.2014.02.002.

Ederer, P., & Leroy, F. (2023). The societal role of meat: What the science says. *Animal Frontiers*, *13*, 3–8. https://doi.org/10.1093/af/vfac098.

Erickson, J., Sadeghirad, B., Lytvyn, L., Slavin, J., & Johnston, B. C. (2017). The scientific basis of guideline recommendations on sugar intake. *Annals of Internal Medicine*, *166*(4), 257–267. https://doi.org/10.7326/M16-2020.

FAO. (2023). *Contribution of terrestrial animal source food to healthy diets for improved nutrition and health outcomes – An evidence and policy overview on the state of knowledge and gaps*. FAO. https://doi.org/10.4060/cc3912en.

Fassler, J. (2023, May 3). Inside big beef's climate messaging machine: confuse, defend and downplay. *The Guardian*. https://www.theguardian.com/environment/2023/may/03/beef-industry-public-relations-messaging-machine.

Garcia, D., Galaz, V., & Daume, S. (2019). EATLancet vs Yes2meat: The digital backlash to the planetary health diet. *Lancet*, *394*(10215), 2153–2154. https://doi.org/10.1016/S0140-6736(19)32526-7.

Herzon, I., Mazac, R., Erkkola, M., Garnett, T., Hansson, H., Kaljonen, M., Kortetmäki, T., Lonkila, A., Jonell, M., Niva, M., Pajari, A. -M., Tribaldos, T., Toivonen, M., Tuomisto, H.L., Koppelmäki, K., & Röös, E. (2023). A rebalanced discussion of the roles of livestock in society. *Nature Food*, *4*, 926–927. https://doi.org/10.1038/s43016-023-00866-y.

Hultman, M. (2017). Exploring industrial, ecomodern, and ecological masculinities. In S. MacGregor (Ed.), *Routledge handbook of gender and environment* (pp. 239–252). Routledge.

Hutcheon, P. (2017, May 28). Food watchdog member linked to confectionary industry advised against sugar tax. *The Herald*. https://www.heraldscotland.com/news/15314027.food-watchdog-member-linked-confectionary-industry-advised-sugar-tax/.

Hutcheon, P. (2019, February 24). Food standards board member in conflict-of-interest row over processed meat. *The Herald*. https://www.heraldscotland.com/news/17455283.food-standards-board-member-conflict-interest-row-processed-meat/.

Jacobs, A. (2019, September 16). A shadowy industry group shapes food policy around the world. *New York Times*. https://www.nytimes.com/2019/09/16/health/ilsi-food-policy-india-brazil-china.html.

Johnston, B. C., Zeraatkar, D., Han, M. A., Vernooij, R., Valli, M., El Dib, C., Marshall, R., Stover, C., Fairweather-Taitt, P. J., Wójcik, S., Bhatia, G., de Souza, F., Brotons, R., Meerpohl, C., Patel, J. J., Djulbegovic, C. J., Alonso-Coello, B., Bala, P., & Guyatt, M. M. (2019). Unprocessed red meat and processed meat consumption: Dietary guideline recommendations from the nutritional recommendations (NutriRECS) consortium. *Annals of Internal Medicine*, *171*(10), 756–764. https://doi.org/10.7326/M19-1621.

Krattenmacher, J., Casal, P., Dutkiewicz, J., Huchard, E., Sanders, E., Treich, N., Wadiwel, D., Williams, A., Bègue, L., Cardilini, A. P. A., Dhont, K., Dugnoille, J., Espinosa, R., Gagliano, M., Lairon, D., Maheta, M., Mendez, L., Nowicki, P., Quinn, T. P., & …Twine, R. (2023). Universities should lead on the plant-based dietary transition. *The Lancet Planetary Health*, *7*(5), e354–e355. https://doi.org/10.1016/S2542-5196(23)00082-7.

Krattenmacher, J., Espinosa, R., Sanders, E., Twine, R., & Ripple, W. (forthcoming 2024). The Dublin declaration: Gain for the meat industry, loss for science. *Environmental Science and Policy.*

Legg, T., Hatchard, J., & Gilmore, A. B. (2021). The Science for Profit Model—How and why corporations influence science and the use of science in policy and practice. *PLoS ONE*, *16*(6), e0253272. https://doi.org/10.1371/journal.pone.0253272.

Leroy, F., Abraini, F., Beal, T., Dominguez-Salas, P., Gregorini, P., Manzano, P., Rowntree, J., & van Vliet, S. (2022). Animal board invited review: Animal source foods in healthy, sustainable, and ethical diets – An argument against drastic limitation of livestock in the food system. *Animal*, *16*(3), 100457. https://doi.org/10.1016/j.animal.2022.100457.

Leroy, F., & Ederer, P. (2023). The Dublin declaration of scientists on the societal role of livestock. *Nature Food*, *4*, 438–439. https://doi.org/10.1038/s43016-023-00784-z.

Leroy, F., Heinrich, F., Lee, M. R. F., & Willems, K. (2023). Meat matters - making the case for a valuable food in a hostile environment. *Italian Journal of Animal Science*, *22*(1), 885–897. https://doi.org/10.1080/1828051X.2023.2221696.

Moore, J. W. (2015). *Capitalism in the web of life – Ecology and the accumulation of capital*. London: Verso.

Morris, V., & Jacquet, J. (2024). The animal agriculture industry, US universities, and the obstruction of climate understanding and policy. *Climatic Change*, *177*, 41. https://doi.org/10.1007/s10584-024-03690-w

National Health and Medical Research Council. (2013). *Australian Dietary Guidelines*. NHMRC.

Neslen, A. (2023a, October 20). The anti-livestock people are a pest': how UN food body played down role of farming in climate change. *The Guardian*. https://www.theguardian.com/environment/2023/oct/20/the-anti-livestock-people-are-a-pest-how-un-fao-played-down-role-of-farming-in-climate-change.

Neslen, A. (2023b, October 20). Ex-officials at UN farming body say work on methane emissions was censored. *The Guardian*. https://www.theguardian.com/environment/2023/oct/20/ex-officials-at-un-farming-fao-say-work-on-methane-emissions-was-censored.

Parker-Pope, T., & O'Connor, A. (2019, October 4). Scientist Who Discredited Meat Guidelines Didn't Report Past Food Industry Ties. *New York Times*. https://www.nytimes.com/2019/10/04/well/eat/scientist-who-discredited-meat-guidelines-didnt-report-past-food-industry-ties.html.

Pethick, D. W., Bryden, W. L., Mann, N. J., Masters, D. G., & Lean, I. J. (2023). The societal role of meat: The Dublin declaration with an Australian perspective. *Animal Production Science*, *63*(18), 1805–1826. https://doi.org/10.1071/AN23061.

Plumwood, V. (2002). *Environmental culture - The ecological crisis of reason*. Routledge.

Popkin, B. M. (2009). Reducing meat consumption has multiple benefits for the world's health. *Archives of Internal Medicine*, *169*(6), 543–545. https://doi.org/10.1001/archinternmed.2009.2.

Sherrington, R., Carlile, C., & Healy, H. (2023, December 9). Big meat and dairy lobbyists turn out in record numbers at Cop28. *The Guardian*. https://www.theguardian.com/environment/2023/dec/09/big-meat-dairy-lobbyists-turn-out-record-numbers-cop28.

Steinfeld, H., Gerber, P., Wassenaar, T.D., Castel, V., Rosales, M., & de Haan, C. (2006). *Livestock's long shadow: Environmental issues and options*. Food and Agriculture Organization of the United Nations.

Twine, R. (2010). *Animals as biotechnology: Ethics, sustainability, and critical animal studies*. Routledge.

Twine, R. (2012). Revealing the 'animal-industrial complex' – A concept & method for critical animal studies? *Journal for Critical Animal Studies*, *10*(1), 12–39.

Twine, R. (2020). Where are the nonhuman animals in the sociology of climate change? *Society and Animals*, *31*, 105–130. https://doi.org/10.1163/15685306-BJA10025

Twine, R. (2024). *The climate crisis and other animals*. Sydney University Press.

Unearthed. (2023). Revealed: The livestock consultants behind the Dublin Declaration of scientists. https://unearthed.greenpeace.org/2023/10/27/dublin-declaration-meat-livestock-industry/

3

DISPLAYING COMPASSION TO HIDE HARMS

An Analysis of the Visual Communication Strategies of the Spanish Animal-Industrial Complex[1]

Laura Fernández, Estela M. Díaz, Olatz Aranceta-Reboredo, and Núria Almiron

Industry Reactions to Rising Compassion

We find ourselves in a context of increased compassion towards nonhuman animals (NHAs) and heightened environmental, and food consciousness. There is a widespread social concern for the welfare and care of NHAs (European Commission, Directorate-General for Health and Food Safety, 2016) and a growing compassionate societal response to their suffering, driven by a recognition of their sentience (Panksepp et al., 2012) and their victimisation within the current socio-economic order (Joy, 2013; Ruby, 2012). In this context, distinct animal-based industry responses are evident. For instance, the Interprofessional Organisation for Beef and Veal in Spain (PROVACUNO; the cow flesh lobby), along with the Belgian meat organisation APAQ-W, launched a controversial campaign in 2020, financed by European Union funds. The campaign's main slogan encouraged the population to become "beefatarian" (Campbell, 2020). Paradoxically, the meat lobbies appropriated the concept of "vegetarian" and "vegan" to promote a new identity, no longer guided by animal and/or environmental ethics but by taste preferences and, above all, the economic benefit of their business. This campaign occurred amid a societal shift towards vegetarian and vegan lifestyles (Lantern, 2020; Proveg, 2021). Interestingly, this case is comparable to the more recent multi-million-pound campaign promoted by the UK government-backed Agriculture and Horticulture Development Board (AHDB), which aimed to promote meat consumption by emphasising the supposed health benefits while omitting the connection between meat consumption and climate change (Carlile, 2024). Furthermore, this campaign was targeted at a younger audience and was launched specifically in the same period as the traditional Veganuary campaign, which aims to promote vegetarianism and veganism during the month of January.

DOI: 10.4324/9781003441908-4

Similarly, in 2018, a citizen initiative in Barcelona called ZOOXXI prompted changes in the ordinance for the protection of animals in the city in May 2019 and involved a reconversion of the zoo. The World Association of Zoos and Aquariums (WAZA) responded by relocating its administrative office from Switzerland to Barcelona (WAZA, 2018) and advocated against it. Loro Parque, a zoological park from the Canary Islands, also sent a dossier to the Spanish Parliament to push against animal rights activists' arguments. Likewise, one of Spain's main animal testing pressure groups, the Society for Laboratory Animal Science (SECAL), organised an exhibition at the Elder Museum of Science and Technology in Las Palmas de Gran Canaria, Canary Islands, in 2017. This exhibition conveyed to visitors – particularly younger audiences – the message that animal experimentation is crucial for human health, simultaneously emphasising animal welfare as a core value in laboratory and breeding practices (SECAL, 2019). Additionally, this prominent lobby conducted informative talks in educational centres and organised a storytelling contest about animal experimentation targeted at children aged 8 to 12 (SECAL, 2016).

These NHA-based industry actions occur in the backdrop of the "Animal-Industrial Complex" (A-IC), a term coined by anthropologist Barbara Noske in 1997 and later defined by Richard Twine as "a partly opaque and multiple set of networks and relationships between the corporate sector, governments, and public and private science" (2013, p. 89). The A-IC's opacity is rooted in promoting collective denial about the harms inflicted by animal exploitation on both nonhuman and human animals, achieved through culturally specific discourses and visual imaginaries. This communication strategies align with Johan Galtung's concept of "cultural violence," defined as "those aspects of culture, the symbolic sphere of our existence - exemplified by religion and ideology, language and art, empirical science and formal science (logic, mathematics) – that can be used to justify or legitimise direct or structural violence" (1990, p. 291).

In this context, we contend that the promotional campaigns and educational materials of NHA exploitation lobbies contribute to creating the symbolic framework that facilitates institutionalised psychological and physical violence towards other animals, that is, "behaviour and practices involving physical force or power that adversely affect, hurt, damage and in particular, kill" them (Cudworth, 2015, p. 5). In the case of Spain, animal exploitation is strongly intertwined economically, politically, and culturally from bullfighting being considered of national cultural interest with constitutional protection according to Law 18/2013, of 12 November, to the enormous monetary worth of its animal-based food industries. Spain is the country that consumes most meat in the European Union (Rodríguez Cánovas & Guzmán Rincón, 2023) and the fifth global exporter of it (Millán, 2022). The size of the Spanish A-IC and its cultural specificity play in favour of its protection and perpetuation, with appalling consequences in terms of violence towards NHAs.

In this context, human compassion towards other animals is influenced by powerful economic forces that promote the consumption of products or services involving direct physical and psychological violence towards them. In modern democracies, interest groups have a big influence on both public opinion and political decision-making (Holyoke, 2020). This is also the case for the A-IC, with its think tanks and lobby groups – actors with a vested interest that employ persuasive techniques to uphold public approval of exploiting animals (Almiron, 2016; Spasser, 2013).

Therefore, society is currently faced with a paradox of compassion characterised by the desire to care for NHAs while, at the same time, participating – directly or indirectly – in the systemic apparatus that facilitates and profits from harming them. The A-IC uses its power and influence to limit our freedom to refrain from engaging in harmful actions, impeding the conditions necessary for making morally decent choices (Boyer et al., 2020).

Our research aims to critically explore the A-IC's use of strategic communication, focusing on their use of visual imagery and iconography in disseminating their messages. Specifically, we analyse the visual rhetoric of three significant Spanish industries within the A-IC: the animal-based *food* industry, the animal-based *entertainment in captivity* industry, and the animal-based *experimentation* industry. Our focus is on how these industries navigate the notion of compassion through strategic communication in public relations (Almiron, 2023). This endeavour seeks to challenge the A-IC's vested interests, potentially transform prevailing societal views regarding animal exploitation and to foster attitudes towards compassion and care, thereby instigating a revaluation of established societal perceptions in these matters.

Methodological Approach

In this chapter, we analyse the visual communication used by animal-based interest groups in their advocacy.[2] Our sample consists of 11 interest groups. For the food industry, we selected seven major Spanish interprofessional groups acknowledged by the Spanish Ministry as principal representatives of the meat industry lobbies: AVIANZA[3] (Spanish Interprofessional Association of Poultry Meat), INTERPORC (Spanish Inter-professional Agri-Food Organization for White Pork), ASICI (Interprofessional Association of Iberian Pork), INTEROVIC (Interprofessional Agri Food Industry for Sheep and Goat Meat), INTERPALM (Interprofessional Association of Fatty Palmipeds), PROVACUNO, and INTERCUN (Interprofessional Organisation for the Promotion of the Rabbit Sector). For the entertainment industry, we chose three groups involved in animal-based entertainment in captivity in Spain: AIZA (Iberian Association of Zoos and Aquaria), Parques Reunidos (Reunited Parks), and Loro Parque Fundación. Lastly, representing the animal experimentation industry, we chose the largest Spanish lobby in defence of animal experimentation in Spain: SECAL.

We conducted a visual content analysis of a variety of materials from those selected interest groups, including publicly available promotional images, videos, posters, scientific journals, and educational notebooks aimed at children as available on their websites. The sample selection criteria were designed to achieve representativeness, and the documents for all three industries were collected between June 2022 and December 2022, with the analysis conducted throughout 2023. We have followed an inductive procedure by first gathering images and then selecting our sample. The results presented in this research are based on a sample of images that explicitly show some sign of "care" towards the animal. With this approach, the sample analysed here consisted of 122 images and 11 videos. As for the origin of the images according to the different industries, 20 images are from the food industry; 71 photographs and 10 videos are from the entertainment industry; and 9 photographs and illustrations and one promotional video are from the animal experimentation industry.

Shaping Animal Exploitation Imagery Through a Pseudo-Care Discourse

We used the notion of pseudo-care to describe a prominent visual frame common to all three industries and that manifested itself in industry-specific ways. Pseudo-care contrasts with the concept of "care" used by ecofeminist and non-speciesist care ethics (Donovan & Adams, 2007; Gruen, 2015); grounded in an embodied, contextual, and mindful commitment to the specific needs of NHAs as unique sentient individuals, this view of care conflicts with and deeply challenges the understanding of care as welfare or well-being, a pseudo-care used in the context of industries that use NHAs for economic purposes (Leth-Espensen, 2023). We contend that these images of care are not just superficial or casual representations; they seem carefully crafted constructions that disseminate specific understandings about the power dynamics exerted by human over NHAs. The pseudo-care frame of our sample is presented through two main themes: (1) Pseudo-Care Dynamic; and (2) Pseudo-Care Context.

(1) Pseudo-Care Dynamic: A Mirage of Mutual Understanding

This theme focuses on the identified types of interactions between humans and NHAs as depicted in the materials of the sample. In the realm of visual communication, images are instrumental in shaping our perceptions and understandings. This holds especially true in the context of how different interest groups represent relationships between humans and NHAs in their respective industries. The representation of these interactions is more than an aesthetic choice; it profoundly reflects the underlying narratives and ideologies that these interest groups aim to convey. Specifically, we highlight three sub-themes, referring respectively to the representation of humans, NHAs, and the relationship between them: (1) the mirage of human compassion; (2) the mirage of subjectivity; and (3) the mirage of intimate contact.

The Mirage of Human Compassion

One of our initial findings is that certain characteristics of the human figure within the images replicate typical human codes of care. These features are essential to project a "caring relationship" between humans and NHAs and include gestures and postures, physical proximity and contact, and the actions being performed. Such elements, constant in the analysed images, are critical to unravel the underlying narratives of "care" and "concern" promoted by the lobbies. However, there are variations among the three sectors that reveal certain divergences in their visual communication strategies.

In the animal-based food industry, processes such as quality control and health surveillance are commonly emphasised in images of human-NHA interaction within sanitised environments. More personal and "caring" interactions – such as those of a worker showing affection to animals – are rare. This form of representation forges a narrative of "professional-care" that emphasises animal welfare as a technical matter; hygienic conditions handled by qualified professionals who are equipped with the necessary knowledge and skills. These images promote a view of care that focuses more on physical health and subsistence than on the emotional well-being or natural behaviours of NHAs. Furthermore, such images are often accompanied by text referring to welfare certificates, which presents care as driven by the goal of assuring consumers of quality and "food safety." This approach omits the emotional dimensions of NHAs and limits the connections between humans and NHAs, minimising care to providing basic physiological needs. This style of visual communication, employed by Spanish meat industry lobbies, reveals a sophisticated strategy that shows "care" without departing from underlying economic interests. It raises critical questions about the authenticity of the portrayed welfare narratives and their strategic alignment with societal concerns about animal rights and ethical treatment.

The animal-based entertainment industry emphasises interpersonal connections between humans and NHAs, particularly caretaking bonds. This "trainer-care" is exemplified in scenes featuring workers nurturing newborn mammals or endangered species like pandas and parrots, notably observed in content from AIZA and Loro Parque Fundación. Across various lobbies, emphasis is placed on relationships with "charismatic megafauna" (Vaughan, 2023) like orcas, dolphins, and seals; often depicted through eye contact between animals and workers, these pictures convey values of friendship and mutual understanding. These glimpses into caretakers' work embody industry values and representations of relationships that extend beyond scientific care to include family care and meeting NHAs' affection needs. Contact is portrayed as mutually beneficial and even therapeutic for children, framed as a wish come true. This narrative aligns with altruistic/salvatory care linked to species conservation and human responsibility, emphasised through the representation of captive-born animals as "ambassadors" and extended to zoo visitors as "animal ambassadors." Young people and families appear engaging in interactions with animals in the advertising content of the lobbies – notably in

Parques Reunidos' – and evoke themes of awe, wonder, shared moments of play and apparent joy, reinforcing the leisure aspect of parks.

Our findings in the animal experimentation industry present a contrasting picture that we have identified as an embodiment of "scientific-care." Technicians are usually depicted in close bodily proximity to NHAs within laboratory settings, and strikingly, one of the recurrent features of the images includes technicians directly gazing into the eyes of an NHA. There is also a significant prominence of hands in interaction; technicians are often shown feeding NHAs in pictures that include offering a jelly to a pig, a seed to a mouse, or a piece of apple to a rabbit. This closeness, together with eye contact, suggests a high level of engagement, interaction, and attention. However, the animal experimentation industry aligns with the animal-based food industry in its scientific and sanitising tone. This approach inherently creates a distance between humans and other animals, emphasising basic needs and assistance rather than emotional connection or understanding of the NHAs' natural behaviour. A critical visual element in these depictions is the attire of the technicians, typically white lab coats or green scrubs, complemented by personal protective equipment such as face masks, latex gloves, and caps. This clothing not only serves practical purposes but also symbolises the scientific and clinical nature of the interaction, further reinforcing the professional barrier between human and nonhuman actors.

The Mirage of Animal Subjectivity

The A-IC's lobbies' visual communication presents a distinctive pattern of representation that – regardless of whether NHAs are shown individually or in groups – points to an acknowledgement of their individuality, collectivity, or collectivises them; we have named it the "Mirage of Animal Subjectivity." In visual representations, collectivisation frequently attributes common characteristics to the group, blurring animal individuality. This mechanism is like that observed in the representation of marginalised human groups, such as migrants (Fernández, 2021). Regardless of the approach, however, the authenticity of this recognition is often called into question.

The lobbies of the meat industry, for instance, usually opt for a collectivised representation of NHAs that turns them into a canvas of forms and movements. This approach seems to emphasise their separation from the human species, presenting nonhumans as wild creatures that are part of nature's food chain. This, in turn, serves to blur their existence as conscious, unique, and irreplaceable beings. In vast meadows of idyllic landscapes, sheeps form part of flocks and geese become harmonious ensembles in extensive farming. Even if they are confined to artificial spaces and intensive enclosures, pigs and chickens are integrated into the fabric of natural life in the same way. Thus, rather than recognising them as individuals with their own stories, this collective representation metamorphoses NHAs into homogeneous components of a mass production machine.

In contrast, the lobbies of zoos, aquariums, and animal theme parks usually depict NHAs as individuals or "specimens." This form of representation seems designed to create the image of a genuine connection between humans and NHAs, whether they are workers or visitors. Some of the NHAs are presented as ambassadors of their species and get an individualised description; however, it is necessary to consider that this approach varies between species. For instance, sharks and eagles might be less collectivised than other fishes and birds, and not shown so much as individual ambassadors, but they especially foster a sense of awe and aesthetic enjoyment as individuals because of how they are presented within the "collections." However, this individualisation can be misleading; the marketing ploy on the illusion of bonding diverts attention from the rest of the animal's welfare and, more importantly, might potentially redirect the public's compassion towards a less critical position.

The animal experimentation industry lies somewhere in between these two industries. Here, NHAs are almost always represented individually. However, their presentation is also framed in a utilitarian way that eclipses their true essence. In the laboratory world, each individual becomes a "model" or specimen, an interchangeable entity whose scientific value, usually in terms of human benefit, is prioritised over their welfare, sentience, and consciousness. Moreover, these NHAs are portrayed as willingly offering their lives on the altar of human medical progress. Although the interest group places them as the protagonists of its images, each one becomes "part of the sample studied," a symbol of a living organism representative of its species. Thus, the individual disappears as part of a larger narrative of collective objectification, which is not explicitly stated.

The Mirage of Intimate Contact

Physical contact is a critical aspect of the interaction dynamics – and a key tactic in the pseudo-care strategy – that conveys empathy and concern for animals. We consider that the type of interaction and intimate contact between humans and NHAs often resembles a "fetishisation of animal contact" (Molloy, 2011). This visual narrative is most prominent in the animal-based entertainment and experimentation interest groups.

Unlike other industries, in zoos, aquariums, and animal theme parks the body contact extends to the public. The photographs and videos in our sample depict direct contact in the form of petting marine and terrestrial animals, facilitated either by trainers or other workers of the industry. These interactions are inherently linked to the opportunities for physical touch offered at these parks; regardless of the physical barriers, children and members of vulnerable communities are often shown swimming with seals, holding birds, petting fishes, starfishes, and feeding giraffes. In aquariums, a more indirect type of interaction predominates with workers pointing to the animals behind the glass of the tanks in an attempt to establish a form of "symbolic-contact." Two photographs from our collection illustrate this point: one with a manatee and one with a shark. Despite the physical barrier of the glass, these images are intended to convey a sense of closeness and emphasise otherness and bodily difference, suggesting proximity within a safe environment.

Physical contact between humans and nonhumans occurs quite frequently in experimentation; we found examples in SECAL's promotional video, on posters, in photographs in its scientific journal and even in illustrations aimed at children's audiences. Notably, the image of human hands with white gloves holding a mouse, either in profile or looking directly at the camera, could be considered paradigmatic of pseudo-care within this industry. It is interesting to note that this specific approach is exclusive to one of the most widely depicted and symbolically significant species in animal experimentation; probably, due to its status as one of the most heavily used species, as evidenced by its presence in the SECAL logo.

Lastly, in our food industry sample, there are only a few occasions in which contact depictions appear, and they are always concerning baby NHAs. One example is a photograph showing a farmer holding two baby ducks, one in each hand, with a "friendly" smile. In the same line, a squatting, uniformed farmer holds a piglet on his knees, gently stroking them while surrounded by a group of twenty other healthy-looking, clean piglets in a covered but barrier-free space. Finally, a black and white photograph of an older man next to a mother sheep and her calf stands out. His kneeling posture and attire are not only intended to convey his know-how, culture, and tradition but also to evoke tenderness and care, by reflecting the human connection with the calf and its mother. However, despite the potential emotionality of these images, they are presented in a frame that commodifies the animals' bodies: turned into products to be proudly displayed in front of the camera. Moreover, symbolically, these representations also seem to convey a message of absolute control over the entire product manufacturing process.

(2) The Context of Pseudo-Care: A Mirage of Animal Welfare

In this second theme of our analysis, we focus on the scenery behind the subjects: the physical spaces and environments in which NHAs and humans appear. Here we pay special attention to absences: the suppression or omission of visual elements as a strategy to build sympathetic social imaginaries around animal "care" in the A-IC. Specifically, we highlight two sub-themes: (1) the mirage of freedom and (2) the mirage of non-violence.

The Mirage of Freedom

To present their facilities, a common tactic of A-IC lobbies is to resort to the idealisation of spaces. This strategy – often involving "humane-washing" and "greenwashing" techniques – tends to disguise the actual settings in which the animals are confined. Sometimes, the images show green enclosures that, resembling naturalised spaces, dilute the real living conditions. Even erasing the idea of captivity, they create the illusion that NHAs enjoy a fulfilling life. Other times, the images are composed in such a way that it is difficult or impossible to identify the spatial location of the NHAs. This type of visual representation places animals in a "spatial limbo" or "non-place," usually on a monochrome and neutral background (e.g., white, or grey), that decontextualises and obscures their reality. Both practices misinform and perpetuate the idea that most NHAs live in good conditions; however, this does not correspond to the reality of their daily existence. Here again, we observe certain differences in the approach of the industries. The ambiguity of the locations is a powerful tool to mitigate the cognitive dissonance the audience might feel when contemplating the true nature – or the whole picture – of those spaces.

The animal-based food industry uses photography to soften the perception of its operations. From farms that appear immaculate and spacious to pastures that evoke the image of pastoral liberty (Holmberg, 2022), these visual depictions often omit any sign of confinement, restraint, or automation – which could challenge the welfare image that the industry seeks to convey. Through these images, the following idea is promoted: NHAs live in harmony with their environment, enjoying an existence that satisfies all their natural and social needs. This depiction is a far-removed version of the conditions typical of large-scale production.

The animal-based experimentation industry oscillates between transparency and concealment. The laboratory prevails as the primary setting, but images are often taken in "playrooms" or neutrally coloured spaces designed to convey an image of clean, controlled science. Informational posters resort to decontextualised images of animals, often opting for stock images showing animals outdoors or against backgrounds that eliminate any association with experimentation. This choice of imagery sanitises the reality of the laboratory, a space where NHAs live in confinement and are subjected to invasive procedures. The laboratory environment is rarely shown in its raw form, thus maintaining the curtain between scientific practice and the conditions experienced by the subjects of study.

As for animal-based entertainment, most of the images use colours and compositions that give the impression of "naturalised" spaces. Carefully selected compositions of zooms and close-ups often obscure the actual captive environment; at the same time, the use of vivid colours creates an environment that simulates their free and natural habitat. However, this visual approach often masks the restrictive enclosures and artificially constructed environments. Close-ups that capture an individual animal's gaze into the camera often blur wires, obscure concrete, conceal glass reflections, or even human presence unless you look closely. It is sometimes unclear whether the images depict individuals in captivity, in semi-freedom, in nature or are simply stock images. Interestingly, some of the children-oriented educational content shows NHAs decontextualised; neither the image nor the text accompanying it places the individual within a group, habitat, or ecosystem.

The Mirage of Non-Violence

A notorious omission in the visual representations of Spanish A-IC's interest groups is violence – or the obvious signs of it. The absence of NHAs in diseased states is particularly notable, rendering invisible the realities of their physical and psychological suffering to the public. For example, utensils and devices designed to manipulate, control, and even kill NHAs are rarely displayed. The only exception is when these tools are presented in the aseptic light of sanitary or research uses.

Likewise, bodily harm resulting from confinement or the invasive and painful procedures of experimentation, intensive breeding, and animal training – such as wounds, mutilations, or tumours – are systematically missing from these images. The only instances of suffering depicted in the content are the ones showing animals in nature that face anthropogenic harms. In other words, industry-specific damages are erased from the visual narrative, and, in an act of deliberate concealment, they present a watered-down version of reality.

Moreover, there is a hyper-representation of births and offspring in the animal-based entertainment and food industries. This choice of images contrasts dramatically with the scarce representation or total absence of death and the act of killing. Slaughterhouses are also absent in the group of images selected following the "care" representations. However, when dead bodies of farmed animals appear in the photographs of the animal-based food Spanish interest groups, they are sufficiently dissociated from the individual animal that once inhabited them. The dead bodies that are displayed show no skin, no face, and no blood, only the body with the muscle tissue that will later be chopped up and commodified as meat. This discrepancy is particularly striking; the life cycle of NHAs appears to be incomplete or distorted in these contexts, presenting only the beginning of life and obviating its end, which is, however, a constant in these industries. As an illustration, there is a tender image of a lioness with her cubs that, accompanied by a text on population management, is used to argue that managing "excedents" is inevitable for proper welfare.

Conclusion: Displaying Compassion to Hide Harms

Our research reveals that Spanish A-IC's interest groups employ pseudo-care narratives in their visual communication, therefore perpetuating a mirage of mutual understanding and animal welfare that conceals the different realities of NHA exploitation.

Presenting a softened, sanitised, and idealised vision of NHA exploitation, the set of visual narratives of these interest groups illustrates the "iconographies of oppression" (Cronin & Kramer, 2018). These "iconographies of oppression" present in marketing strategies of the A-IC business favour the progressive invisibilisation of animals in a chain of commodification of their bodies that occurs at the different stages of NHA exploitation, from breeding, confinement, or the invasive alteration of their bodies in the case of animal experimentation, to the packaging and labelling of certain animal-sourced food products (Capilé et al., 2021). We find a speciesist hierarchisation where some animal species are underrepresented and others overrepresented, such as the so-called "charismatic megafauna" (Vaughan, 2023) and the "good animals" that "have high moral status because they willingly accept their subordinate place in society, with some able to enjoy their niche while others only dutifully comply with it" (Arluke & Sanders, 1996, p. 169). We consider that, as Linné identified for the Swedish dairy industry (Linné, 2016), this strategy could be due to the potential exploitation of the emotional labour of NHAs for capital accumulation of these industries. However, as the case of the mouse and the animal experimentation lobby illustrate, their representative and even "iconic" status does not provide them a better life within the industry.

We also identified a simultaneous appeal to traditional modes of production and an emphasis on the idea of the progress of technology. A pastoral narrative of care and bucolic representation (Holmberg, 2022), natural spaces, and rurality contrast with constant improvement and milestones in ways of taking care of animals and the science of managing them (Almiron et al., 2024). Sometimes these are combined, appealing to an ethical and improved way of doing things that maintains its original spirit; other times, these images seem contradictory.

The A-IC interest groups find in visual communication a powerful tool for appropriating the imaginary of care. We argue that they mislead audiences between the type of care they can really offer as a business and a more ethical, non-speciesist care of individuals. The first one, the "care through regulation," refers to "the way in which the state governs human-animal relations by means of monitoring care to ensure 'well-cared-for' animals" (Leth-Espensen, 2023, p. 39); the second one would imply the end of their exploitation and use, as embodied by the anti-speciesist ecofeminist tradition (Donovan & Adams, 2007; Gruen, 2015) and the care promoted by animal sanctuaries (Leth-Espensen, 2023). Moreover, we consider that, through the pseudo-care visual frame, industries generate harm to NHAs through cultural violence (Galtung, 1990) and harm human society through various mirages; these produce concealment and confusion that limit collective compassion towards other animals.

The economic and political influence of interest groups' visual discourses can distort NHA suffering and oppression and, simultaneously, reinforce already existing systematic barriers to recognising and restoring injustices done to NHAs (Boyer et al., 2020). The Spanish A-IC lobbies have employed an "ethics-washing strategy" (Breakey, 2023) that, applied to our study, is characterised by strategic visual communication that instrumentalises public compassion to justify the exploitation of NHAs, all while commodifying them under the guise of NHA care and welfare. Ethics-washing is strategically used to deceive the government and the public into believing that the industry has greater support than it does and that it takes its social responsibilities more seriously than it does.

From our analysis of the Spanish A-IC interest groups' visual communication, we infer that they strategically use visual rhetoric that distorts the harm inherent in the captivity and exposition, farming and experimentation carried out by these industries. These communication strategies can be contested from the critical public relations field and the critical animal and media studies fields, that are useful frameworks to detect in which cases the interest groups' persuasion follows ethical principles within the public relations and advertising practices or, alternatively, provide critical inquiry to question the lack of it (Almiron & Fernández, 2021). The Spanish A-IC lobbies aim to display an image of care that is a mirage on several levels: a mirage of human compassion, NHA subjectivity, and intimate contact – that is, a mirage of mutual understanding between human and NHAs that distorts human supremacy and NHA exploitation as consented – and a mirage of freedom and non-violence – one of animal welfare, which creates confusion and doubt about living conditions, deprivation of freedom and the exercise of physical and psychological violence over nonhuman individuals trapped in the A-IC.

Funding

This work is part of the COMPASS Research Project (Lobbying and Compassion: Interest Groups, Discourse and Nonhuman Animals in Spain), PID2020-118926RB-100 funded by MICIU/AEI/10.13039/501100011033.

The grant for predoctoral contracts for the training of doctors PRE2021-099988, supported by the Ministry of Science, Innovation and Universities (Spain), and funded by MCIN/AEI/10.13039/501100011033 and the European Social Fund Plus (ESF+).

The Juan de la Cierva Postdoctoral contract is part of the grant FJC2021-046861-1, funded by MCIN/AEI/10.13039/501100011033 and by the European Union NextGenerationEU/PRTR.

Notes

1 A previous version of this chapter was presented as a panel at the conference *The Factual Animal: Audiovisual Representations of Real Other-than-Human Animals* that took place at University of Valencia, November 29–December 1, 2023.

2 This work is part of the COMPASS Research Project (Lobbying and Compassion: Interest Groups, Discourse and Nonhuman Animals in Spain), PID2020-118926RB-100 funded by MICIU/AEI/ 10.13039/501100011033.

3 If the name of the organisation was not publicly available in English, the translations are provided by the authors. The following are the acronyms and full names in Spanish of the organisations mentioned: AVIANZA (Asociación Interprofesional Española de Carne Avícola), INTERPORC (Organización Interprofesional Agroalimentaria del Porcino de Capa Blanca), ASICI (Asociación Interprofesional del Cerdo Ibérico), INTEROVIC (Interprofesional Agroalimentaria del Ovino y el Caprino de Carne), INTERPALM (Asociación Interprofesional de las Palmípedas Grasas), PROVACUNO (Organización Interprofesional Agraria de Vacuno de Carne), INTERCUN (Organización Interprofesional para Impulsar el Sector Cunícola), AIZA (Asociación Ibérica de Zoos y Acuarios), Parques Reunidos, Loro Parque Fundación and SECAL (Sociedad Española para las Ciencias del Animal de Laboratorio).

References

Almiron, N. (2016). Beyond anthropocentrism: Critical animal studies and the political economy of communication. *Political Economy of Communication*, *4*(2), 54–72.
Almiron, N. (Ed.). (2023). *Animal suffering and public relations: The ethics of persuasion in the animal-industrial complex* (1st ed.). Routledge. https://doi.org/10.4324/9781003324065.
Almiron, N., & Fernández, L. (2021). Including the animal standpoint in critical public relations research. *Public Relations Inquiry*, *10*(2), 139–161.
Almiron, N., Fernández, L., & Rodrigo-Alsina, M. (2024). Illusory authenticity: Negotiating compassion in animal experimentation discourse. *Discourse Studies*, *26(2),* 153–172. https://doi.org/10.1177/14614456231219633.
Arluke, A., & Sanders, C. R. (1996). *Regarding animals*. Temple University Press.
Boyer, K., Scotton, G., & Wayne, K. (2020). Beyond complicity and denial: Nonhuman animal advocacy and the right to live justly. In G. Garmendia da Trindade & A. Woodhall (Eds.), *Intervention or protest. Acting for nonhuman animals* (pp. 159–182). Vernon Press.
Breakey, H. (2023). The social licence to operate: Activist weapon, industry shield, empty buzzword, or vital ethical tool? In H. Breakey (Ed.), *Social licence and ethical practice* (pp. 1–10). Emerald Publishing Limited. https://doi.org/10.1108/S1529-209620230000027001.
Campbell, M. (2020, November 25). 'Become a beefatarian' says controversial EU-funded Red Meat Campaign. *Euronews*. https://www.euronews.com/green/2020/11/25/become-a-beefatarian-says-controversial-eu-funded-red-meat-campaign.
Capilé, K., Parkinson, C., Twine, R., Kovalski, E. L., & Paixão, R. L. (2021). Exploring the representation of cows on dairy product packaging in Brazil and the United Kingdom. *Sustainability*, *13*, 8418.
Carlile, C. (2024, January 19). Government-Backed Meat Ad Campaign Targets Gen Z in Veganuary. DeSmog. https://www.desmog.com/2024/01/19/government-backed-meat-promo-campaign-targets-gen-z-over-veganuary//.
Cronin, J. K., & Kramer, L. A. (2018). Challenging the iconography of oppression in marketing: Confronting speciesism through art and visual culture. *Journal of Animal Ethics*, *8*(1), 80–92. https://doi.org/10.5406/janimalethics.8.1.0080.
Cudworth, E. (2015). Killing animals: Sociology, species relations and institutionalized violence. *Sociological Review*, *63*(1), 1–18. https://doi.org/10.1111/1467-954X.12222.
Donovan, J., & Adams, C. J. (Eds.). (2007). *The feminist care tradition in animal ethics*. Columbia University Press.
European Commission, Directorate-General for Health and Food Safety. (2016). *Attitudes of Europeans towards animal welfare* – Report, European Commission. Retrieved January 8, 2024, from https://data.europa.eu/doi/10.2875/884639.
Fernández, L. (2021). Seeking a place to live. Visual representations of human and anymal migrants in images from the international daily press. In N. Khazaal & N. Almiron (Eds.), *"Like an Animal". Critical animal studies approaches to borders, displacement, and othering* (pp. 341–385). Brill. https://doi.org/10.1163/9789004440654_012.
Galtung, J. (1990). Cultural violence. *Journal of Peace Research*, *27*(3), 291–305.
Gruen, L. (2015). *Entangled empathy: An alternative ethic for our relationship to animals*. Lantern Books.
Holmberg, M. (2022). Constructing captive ecology at the aquarium: Hierarchy, care, violence, and the limits of control. *Environment and Planning E: Nature and Space*, *5*(2), 861–880. https://doi.org/10.1177/25148486211014508.
Holyoke, T. T. (2020). *Interest groups and lobbying: Pursuing political interests in America* (2nd ed.). Routledge. https://doi.org/10.4324/9781003041795.
Joy, M. (2013). *Por qué amamos a los perros, nos comemos a los cerdos y nos vestimos con las vacas: una introducción al carnismo* (1st ed.). Plaza y Valdés.

Lantern. (2020). *The green revolution 2020. Entendiendo el boom de los 'vegetarianos'*. https://www.lantern.es/papers/the-green-revolution-entendiendo-el-auge-del-mundo-veggie.

Leth-Espensen, M. (2023). *Animals and the politics of suffering: Essays on law, care and interspecies relations*. [Doctoral dissertation, Lund University]. https://lucris.lub.lu.se/ws/portalfiles/portal/146374393/e_spik_ex_Marie.pdf.

Linné, T. (2016). Cows on Facebook and Instagram: Interspecies intimacy in the social media spaces of the Swedish Dairy Industry. *Television & New Media*, *17*(8), 719–733. https://doi.org/10.1177/1527476416653811.

Millán, J. (2022, January 7). España es el quinto exportador mundial de carne y el país que más cerdo vende en el extranjero. 20 Minutos. https://www.20minutos.es/noticia/4936972/0/espana-nacion-jamon-es-el-quinto-exportador-mundial-de-carne-y-el-pais-que-mas-cerdo-vende-en-el-extranjero/.

Molloy, C. (2011). *Popular media and animals*. Palgrave Macmillan.

Noske, B. (1997). *Beyond boundaries: Humans and animals*. Black Rose Books.

Panksepp, J., Reiss, D., Edelman, D., Van Swinderen, B., Low, P., & Koch, C. (Eds.). (2012). *The Cambridge declaration on consciousness*. University of Cambridge, UK. http://fcm-conference.org/img/CambridgeDeclarationOnConsciousness.pdf.

Proveg. (2021, February 25). *El Mercado de los productos plant-based Registra un crecimiento récord del 48% en España en solo dos años*. ProVeg Spain. https://proveg.com/es/press-releases/el-mercado-de-los-productos-plant-based-registra-un-crecimiento-record-del-48-en-espana-en-solo-dos-anos/.

Rodríguez Cánovas, B., & Guzmán Rincón, A. (2023). Consumo de carne y sostenibilidad. *Actitudes de los jóvenes en España. HUMAN REVIEW. International Humanities Review/Revista Internacional de Humanidades*, *19*(3), 1–9. https://doi.org/10.37467/revhuman.v19.4924.

Ruby, M. B. (2012). Vegetarianism. A blossoming field of study. *Appetite*, *58*(1), 141–150.

SECAL. (2016). *Bases del concurso. Concurso de cuento infantil: la importancia de ser un animal de laboratorio*. Secal. https://secal.es/wp-content/uploads/2016/10/02-BASES-CONCURSO.pdf.

SECAL. (2019, March 11). *Reportaje - Área investigación biomédica con modelos animales del Museo Elder*. Secal. http://secal.es/reportaje-area-investigacion-biomedica-con-modelos-animales-del-museo-elder/.

Spasser, A. J. (2013). *Winning hearts and minds: Using "Ag-Gag" outrage and corporate rebranding to achieve a public image makeover for the animal rights movement*. [Master's Projects and Capstones. 4, University of San Francisco]. https://repository.usfca.edu/capstone/4

Twine, R. (2013). Addressing the animal industrial complex. In R. Corbey & A. Lanjouw (Eds.), *The politics of species: Reshaping our relationships with other animals* (pp. 205–240). Cambridge University Press.

Vaughan, D. (2023). *Charismatic megafauna*. Encyclopaedia Britannica. https://www.britannica.com/science/charismatic-megafauna.

WAZA. (2018, September 6). *Waza executive office relocation*. WAZA. Retrieved April 7, 2023, from https://www.waza.org/news/waza-executive-office-relocation/.

4

"BUT BACON!" THE PERFORMATIVE VIOLENCE OF ANTI-VEGAN TROLLING[1]

Jason Hannan

The spectacle of anti-vegan trolling is among the more peculiar aspects of our never-ending culture wars. Of the many anti-vegan trolls who have proliferated in our digital public sphere, by far the most extreme and disturbing is Gatis Lagzdins, better known by his online sobriquet Sv3rige. Lagzdins is a YouTuber who has built an infamous public persona entirely around his twisted delight in shocking and offending vegans. For years, Lagzdins has made headlines for himself through his antics at vegan festivals, where he scandalizes visitors by devouring raw meat, in some instances even biting into whole animal carcasses. In 2018, for example, Lagzdins appeared at Amsterdam's Vegan Food Festival, where he created a scene by biting into a bleeding slab of raw animal flesh. Wearing a black tank top emblazoned with the words, "Go Vegan and Die," Lagzdins claimed to be protesting the supposed dangers of a vegan diet (Barr, 2018). In 2019, he showed up at Veg-Fest Brighton, where he bit off chunks of a raw pig's head. Veganism, he said, is "human cruelty" and "malnutrition," because "you can't get over 15 nutrients from plants" (Cockburn, 2019). That same year, Lagzdins and a friend appeared at a vegan food market in Soho, London, this time gnawing at the carcasses of decapitated squirrels, their fur still intact. For added shock value, he wore a dead woodcock around his neck. This stunt brought an onlooking child to tears, leading to the duo's arrest (Ackerman, 2019). Around this time, disturbing details of Lagzdins's past and present came to light. It was revealed, for example, that he had stabbed four of his high school classmates with a knife, only to be later found sitting calmly in the computer lab, as if nothing had happened. Apart from his grisly and nauseating antics, Lagzdins has also posted videos revealing his unapologetic racism. He only "feels at home," he says, around "white people" (Robert, 2019).

To be sure, Lagzdins is an unusual and excessive case of anti-vegan trolling, though the sadism and perversity of his antics are not uncommon among

DOI: 10.4324/9781003441908-5

anti-vegans. An entire army of anti-vegan trolls has formed a distinct and recognizable subculture in response to the rapid growth of the global animal liberation movement. Vegans know these trolls best from their signature line, "But bacon!," a blunt and pithy retort to animal rights activism, often accompanied by an explicit meme featuring heaps of bacon for gratuitous offense. Unfortunately, what began as an online phenomenon has since become a popular gimmick among prominent chefs, fast food companies, politicians, and more than a few rogue physicians (Kelly & Persons, 2019; Todisco, 2019; Vanderhoof, 2018).[2] Like Lagzdins, these deanonymized trolls have built their brands and personas around the deliberate mockery and harassment of vegans, which suggests that this form of sociopathic behaviour is becoming increasingly normalized in our culture.

Why do so many carnivores turn to such childish and violent spectacles of harassment and abuse of vegans? Is it merely a case of individual psychopathology coincidentally multiplied many times over? Or is anti-vegan trolling part of a distinctly social pattern explicable through general analysis?[3] If the latter, what is it about our political culture and our online platforms that encourage this kind of abusive behaviour? This chapter explores the disturbing phenomenon of anti-vegan trolling. It attempts not just to document the nature and content of this unique style of hateful conduct, but also seeks to make sense of it as a violent symptom of our increasingly toxic political environment.[4] As I argue below, anti-vegan trolling is perfectly at home in a culture of possessive individualism, in which morality is a private affair and public discourse necessarily takes the form of a violent culture war.[5] But first, it's necessary to define basic terms. What exactly is trolling?

Making Sense of Trolling

The phenomenon of trolling has a long and ignoble history (Graham, 2019; Tepper, 2013). The first trolls emerged in the 1990s on Usenet, the still-operative, decentralized digital discussion forum that dominated the early years of the digital era. By connecting people from around the world through dial-up technology, Usenet fast developed into a truly global network. During its peak, Usenet competed with telephones, fax machines, and postal mail as a medium of global communication. Usenet was best known for its newsgroups—online message boards consisting of a vast range of topics for public discussions open to anyone with a computer. These message boards enabled users to share research, discuss science, debate politics, and build all manner of lifestyle-based communities. The key feature of Usenet participation was anonymity (Donath, 2002). As standard practice, newsgroup users created imaginative screennames that obscured their true identities. While this feature enabled a certain freedom to discuss controversial topics openly and unapologetically, it also had a notable downside: it encouraged the more juvenile participants on newsgroup forums to indulge in childish and offensive behaviour.

These early digital delinquents revelled in the freedom to shock and offend others without accountability. They took advantage of the trust of newsgroup communities by masquerading as good faith users, only then to sow discord, create controversy, and drag whole communities into bitter conflict. They pretended to be naïve. They asked silly questions. They made outrageous comments. Like Borat or the fake news reporters of *The Daily Show*, they went out of their way to create awkward scenarios that left others cringing with embarrassment and horror. They waited to see how long it would take before everyone would notice. Other delinquents were far more aggressive, deliberately targeting vulnerable users for special harassment and abuse. Newsgroup communities quickly became savvy to this menacing phenomenon, dubbing the troublemakers "trolls" and their obnoxious and abusive behaviour "trolling." The precise etymology of these terms is unclear. They may have originally referred to the technique of pulling a fishing line back and forth along the side of a boat to lure and catch unlucky fish. Alternatively, they may have referred to the hideous creatures of Norse mythology, who in the legends were known for accosting unsuspecting human travellers and even gobbling them up (Phillips, 2015, p. 15). Regardless of their precise origins, these terms soon became the most common means of naming and scrutinizing a unique kind of online menace.

While the first trolls can be traced back as far as the 1980s, the subculture of trolling as it has come to be popularly known emerged on 4chan in the early 2000s (Phillips, 2015, pp. 13–36). Like Usenet, 4chan is a digital discussion forum. Unlike Usenet, however, its message boards are image-based. That is, users primarily post images—sometimes benign, often depraved. 4chan users were notorious for posting pictures of blood, gore, death, bodily waste, gratuitous racism, morbid suicidal fantasy, and indescribable sexual perversion—acts of visual provocation ultimately earning the platform the infamous moniker "the asshole of the internet" (Bartlett, 2013). It was on 4chan that trolling first developed into a self-conscious community. On 4chan, the new subculture produced its signature lingo, imagery, and disruptive tactics. It was also here that memes as we know them today first arose (Coleman, 2015, pp. 4–42). The trolls of 4chan were notorious for orchestrating both non-violent pranks as well as dangerous harassment campaigns that caused material harm to their hapless victims. Whether harmless or malicious, the motive behind both pranks and harassment campaigns was always the same. The trolls did it for the "lulz," a term connoting sociopathic glee without any regard for principle or consequence (Bishop, 2014). 4chan was then, and is now, by no means a homogeneous space. On the one hand, it gave rise to Anonymous, the anarchist hacktivist group known for its distributed denial-of-service attacks on the Church of Scientology, the Israeli Knesset, the Minneapolis Police Department, and Gene Simmons. On the other hand, 4chan was also the site on which the alt-right came into being, the reactionary movement that strategically manoeuvred their way through America's evolving political landscape and helped catapult Donald Trump to the White House in 2016.

Because trolling has proven to be such a pervasive phenomenon in our digital culture, we now have a sizable literature devoted to the topic. Some of the most prominent media theorists who have sought to make sense of trolling include Judith

Donath, Gabriella Coleman, Whitney Phillips, Alice Marwick, and Rebecca Lewis (Coleman, 2015; Donath, 2002; Marwick & Lewis, 2017; Phillips, 2015). These scholars have conducted detailed studies of the subculture of trolling, providing a poignant portrait of their history, nature, psychology, volatility, and limitations. A common theme that emerges in their work is the element of anonymity. Trolling as it has been understood in the scholarly literature is a distinctively anonymous practice, since anonymity provides trolls with a certain strategic advantage over other users. Trolls characteristically prey on earnest targets whose conspicuous visibility and vulnerability leave them ripe for harassment: women, racialized minorities, queer and trans people, and conservative Christians. Anonymity thus ensures a fundamental power asymmetry between the trolls and the trolled. If you know nothing about a troll, then you can't find their weak spot, which keeps them invulnerable to retaliatory harm.

This standard scholarly understanding of the troll as an anonymous digital vandal is useful up to a point. Where it loses explanatory power is in making sense of the phenomenon of deanonymized trolls: those who may not necessarily think of themselves as trolls, yet who nonetheless engage in the behaviours typically associated with trolling and even worse. In some respects, we have entered a new era of trolling, in which a new generation of trolls has abandoned the need for anonymity and now happily trolls out in the open, unconcerned about personal repercussions. This new generation of trolls feels no need to hide their faces or obscure their names. Worse, they make the mockery and abuse of others a part of their public identity and persona. The new generation includes heads of state, business leaders, celebrities, and even academics. For many of us, it includes co-workers, neighbours, and even family members. Whether or not deanonymized trolls consciously think of themselves as trolls, something about the platforms in particular and our political culture more generally appears to bring out their inner monsters. This requires explanation.

Moral Rhetoric in the Culture Wars

Democracy has long been envisioned by its more idealistic theorists as a system of open and accountable government. According to a certain school of democratic thought, the core of this system is the social practice of public reasoning (Cohen, 2007; Bohman & Rehg, 1997; Gutmann & Thompson, 2004). In principle, rather than turning to violence and coercion, we turn to words, speech, argument, and persuasion to settle our differences and collectively govern a shared society. While this conception of democracy has always seemed nice in theory, in practice democracy is a far messier and more dispiriting thing. In his classic book, *Culture Wars: The Struggle to Define America* (1992), the sociologist James Davison Hunter makes explicit just how messy and dispiriting democratic culture has become. Davison conceives of the culture wars as the ongoing battle for the heart and soul of America. Focusing on such divisive issues as abortion, affirmative action, gender equality, and prayer in public schools, Hunter describes a virtual civil war over what it means to be American. He finds a deep and unbridgeable chasm pitting American citizens against each other on just about every major issue of public concern.

The culture wars are hardly limited to the United States. They can be found in almost every society, democratic or otherwise, in which citizens have some stake defining their political identity. Typically, the divide is between conservatives, who wish to restore some allegedly golden past, and progressives, who wish to depart from that past in favour of a new and different future. The two sides therefore battle it out in the public arena, seeking to persuade and entice each other and the undecided, sometimes through a message of hope, though quite often through a rhetoric of fear. The culture wars are very much a semantic struggle, in which rival and competing factions seek to define the essence of national identity, what it means to be American, Canadian, British, Indian, and so forth. The culture wars are also very much a political struggle for what Ernesto Laclau and Chantal Mouffe, following Antonio Gramsci, call hegemony—that is, the contingent institutional power to determine material outcomes for a political society (Laclau & Mouffe, 2014).

Aside from the sociological fact of our culture wars, there is also the historical question of why we are in a culture war in the first place. Why have modern liberal democracies tended to veer towards heated and increasingly violent battles over national identity? These are complex questions for which political theorists have offered various clues. Charles Taylor, for example, has argued that the transition from a premodern world to a modern one entailed a decentring of tradition and the centring of the economic sphere (Taylor, 2004). This transition marks a shift from communal bonds, however hierarchical and oppressive, to a moral and political order revolving around the exchanges of the marketplace. The moral centrepiece of the new moral and political order was the individual, a political creature conceived as the independent sovereign and master of his own private affairs. Taylor refers to a society composed of a collection of disconnected and independent individuals as political atomism (Taylor, 1979). In an atomistic society, individuals live for personal gain. Human relationships are therefore fundamentally instrumental and transactional. Without the social bonds of tradition, the individual necessarily turns inward, rather than outward to community, to find personal meaning, purpose, and the sources of the self. While not inherently problematic, this inward turn can lead, as Taylor has argued, to a disaffection with modern life. Following Taylor, we can see why atomized and alienated individuals might feel compelled to fashion their sense of self in evermore extreme ways. Without inner resources for the self, they might seek to define themselves, not by some personal essence, but rather by what they are against, such as veganism.

Alasdair MacIntyre (1981) has made a similar argument about human relationships under liberal modernity. He, too, sees those relationships as fundamentally instrumental and transactional. However, MacIntyre paints a much bleaker portrait of liberal societies. He sees the centring of the economic sphere, our market-based political order, as profoundly corrosive to moral and political rhetoric. In a vivid description of the culture of marketplace exchange, he proposes that the very idea of the individual erases the distinction between manipulative and non-manipulative human relationships (MacIntyre, 1981, pp. 22–35). In a social world where each individual seeks personal gain, we by definition lack a basic social solidarity and common ethical ground. That is, we have no shared standards by which to engage each other on matters of moral and political concern. This is why, MacIntyre argues, our moral rhetoric is so heated. We shout our opinions across the chasm, failing to persuade those on the other side. And those others shout their opinions back at us, failing to persuade us in turn. At some point, our conversations will likely break down, at which point we resort to non-rational means of persuasion and influence, such as unmasking each other's ulterior motives and conducting duelling protests and counterprotests.

Drawing from MacIntyre, we might understand political trolling as one of these non-rational means of persuasion and influence. Political trolling is a symptom of the breakdown of public communication, a sign of the limits of rational argument. It signals a loss of confidence in the Other. Worse, it marks an abandonment of civility, a turn to the malicious, the vindictive, the sadistic. The climate activist Greta Thunberg has described breakdown in her dealings with her critics (Klein, 2019). Thunberg has become one of the most prominent and influential voices for climate action today. Unsurprisingly, she has attracted more than her fair share of detractors, who either dispute climate science or oppose climate action, or both. Where those detractors bother to challenge her with actual arguments, Thunberg consistently points to the science, which is basically irrefutable. At that point, their initial pretence of reasonableness runs its course. Out of arguments, they resort to verbal abuse and harassment. They mock her looks, her voice, her intelligence, and her mental state. They insist she's too young and ignorant to understand the science. They say her parents have brainwashed her. She's merely a puppet for corrupt entities operating behind the scenes. She's the victim of child abuse. Her speeches and social media posts are even the stuff of Nazi propaganda. The trolls have thus unleashed their violent hatred and abusive contempt for a young woman merely for raising awareness about our climate crisis. While many of these trolls maintain their anonymity, a disturbing number do not. This latter group is perfectly happy to abuse and humiliate her out in the open. The most notorious of them is Donald Trump, who has famously developed a twisted obsession with the young Thunberg, publicly mocking her without remorse (Levin, 2020). Given this new political environment for trolling out in the open, we can see why so many belligerent social media users have no qualms trolling vegans without the veil of anonymity.

The Rise of Troll Influencers

Trolling is now a mainstream part of our public culture. The story of the gradual migration of trolling from an anonymous practice on Usenet and 4chan to an open and defiant rhetorical style on just about every social media platform today is instructive for understanding anti-vegan trolls. This transformation reflects the emergence of digital platforms as the dominant media of our age and the consequent high-schoolification of public culture. Facebook played a critical role in this massive cultural shift. Originally designed as a digital yearbook to keep student cohorts connected, Facebook recreated high school culture in a digital environment, feeding narcissistic self-regard and delayed adulthood. To play the Facebook game, as it were, was to compete for popularity, measured by the number of friends one could accumulate. The rapid speed with which Facebook attracted an entire generation of young people was not lost on either political observers or business analysts. Just as television had become a standard medium for political branding and commercial advertising, Facebook fast became a *de rigueur* medium for both politicians seeking to establish their political credibility and businesses seeking to establish their commercial credibility. On the political front, the unexpected candidacy and historic presidency of Barack Obama is a testament to this profound change in our media environment. Obama has rightly been called the first social media president (Katz et al., 2013). He built a political persona attuned to the younger sensibility of the Facebook generation, wielding a spell-binding charm that played well on viral videos and attracted a loyal following. But Obama's success on Facebook was the feel-good story of social media.

Facebook's basic social network structure was later replicated by YouTube, Twitter, and Instagram. This time, users competed to amass "followers," instead of "friends." Those who learned to play the social media game well could accumulate truly colossal followings, which translated into considerable social clout. Popular politicians, musical artists, and Hollywood celebrities automatically commanded formidable followings online, a feat owing to their sheer star power. However, non-celebrities seeking to build that clout from scratch, but who lacked any prior claim to fame, were forced to engineer a social media brand, offering something unique that could attract a coveted cult following. The range of niche areas for cultivating these brands was staggering: health and wellness, food, fashion, beauty, home décor, fitness, travel, cats, dogs, politics, and couples' pranks, to name just a few. Those who succeeded in attracting a veritable army of followers earned themselves the title of "influencer" (Khamis et al., 2017).

Sadly, trolling was one more niche area in which to become an influencer. The formative and paradigmatic example of the influencer-troll was Milo Yiannapolous, the now-disgraced polemical "journalist" who earned global notoriety by writing inflammatory articles with shocking headlines for the far-right opinion outlet Breitbart News. Milo earned rockstar status in the conservative movement through his cheeky form of reactionary commentary attacking feminists, trans people, fat people, Muslims, and "social justice warriors" (Stein, 2016). He became a hero to the Alt-Right after the infamous Gamer Gate controversy, in which he ridiculed female gaming critics who sought more equitable representation in video games. Milo showed how to become a social media influencer simply by being a jerk.

If you selected the right targets, if you provided the right rationale for bullying easy prey, then you, too, could achieve instant popularity and enjoy social and political clout. This had profound implications for the culture of social media. During the 2016 US general election, Milo aggressively supported Donald Trump, who made trolling part of his political brand. Although Milo has since long faded into political oblivion, he will be forever remembered for his critical role in degrading social media into a poisonous space for the new breed of online trolls (Lynskey, 2017). It is into this twisted niche that anti-vegan trolls were to forge their identity and find their place in the online moral universe.

A Profile in Anti-Vegan Trolling: Sama Hoole

Sama Hoole is a pertinent example of a troll who uses hostility to veganism as part of his public persona and professional brand. Hoole works as a fitness trainer in Wrexham, Wales. His services include in-person training and online coaching. Hoole's website invites visitors to "train with the biohacker chief" and "get back to the top of the food chain," suggesting that he has cracked the code to optimal living. To climb atop the food chain, he says, requires a radical approach to "muscle," "metabolism," and "mind" (Hoole, n.d.). Hoole's "coaching blueprint" provides a window into his dietary and psycho-calisthenic worldview. The centrepiece of his program is meat consumption. Hoole instructs us to "forget everything you know about nutrition" in favour of "the apex diet" (Hoole, 2020, July 12). Against the basic principles of conventional nutritional science, Hoole recommends minimizing or eliminating plants. He tells us that plants are "foreign invaders" that lead to "brain fog" and "imbalanced hormones." Berries are bad for you. Broccoli is a low-nutrient food. Beans lack nutrition value. Fibre is "not as essential as you might think" (Hoole, 2020, November 14). Instead of fruits, vegetables, and grains, Hoole recommends massively increasing our consumption of red meat, organ meat, and saturated fats. Hoole advocates a full-on carnivore diet, which promises not only to build muscle, but also to increase our brain power, reduce our stress, sharpen our focus, boost our immunity, heal our gut, clear our skin, reduce autoimmune issues, resolve all metabolic dysfunction, and enable us finally to enjoy "deeper" sleep. Hoole even goes so far as to claim that a carnivore diet provides "The Best Natural Dental Care." By sticking exclusively to meat and butter, you can "avoid cavities without having to brush your teeth for fifteen minutes a day" and even enjoy "perfect rows of shiny teeth" (Hoole, 2020, January 16).

Hoole's Twitter account promotes his apex fitness and dietary lifestyle. It also showcases his pathological obsession with vegans and veganism. Even a cursory glance through his tweets reveals daily, almost hourly, ridicule of vegans. One tweet, for example, features a stock photo of a burger and fries made of cardboard, implying that vegan food is necessarily junk food devoid of nutritional value, a common trope in anti-vegan trolling. Hoole's caption sarcastically reads, "Vegan burger and fries. Low carb, gluten free, low fat, gut friendly" (Hoole, 2024,

February 18). Another tweet, posted without an image, reads, "Vegans will really look animals in the eye and say 'my monocrop foods, phone, computer, TV, camera, car, and fleece jacket are important [sic] than your suffering'" (Hoole, 2024, February 17). Somehow, veganism, an ethical outlook based on the principle of compassion and non-violence towards all living creatures, has been redefined as the consumption of wasteful and environmentally damaging commodities. Hoole here also presents veganism as more harmful to animals than a diet based on the maximal consumption of meat, another common trope in anti-vegan trolling. In another tweet, we read, "For 99.4% of our evolutionary history, we ate almost exclusively meat. We've been eating plants in quantity for only 0.6%. So tell me how exactly meat can be the problem for our modern woes, and plants are the solution?" (Hoole, 2024, February 17). This tweet invokes the recurring idea, popular among anti-vegan trolls, that veganism somehow violates human biology and that a carnivorous diet, far from being a fringe trend, in fact marks a return to the historical human diet. Of course, Hoole provides no evidence for this claim.

Another recurring theme in Hoole's tweets is a tangible contempt for vegan arguments for protecting the environment. One such tweet features a stock photo of green hills covered with trees and meadows. The caption reads, "Vegans love saying how 80% of the worlds [sic] agricultural land is used for livestock without having the slightest clue what they are actually talking about. The vast majority of that 80% is too dry, or too steep or rocky to grow anything but grass. Good luck trying to repurpose this land for farming" (Hoole, 2024, February 17). Here, Hoole implies that vegans are ecologically illiterate and that sloping grasslands can serve no agricultural function other than animal farming. Hoole falsely assumes that vegans wish to convert all animal farms to crop farms, apparently failing to realize that a plant-based food system would require only a small fraction of the land currently devoted to animal agriculture.

In another tweet, Hoole posts a screenshot of a *Men's Health* article with the catchy headline, "Vegan Men Fart Seven Times More Than Non-Vegans, Study Finds." Hoole's caption reads, "If cow farts are so catastrophic for the environment: Why aren't we addressing the problem of vegans?" (Hoole, 2024, February 16). This caption reveals Hoole's touchiness regarding environmental concerns over livestock farming. As we now know, animal agriculture is a major source of methane emissions. Hoole's tweet suggests that vegans play a greater role than livestock animals in exacerbating climate change. Once again, Hoole provides no evidence to support this wild suggestion, feeling contented with a vast leap of logic from a men's magazine story about the social consequences of flatulence to an outlandish suggestion about a primary driver of global warming. In another tweet, Hoole shares a vertically split photo, the top half featuring an image of aerial crop dusting, the bottom half featuring a rifle next to a pile of dead rabbits. Hoole writes, "Vegans hate hearing this. Whether you eat meat or plants, you are complicit in the killing of animals. Field deaths aren't "accidental." They are a very intentional part of protecting those crops" (Hoole, 2024, February 10). Hoole recycles the discredited argument that crop farming kills more animals than animal farming. The gratuitous image of dead rabbits here serves no purpose other than to dish out tough love, as if educating vegans out of their supposed ecological delusions entails forcing them to witness animal violence.

In fact, just about every other tweet on Hoole's account is about vegans: their supposed ignorance about human health, their naïveté about the environment, their hypocrisy about protecting animals, their weak and enfeebled physical appearance, their dwindling numbers, their vile personalities, their self-defeating activism. In one tweet, Hoole writes, "Vegans: 'Lets [sic] also pretend to be the morally superior, compassionate, good guy protagonist while we mock, slander, insult, berate, disrespect and name call the ones who don't follow our cult's diet.' Funny why it puts people off" (Hoole, 2024, February 16). Hoole wrote these words apparently without considering that he was projecting his personal insecurities upon vegans. An especially striking example of Hoole's unceasing preoccupation with vegans is a 2023 photo from Chester, England. In it, Hoole sits on an outdoor step, dressed in fitness attire and sporting minimalist shoes, the latter a staple of carnivore-ancestral-caveman-all-natural-CrossFit chic. The historic backdrop and the high-saturation photo suggest that Hoole is on vacation. Still, Hoole seems unable to help himself from unleashing his hatred upon vegans. The caption reads: "Vegans desperately try to make their vegetables look like meat. But nobody tries to make steak look like a vegetable. Makes you think" (Hoole, 2023, August 18). Even while apparently on vacation, Hoole cannot seem to enjoy himself without broadcasting his mania for vegans.

Hoole's tweets are not unusual for anti-vegan troll influencers. On the contrary, they are rather tame by comparison. Yet they capture a general pattern in the subculture of anti-vegan trolling. On the one hand, Hoole's tweets showcase extreme pettiness, immaturity, vindictiveness, bad faith, envy, and sadism. On the other hand, they reveal the psychology of an addict, a profoundly alienated, lonely, unhappy, and desperate individual for whom the thrill of cruelty towards vegans functions as a kind of narcotic. Judging by the sheer number of anti-vegan comments he tweets each day, it would seem that Hoole requires several doses of this narcotic just to help him get through the day.

The Reactionary Politics of Anti-Vegan Trolling

Hoole's daily trolling of vegans is symptomatic of a broader reactionary politics in which carnivory is identified with manhood, heritage, faith, and patriotism, and veganism is identified with femininity, weakness, decadence, and a generally liberal or left-wing agenda. Culture warriors such as Piers Morgan, Tucker Carlson, and Jordan Peterson have all taken up anti-veganism as part of their reactionary platforms. As Jan Dutkiewicz and Gabriel N. Rosenberg have argued, we are in the midst of a full-blown culture war over meat (Dutkiewicz & Rosenberg, 2023). The right-wing hostility to veganism is about far more than a mere dislike of vegetables or a misguided understanding of nutrition. Rather, it reflects a profoundly paranoid fantasy, in which vegans are part of a global conspiracy to undermine basic freedoms, liberties, and masculine ideals. Anti-vegan trolling is not like the juvenile antics, pranks, and harassment campaigns of the old breed of anonymous trolls on Usenet or 4chan. Anti-vegan trolls are not in it for the "lulz." Rather, they see themselves as soldiers, even holy warriors, in a historic battle for traditional values.

The reactionary psychology of anti-vegan trolling is driven by the defensive sense of male victimhood. They see vegans as tyrants and conspirators, working behind the scenes to undermine Western society, the Christian faith, and wholesome family values. Vegans are lumped together with a motley and amorphous cadre of CEOs, deep state operatives, godless heathens, and sexual perverts. The moral panic over "soy boys," for example, reflects the deeply homophobic and transphobic fears of anti-vegan trolls, their nagging nightmare that tofu is giving men "soy boobs," turning kids gay, and causing them to become transgender (Amidor, n.d.; Clemens & Flannery, 2022; Villarreal, 2019). The anti-vegan worldview is nothing short of a conspiracy theory, a black-and-white historical drama of good vs. evil. Seen in this light, the trolling of vegans is not unlike the harassment of suspected communists during the 1950s or the racial profiling of Muslims after 9/11. Vegans have become one more spectre haunting the extremely fragile and sensitive reactionary mind.

Although some vegans do admittedly succumb to the anti-social temptations of online culture and themselves misguidedly engage in trolling, veganism has the merit at least of resting on genuine principles of concern for animals and the planet. Veganism is a positive philosophy. It presents a practical vision of a better world. The same cannot be said for anti-vegan trolls, who desperately need to target and demonize vegans, thereby giving their lives meaning and purpose. The proper response to anti-vegan trolling is not to troll the trolls, a never-ending tit-for-tat that only perpetuates the culture wars, but rather to understand the profound alienation at the heart of the reactionary mind and to build a culture of trust, understanding, dialogue, and communication. As I have argued elsewhere, "Either we can work to repair our civic culture by building a spirit of trust…or we can relinquish our politics to the pathogen of unreason and continue trolling ourselves to death" (Hannan, 2023, p. 130). For the sake of animals and the planet, too, we can afford nothing less.

Notes

1 I wish to thank Benny Malone for sharing numerous examples of anti-vegan trolling. I am grateful to Benny and also to Emilia Leese for reading earlier versions of this chapter.
2 Perhaps the most extreme example of an anti-vegan physician troll is Shawn Baker, whose Twitter account @SBakerMD promotes his "carnivore diet" and routinely mocks vegans and veganism.
3 For a social psychological approach to this question, see Gregson et al. (2022).
4 For a general historical materialist account of trolling, see Hannan (2023).
5 The concept of possessive individualism is taken from Macpherson (2010).

References

Ackerman, N. (2019, July 23). Anti-vegan protesters who ate squirrels at food fair guilty of causing public distress. *The Standard*. https://www.standard.co.uk/news/crime/protesters-who-ate-squirrels-at-vegan-food-fair-guilty-of-causing-public-distress-a4196221.html.

Amidor, T. (n.d.). Does eating soy give you man boobs? *Muscle & Fitness* https://www.muscleandfitness.com/nutrition/healthy-eating/does-eating-soy-give-you-man-boobs/.

Barr, S. (2018, August 14). Man eats bloody raw meat outside vegan food festival. *The Independent*. https://www.independent.co.uk/life-style/food-and-drink/vegan-food-festival-man-eats-raw-meat-video-amsterdam-a8491656.html.

Bartlett, J. (2013, January 10). 4chan: The role of anonymity in the meme-generating cesspool of the web. *WIRED*. https://www.wired.co.uk/article/4chan-happy-birthday.

Bishop, J. (2014). Trolling for the lulz?: Using media theory to understand transgressive humour and other internet trolling in online communities. In *Transforming politics and policy in the digital age* (pp. 155–172). IGI Global.

Bohman, J., & Rehg, W. (Eds.). (1997). *Deliberative democracy: Essays on reason and politics*. MIT Press.

Clemens, R., & Flannery, B. (2022). 'Soy Boy', ecology and the fascist imaginary. R. Dolphijn & R. Braidotti (Eds.), *Deleuze and Guattari and fascism, Deleuze connections* (pp. 341–361). Edinburgh University Press.

Cockburn, H. (2019, March 25). Man eats raw pig's head outside vegan festival in 'protest'. *The Independent*. https://www.independent.co.uk/news/uk/home-news/raw-pigs-head-brighton-vegan-festival-protest-a8838656.html.

Cohen, J. (2007). Deliberative democracy. In *Deliberation, participation and democracy: Can the people govern?* (pp. 219–236). Palgrave Macmillan.

Coleman, G. (2015). *Hacker, hoaxer, whistleblower, spy: The many faces of anonymous* (pp. 41–43). Verso.

Donath, J. S. (2002). Identity and deception in the virtual community. In *Communities in cyberspace* (pp. 37–68). Routledge.

Dutkiewicz, J., & Rosenberg, G. (2023, April 17). Why right-wingers are so afraid of men eating vegetables. *The New Republic*. https://newrepublic.com/article/171781/meat-culture-war-crickets.

Graham, E. (2019). Boundary maintenance and the origins of trolling. *New Media & Society*, *21*(9), 2029–2047.

Gregson, R., Piazza, J., & Boyd, R. L. (2022). Against the cult of veganism': Unpacking the social psychology and ideology of anti-vegans. *Appetite*, *178*, 106143.

Gutmann, A., & Thompson, D.F. (2004). *Why deliberative democracy?* Princeton University Press.

Hannan, J. (2023). *Trolling ourselves to death: Democracy in the age of social media*. Oxford University Press.

Hoole, S. (n.d.). *Fit Awakening*. https://fitawakening.co.uk/.

Hoole, S. (2020, November 14). The hidden dangers of plant toxins. *Fit Awakening*. https://fitawakening.co.uk/2020/11/14/plant-toxins/.

Hoole, S. (2020, July 12). Carnivore is the apex diet. *Fit Awakening*. https://fitawakening.co.uk/2020/07/12/the-apex-diet/.

Hoole, S. (2020, January 16). The complete list of benefits on the carnivore diet. *Fit Awakening*. https://fitawakening.co.uk/2021/01/16/carnivore-diet-benefits/.

Hoole, S. [@SamaHoole]. (2023, August 18). Vegans desperately try to make their vegetables look like meat. But nobody tries to make steak look like a vegetable. Makes you think. [Tweet]. Twitter. https://x.com/SamaHoole/status/1692611497002697207?s=20.

Hoole, S. [@SamaHoole]. (2024, February 18). *Vegan burger and fries. Low carb, gluten free, low fat, gut friendly*. [Image attached] [Tweet]. Twitter. https://twitter.com/SamaHoole/status/1759217582186320345.

Hoole, S. [@SamaHoole]. (2024, February 17). Vegans will really look animals in the eye and say "my monocrop foods, phone, computer, TV, camera, car, and fleece jacket are important than your suffering". [Tweet]. Twitter. https://x.com/SamaHoole/status/1758942970751881231.

Hoole, S. [@SamaHoole]. (2024, February 17). For 99.4% of our evolutionary history, we ate almost exclusively meat. We've been eating plants in quantity for only 0.6%. So tell me how exactly meat can be the problem for our modern woes, and plants are the solution?. [Tweet]. Twitter. https://x.com/SamaHoole/status/1758891636623311000.

Hoole, S. [@SamaHoole]. (2024, February 17). Vegans love saying how 80% of the worlds agricultural land is used for livestock without having the slightest clue what they are actually talking about. The vast majority of that 80% is too dry, or too steep or rocky to grow anything but grass. Good luck trying to repurpose this land for farming. [Tweet]. Twitter. https://twitter.com/SamaHoole/status/1758738238691295276.

Hoole, S. [@SamaHoole]. (Hoole, 2024, February 16). If cow farts are so catastrophic for the environment Why aren't we addressing the problem of vegans? [Tweet]. Twitter. https://twitter.com/SamaHoole/status/1758443668665405844.

Hoole, S. [@SamaHoole]. (Hoole, 2024, February 16). Vegans: "Lets also pretend to be the morally superior, compassionate, good guy protagonist while we mock, slander, insult, berate, disrespect and name call the ones who don't follow our cult's diet." [Tweet]. Twitter. https://twitter.com/SamaHoole/status/1758548910182719798.

Hunter, J. D. (1992). *Culture wars: The struggle to define America*. Basic Books.

Katz, J. E., Barris, M., & Jain, A. (2013). *The social media president: Barack Obama and the politics of digital engagement*. Palgrave Macmillan.

Kelly, D. B., & Persons, B. (2019, September 25). The untold truth of Food Network star Carl Ruiz. *Mashed*. https://www.mashed.com/167582/the-untold-truth-of-food-network-star-carl-ruiz/.

Khamis, S., Ang, L., & Welling, R. (2017). Self-branding, 'micro-celebrity'and the rise of social media influencers. *Celebrity Studies*, *8*(2), 191–208.

Klein, N. (2019, September 13). Greta Thunberg on the climate fight: 'If we can save the banks, then we can save the world. *The Intercept*. https://theintercept.com/2019/09/13/greta-thunberg-naomi-klein-climate/.

Laclau, E., & Mouffe, C. (2014). *Hegemony and socialist strategy: Towards a radical democratic politics*. Verso.

Levin, B. (2020, January 21). Trump attacks Greta Thunberg at Davos, then claims he doesn't 'know anything about her. *Vanity Fair*. https://www.vanityfair.com/news/2020/01/donald-trump-greta-thunberg-davos.

Lynskey, D. (2017, February 21). The rise and fall of Milo Yiannopoulos – how a shallow actor played the bad guy for money. *The Guardian*. https://www.theguardian.com/world/2017/feb/21/milo-yiannopoulos-rise-and-fall-shallow-actor-bad-guy-hate-speech.

MacIntyre, A. (1981). *After virtue: A study in moral theory*. University of Notre Dame Press.

Macpherson, C. B. (2010). *The political theory of possessive individualism: Hobbes to Locke*. Oxford University Press.

Marwick, A., & Lewis, R. (2017, May 15). Media manipulation and disinformation online. *Data & Society*. https://datasociety.net/wp-content/uploads/2017/05/DataAndSociety_MediaManipulationAndDisinformationOnline-1.pdf

Phillips, W. (2015). *This is why we can't have nice things: Mapping the relationship between online trolling and mainstream culture* (p. 15). MIT Press.

Robert, J. (2019, April 13). Man who ate raw pig's head at vegan festival stabbed four classmates at school. *Metro*. https://metro.co.uk/2019/04/13/man-who-ate-raw-pigs-head-at-vegan-festival-stabbed-four-classmates-at-school-9184229/.

Stein, J. (2016, September 15). Milo Yiannapolous is the pretty, monstrous face of the Alt-Right. *Bloomberg*. https://www.bloomberg.com/features/2016-america-divided/milo-yiannopoulos/

Taylor, C. (1979). Atomism. In A. Kontos (Ed.), *Essays in honour of C. B. Macpherson* (pp. 39–61). University of Toronto Press.
Taylor, C. (2004). *Modern social imaginaries*. Duke University Press.
Tepper, M. (2013). Usenet communities and the cultural politics of information. In *Internet culture* (pp. 39–54). Routledge.
Todisco, E. (2019, June 26). Arby's trolls vegans and Impossible Burger fans by making 'carrots' made entirely of meats. *People*. https://people.com/food/arbys-trolls-vegans-making-carrots-made-meat.
Vanderhoof, E. (2018, September 18). Why is Ted Cruz so obsessed with tofu? *Vanity Fair*. https://www.vanityfair.com/style/2018/09/why-is-ted-cruz-so-obsessed-with-tofu.
Villarreal, D. (2019, December 30). The far-right thinks Burger King's Impossible Whoppers will literally turn men into women. *LGBTQ Nation*. https://www.lgbtqnation.com/2019/12/far-right-thinks-burger-kings-impossible-whoppers-will-literally-turn-men-women/.

PART II

The Trouble with "Visibility" in Unveiling Commodity Fetishism

5

THE POLITICS OF SMELL AND THE MORALITY OF SIGHT

Challenging "Slaughterhouses with Glass Walls" in Animal Advocacy

Chiara Stefanoni

The connection between violence against non-human animals and the possibility of overcoming it has long been linked to the conceptual framework of "visibility/concealment". The *Animal Activists' Handbook*, for instance, states: "[…] if the realities of factory farms and slaughterhouses were as visible as the meat they produce, all thoughtful compassionate individuals would be vegetarian advocates" (Ball & Friedrich 2009, p. 17). The extremely widespread activist format of the "Cube of Truth" also reinforces this perspective.

Nowadays, however, there are many easily accessible ways to look at every stage of non-human animal exploitation, thus, supposedly, knowing the truth behind the scenes, via TV shows, documentaries, online videos, and more. It does seem that we potentially can see through the "glass walls" and the public awareness about issues associated with the capitalist animal agriculture attests to this. However, despite this increased visibility and awareness, meat and animal source food (ASF) production and consumption continue to rise, maintaining a central role in the global food system (Our World in Data, 2023).

Certainly, through this juxtaposition (growing visibility, yet growing consumption), I do not contend that meat consumption increases *because of* the failure of the equation "visibility equals going vegan", indeed there are many other factors at play. Rather, I posit that the predominant focus on this thesis within animal advocacy discourses and related activist strategies, that goes under the label of "politics of sight" (English & Zacka, 2022; Pachirat, 2011),[1] is both theoretically problematic and practically ineffective. In relation to the latter point, Arcari (2017) persuasively argues that emphasizing *how* ASF is violently produced, rather than questioning *why* non-human animals are used as food in the first place, carries certain risks. According to Arcari, such an approach "risks perverting the goal of ending the use of non-human animals for food by highlighting areas of

DOI: 10.4324/9781003441908-7

concern that the industry can address to maintain its legitimacy in the public eye" (p. 43), just like the happy/sustainable/ethical/humane – in a word "better" – meat market.[2]

On the theoretical level, the idea of the "slaughterhouse with glass walls" rests on the equation between vision and access to truth, based on the assumption of an immediate intelligibility of what violence is and consequent moral shock. I argue that this equation is inherent to the moral approach to the animal question that also shapes mainstream animal advocacy. Such approach encompasses a diverse array of theories and practices centred on the normative inquiry into the treatment of non-human animals and characterized by a definition of speciesism as prejudice (Singer, 2002), mental attitude, or mindset. A crucial limitation of this approach, as illuminated by scholars like Calarco (2015), is its logocentrism or rationalism. I term this logocentric inclination "idealism" to highlight the primacy given to the sphere of ideas, conceptions, consciousness, and the complementary neglecting of material life-processes and, thus, the socio-historical dimension and practices (Marx & Engels, 1932/1976).

Recognizing this idealistic moral approach as a potential constraint, I propose a shift towards a materialist perspective. In adopting this term within the broad tradition of Marx, I advocate for an approach that prioritizes material practices and seeks to deconstruct ideological aspects. Such materialist stance is shared with the method and frame of Animal-industrial complex (A-IC) in critical animal studies (CAS), Noske herself being influenced by Marxism when first coining the term and, more prominently, in the further development of the concept undertaken by Twine (2012) in which the emphasis given to economic structures and power relations is explicitly understood as a move beyond that psychological individualism and mere moral approach to violence against non-human animals.

Within the materialist framework I am proposing, it is pertinent to incorporate approaches rooted in Foucauldian critical epistemology, which offer a significant analytical advancement. Foucault (1966/1994) characterizes epistemology as constituted by ongoing social and political contestations that reframe "knowledge/science", first, in terms of "epistemological field/knowledge regime" and, later, in terms of *dispositif* or apparatus, where an apparatus can be defined as: a network of institutions and mixed practices, authorized by correlated scientific knowledges, with subjectivation effects (1977/1980). Again, the notion of *dispositif* could be seen akin to that of "complex" in the frame of A-IC, one meaning of "complex" being precisely "*network* [emphasis added], conglomerate or system" (Twine, 2012, p. 23). Moreover, the idea of "complex" and of *dispositif* shares an attention for how these networks orient and shape form of subjectivities and identities.

Incorporating these Foucauldian concepts introduces a crucial third element into the equation concerning vision and truth: power. In this context, what qualifies as violence is not merely there to be seen; rather, it is produced within a given *dispositif* and social form. CAS scholar Wadiwel (2016) summarizes this perspective in relation to non-human animals, stating: "Violence, as it is rendered within the public space [...], can only be made visible within the context of available knowledge systems. It is only possible to see violence towards animals when we conceptualise this as possible" (p. 204).

In the following, building upon the outlined "materialist critique of violence", I question the significance of the visibility/concealment pair, primarily by conducting a genealogical reconstruction of the nineteenth-century epistemological field linked to the rise of the industrial slaughterhouse, indeed one crucial institution in the A-IC. The choice of this historical examination seeks to challenge a deeply ingrained notion within the broader field of animal studies: that the concealment of slaughter in modern, secluded abattoirs at the outskirts of the city is historically and conceptually linked to the belief that violence against non-human animal is unbearable to see. In the first section, I demonstrate that this concealment was primarily driven by hygienic discourses of the time, rooted in miasmatic theory – thus, based on smell rather than sight.[3] Surely, that gaze-centred idea is not based on nothing: it finds its roots in the gaze-centred moral discourse of the early animal advocacy associations of the Victorian era. This discourse, characterized by a utilitarian, anthropocentric, and classist perspective on violence against non-human animals – a perspective inherited by the contemporary mainstream animal advocacy movements, which must grapple with its legacy – will be reconstructed in the second section. Finally, in the third section, I briefly examine slaughterhouse tours, both in nineteenth-century and contemporary settings (e.g., transparency campaigns by the meat industry). These tours represent a case study wherein the equation between violence and visibility/concealment reaches a highly problematic juncture.

Hygiene and Miasma in the Rise of the Modern Slaughterhouse: Smell

As mentioned earlier, it is a common trope in the field of animal studies (critical or not) that the concealment of meat production in the modern slaughterhouse, typically situated on the outskirts of cities, is related to the repugnance at the sight of animal suffering. Here the notion of "civilizing process" theorized by historical sociologist Norbert Elias has set a trend. According to Elias (1939/1978), the Western society undertook a "civilizing process" between the sixteenth and nineteenth centuries, which was characterized by the social and cultural construction of a new image of "Man": a Man educated in "good manners", equipped with a new sensitivity and defined by means of binary opposition to barbarism and animality. Elias posits that the civilizing process was largely centred around removing from the public view those repulsive objects or spectacles that, instead, were previously openly visible in daily life. Such repugnant spectacles certainly included the violence of slaughter or cruelty to live non-human animals in city markets: involving animals being beaten, dragged, whipped, and killed.

However, I claim that a more in-depth examination of the dynamics contributing to the emergence of the modern slaughterhouse in the latter half of the nineteenth century reveals that this process is primarily associated with a different sense: not sight but smell. The olfactory sense plays a central role in the power-knowledge apparatus that performs the qualitative change of purpose and function undertaken by non-human animals and meat with the transition to capitalist modernity. This apparatus can be called "dietary *dispositif*" and the change process can be labelled with the expression "hygienizing of meat" (Aloe & Stefanoni, 2021). As will become clear from the following discussion, under the knowledge regime shaped by the theory of miasma, predominant in the medicine of that time, the trickles of blood on the sides of the streets due to the hundreds of small slaughterhouses scattered in the cities or the presence of live animals in the markets were not so "repugnant at the sight", as Elias writes, and therefore to be hidden, but rather repulsive to the sense of smell and therefore to be *deodorized*, through the centralization and segregation of slaughtering processes in large facilities on the outskirts of cities.

Animal Nuisance

In the nineteenth century, protracted debates among various social actors, including municipal health councils, animal welfare associations, butchers, and entrepreneurs, culminated in substantial reforms to the slaughterhouse system. These reforms, integral to the hygienist agenda (Coleman, 1982; La Berge, 1992), resulted in the establishment of industrial slaughterhouses across Europe and its settler colonies, such as the Americas and Australia. A key conceptual driver behind these reforms was the notion of "animal nuisance".

The *nuisance* was a technical legal conception that referred to something injurious or obnoxious to the community due to an "environmental wrong" (Atkins, 2012b, p. 27). The term has been in use since the fifteenth century, and by the late Middle Ages, it had become a subject of calculation and resolution in the adversarial setting of the magistrate's court. Starting from the 1830s, it evolved into one of the major themes within legal systems and was transformed into a principal tool of the public health movement. Such transformation welded nuisance and health into a binomial, in the sense that a nuisance came to be viewed as "injurious to *life*" of the community, a hazard to health blurring the distinction between its purely legal meaning and its health-related implications. It was precisely the so-called animal nuisance that acted as "a catalyst for both medical theories and sanitary theories of the environment" (Atkins, 2012a, p. 14) and gave a major boost to the great season of public health councils in the nineteenth century.

What was injurious to life with regard to non-human animals? To fully grasp the centrality of the concept of "animal nuisance" it is essential to briefly describe the pre-reforms system of institutionalized killing of animals for food production. This system was based on private artisanal slaughterhouses and was characterized by two fundamental attributes: dispersal and privacy (Otter, 2008).

Dispersal in this context implies that there was no single, easily identifiable, or monofunctional space designated for animal slaughtering and carcass dressing.

Instead, these activities took place in numerous scattered locations, ranging from household backyards, predominantly for chickens and hogs, to butcher's sheds, primarily used for slaughtering cattle. Slaughterhouses were typically small, comprising a single room that was dark, almost windowless, poorly ventilated, and often lacked a tap water supply.

Privacy, in this context, had a dual meaning: freedom from public attention and private property rights. Inside slaughterhouses, only the master butcher and his assistants (if present) were allowed, along with the non-human animals. It was challenging for undesired observers, such as public inspectors, to enter these places. This slaughtering system, whether at the family level or among trade professionals, operated under a kind of "one-to-one" ownership principle: non-human animals were the private property of who was going to kill them. Individuals would personally slaughter their own poultry and hogs raised in their backyard, while butchers, having purchased animals at the market, owned the cattle they personally slaughtered and subsequently sold as meat. In addition to butchers, there were various meat vendors in early nineteenth-century cities and markets, such as peddlers and meat-sellers with mobile stands. All these figures animating the meat supply scenario often sold their own illegally butchered meat as well as recycled meat scraps (Horowitz, 2008).

With this picture in mind, it is not surprising that most chronicles of the time about animal nuisance harshly denounced slaughterhouses, often referring to them as "the greatest of all nuisances in a large city" (as cited in Otter, 2008, p. 91) and that these accounts abounded with references to the blood that flows copiously in the streets, almost turning it into a literary trope through repetitive and obsessive mentions.

Contemporary medical reports, newspapers' articles and public health councils' acts provide clear indications on the connection between non-human animals and health, specifically on what rendered all animal nuisances "injurious to life": it is "intolerable stench", "deadly exhalations", "the most horrid, offensive, and disgusting concentration of putrescent odours" (as cited in Atkins, 2012b, p. 27, 30). In a single word: smell. Blood, guts, and manure are the worst nuisances because they stink the most, because of their bad smell, rather than their colour, flavour, or consistency.

Theory of Miasmas and Medicine

What is so special about smell? What made it dangerous? The answer is to be found in the process of "miasmification", which influenced medicine from the late eighteenth century until at least the 1890s, coinciding with the triumph of Pasteur's and Koch's discoveries in microbiology. It is on the basis of miasma theory of disease that, for instance, Edwin Chadwick, a prominent English health and social reformer could state that "all smell is disease" (as cited in MacLachlan, 2007, p. 240). Thus, it is miasma theory the knowledge within which the concept of animal nuisance acquires meaning. To get a sense of how highly hazardous urban animals and their by-products in this framework were, we can refer to a key textbook of the time the *Copland's Dictionary of Practical Medicine* (1834–1856):

> The putrefaction of animal substances has been supposed by many to occasion disease in those who come within the sphere of the exhalations thus produced, and even to generate a malady which has become infectious, and has, partly thereby, and partly from other concurring causes, prevailed to an epidemic, or even pestilential, extent.
>
> *(as cited by Atkins, 2012b, p. 29)*

While these explanations may seem vague and non-specialized to contemporary readers, they spoke the language of science, specifically to the eyes – or perhaps more aptly, the nose – of doctors and public health officers of the time. This is the language of the aerist theory, based on the sense of smell, whose study on miasma is the main branch in the nineteenth century.

As the French historian Alain Corbin (1982/1986) points out in what is probably the most complete historical-social study of olfactory theories and odours in the eighteenth and nineteenth centuries[4]:

> The nose, as the vanguard of the sense of taste, warns us against poisonous substances. Even more important, the sense of smell locates hidden dangers in the atmosphere. Its capacity to test the properties of air is unmatched. [….] The nose anticipates dangers; it recognizes from a distance both harmful mold and the presence of miasmas. […] Increased recognition of the importance of the air led to increased acknowledgment of the importance of the sense of smell as an instrument of vigilance.
>
> *(p. 7)*

This "olfactory vigilance" was based on the new science of osphresiology (literally, the science of smells), i.e. a renewed interest in the principles of ancient Hippocratic medicine, mixed with epidemiology and boosted by the work of nineteenth-century chemists dedicated to translating it into a scientific language based entirely on smell. These scientists endeavoured to develop a nose-based lexicon to define the mixtures and to identify the stages of putrefaction with the aim of eliminating "the vagueness of the putrid" and achieving a complete understanding of the mechanisms of infection.

"Henceforth this vigilance had manifold aims: to detect irrespirable gases and particularly 'airs', and to discern and describe hitherto imperceptible viruses, miasmas, and poisons" (pp. 14–15). It had to be a 360° grade vigilance. The task, in itself worthy of Sisyphus, was aggravated by a convoluted, tricky, and ambiguous classification, with only a handful of indisputable key elements: fixed air, sulfuric acid, inflammable air, volatile alkali, and liver of sulphur. Among these, fixed air held a central role in the theory of putrefaction.

According to this theory, decomposition was believed to be an internal ceaseless movement in perpetual conflict with the principle of the natural cohesion of bodily parts, represented precisely by fixed air transmitted through the blood. Decomposing or diseased parts emitted a fetid odour accompanied by humidity – the released odours of once-fixed air now seeking new combinations. If accidentally someone inhaled these putrid miasmas, their equilibrium of internal forces (decomposition-cohesion) was altered and compromised in favour of decomposition, putrefaction, making the spread of plagues, fevers, gangrene, and other maladies possible (pp. 16–34).

Given the belief that blood carried fixed air, making it the most prominently putrid substance in non-human animal remains, it becomes clear why urban slaughterhouses had to be under "special surveillance" within the smellscape of the city, the reasons why they were considered "the greatest of all nuisances".

With the hygienic "revolution", animal stench also came to connote the malodorous people. The infected and plagued, thus, were to be carefully monitored through the social reading of the olfactory plot of the city, oriented towards localizing its miasmatic networks. No surprise, then, that knackers, gut dressers, butchers, cattle drivers, and urban cowkeepers, along with sewermen and drain cleaners, were at the top of the list of the "stinkers", bearing a dual meaning – both olfactory and, as we explore in the next section, moral.

Animal Protection and Morality in the Rise of Modern Slaughterhouse: Sight

The question of sight and the conceptual pair visibility/concealment do play a role in this *dispositif*, albeit a minor one. It is linked, from the outset, to the discourse of the societies for the prevention of animal cruelty, of the humanitarian leagues, of the anti-vivisection organizations of the vegetarian associations, etc., in a word that protectionist movement which in those same years developed in Europe and overseas with the aim of enacting animal protection laws and instilling good manners in the population through education.

This discourse insisted on the fear of moral degradation (brutalization) of humans, especially children, by means of cruelty to animals through both direct actions and mere observation of the spectacle of violence. Consequently, proponents of this perspective advocate for the concealment of such acts. The most important theory embraced by animal welfare associations, amalgamating new (Bentham's utilitarianism) and older arguments for a fairer treatment of animals (e.g., biblical or based on the idea of animal souls) (Maehle, 2002), is that of indirect moral obligations in particular in its Kantian version. According to this view, we must refrain from cruelty and violence against non-human animals to prevent cruelty and violence against humans. Recurrent were the tirades of humanitarians on the moral damage caused by the sight of cruelty in private slaughterhouses on "'the lad and little boys' that they attracted as onlookers" (MacLachlan, 2008, p. 110).

Humanitarians' pamphlets insisted on an iconography of the slaughterhouse where children peek in the doorway, or "peer through cracks in the fence, with the usual juvenile delight in sensational developments" (as cited in MacLachlan, 2008, p. 110), thus exposing themselves to a dangerous gradual de-moralization. It is essential to note that the focus of concern here is anthropocentric. As Tester (1991) argues, the moral push behind nineteenth-century pro-animal advocacy "is not a morality founded on the reality of animals, it is a morality about what it is to be an individual human who lives a social life" (p. 16).

Indeed, such a morality was not only anthropocentric but also deeply ingrained with classist and racist undertones. The lower classes, to which people employed in the meat production sector, from cattle transport to slaughter, belonged, were the primary target of fear and repulsion. A striking example can be found in Henry Salt's *Animals' Rights: Considered in Relation to Social Progress* essay, deemed the apex of radicalism in the nineteenth-century reflections on animal treatment. In this text, Salt asserts that "the sight, or the mention, or even the very thought" of the slaughtering process is so repugnant "to all people of sensitivity and refinement" that it can only be delegated to a "pariah class" (1892/1980, p. 61). This fear is, thus, mainly to be read as classist and racist dread of social disorder, i.e. fear of the "low moral quality" of the poor and the marginalized, which was, once again, a hygiene issue.

From the 1830s onwards, the racist view associated with animal welfare began to be directed against the Mediterranean region and the "inhumanity of southern European races" – to use the words of a British humanitarian in 1911 (as cited in MacLachlan, 2008, p. 110) – as cruelty to animals came to be seen as peculiarly Latin (e.g., attention was drawn to French vivisection, Italian brutality, Spanish bullfighting). Additionally, intra-national racist sentiments were evident in criticism of Jewish ritual slaughter (*shehitah*), reflecting a broader level of anti-Semitism, particularly in Germany (Brantz, 2002) and Britain (MacLachlan, 2008, pp. 115–117, 123–124).

In summary, as Ritvo (2002) observes regarding the British main protectionist association, the Royal Society for the Prevention of Cruelty to Animals (RSPCA), but the same is true of all the other animal welfare organisations that followed the British example and spread throughout Europe and the United States, it "feared social chaos and tended to focus on what it viewed as the disturbingly irrational behaviour of the uneducated and insufficiently disciplined segments of society" (p. 109).

This irrational behaviour was perceived as such in the framework of modern rationalization and its new processes and practices under the conditions of industrial capitalist society, which made the previous organization of non-human animal use painful, ineffective, and obsolete. The principle that guided animal advocacy groups in imposing "their 'bourgeois moral sensibilities' as a corrective to lower class cruelty", was "that it was wrong to inflict *avoidable* suffering on any animal" (MacLachlan, 2008, p. 110). "Avoidable" means no longer functional within the conditions of this framework and therefore no longer legitimate. For example, the beating of a cow while driving her to a private urban slaughterhouse became obsolete in a context where cows could be easily directed to the kill floor through a precisely designed path of pens and corridors. In this new context, beating a cow would be considered abuse, gratuitous violence, or a sadistic action. Conversely, practices that might be deemed abusive according to other parameters but were productive within this framework (e.g., the imprisoning of non-human animals in confined spaces), were functional and therefore considered legitimate. Hence, the anti-cruelty reforms stood against an outdated, and therefore perceived as uncivilized, system of organization of animal exploitation (the pre-reforms one), its institutions (decentralized slaughterhouses, city livestock markets), and its representatives (butchers, urban cattle owners, cattle drivers, meat traders, street vendors, etc.).

Indeed, as far as slaughterhouse struggles were concerned, their main aim was to abolish the old private slaughterhouses and to introduce municipal, centralized, and licensed abattoirs where humane slaughter could be implemented and surveilled. The fundamental principle of humane slaughter was that animals should be stunned before their blood was drowned, and to achieve this principle innovative science-based technologies were developed from the 1860s onwards, under the impetus of animal advocates. These demands aligned seamlessly with the hygienists, both in terms of reasons and objectives, reflecting a complete confluence with the hygienists' discourse of civilization and sanitation.

Hygeia, the City of Health: Old and New Transparencies

The presence of non-human animals with their smells and their dangerous miasmas, together with the spectacle of brutal cruelty, were no longer acceptable in the hygienist dream of the utopian healthy city of the future *Hygeia*, vividly depicted in 1876 by Benjamin Ward Richardson, an English doctor, temperance movement leader, and advocate of health reforms. Richardson's utopia outlined a meticulously planned city, featuring factories located outside urban areas, underground railroads and sewage systems, paved roads, model hospitals, public street cleaning initiatives, and strict bans on smoking and alcohol. In this envisioned utopia, the abattoir played a prominent role:

> The slaughter-houses of the city are all public, and are separated by a distance of a quarter of a mile from the city. They are easily removable edifices, and are under the supervision of the sanitary staff [...] All animals used for food [...] are subjected to examination in the slaughter-house, or in the market, if they be brought into the city from other depots. The slaughter-houses are so constructed that the animals killed are relieved from the pain of death. They pass through a narcotic chamber, and are brought to the slaughterer oblivious of their fate. The slaughter-houses drain into the sewers of the city, and their complete purification daily, from all offal and refuse, is rigidly enforced.
>
> *(As cited in Atkins, 2012c, p. 87)*

Hygieia represents the city where "progress" and "civilization" have been perfectly realized, and its new slaughterhouses are its flagship. From a moral point of view, there is nothing more to hide, to invisibilize in these facilities designed specifically to give adequate response to the miasmatic concerns initially and, subsequently, the bacteriological ones of hygienists: centralization, surveillance, inspection, and controls that are guaranteed, managed, and standardized by public authorities, either directly or indirectly. Moreover, it is not merely a question of being under the "public eye" – to use the words of the same humanitarian who stigmatized the presence of curious children at the doors of private slaughterhouses – in fact, he continues, "an abattoir [...] ought to be capable of having surprise visits paid to it by the public, as amateur inspectors of nuisances" (as cited in MacLachlan, 2008, p. 110). The abattoir is no longer afraid of "prying eyes", nor is it now seen as a moral monstrosity to be feared; it is now transparent. All are therefore welcome, visitors, families, and even children to witness what is now an edifying spectacle, as it is sanitized and hygienized, a spectacle to be exhibited with pride as a showcase of the progress of modernity, a testament to the achievements of technology and science.

In such a context, slaughterhouse tours became popular, with notable examples such as the family tours at the Union Stock Yard in Chicago. The leaflet produced by the Swift and Company, a prominent player in the industry, to promote these tours focused on the figure of a white girl on visit to slaughterhouse. The message is that tours of slaughter are not disturbing, that there is no reason to be haunted by the sights seen. This is reinforced at the end of the souvenir booklet. In this concluding section, the company states that the booklet serves as a "reminder of the sights of the Stock Yards", one enabling visitors "to see those sights again in memory" (Shukin, 2009, pp. 94–100).

Certainly, Hygeia is a utopia, and the reality was and still is very different: inadequate facilities, violations of health and animal welfare laws, violations of quality standards, abuse and mistreatment, etc. All this must not be seen. The places of animal exploitation have now become almost totally impenetrable to outsiders, making it essential to expose this obscene reality through undercover operations.

And yet, is it not still the model of the City of Health that seems to be working here?

"Video: The abusive slaughterhouse of horrors", "Shock: Pigs clubbed in a slaughterhouse". This is the tenor of the results of investigations by major animal welfare associations around the world. The problem is abuse, i.e. use which is not in accordance with the law – "ab-use", mis-treatment: the problem seems to be a question of degree of compliance with quality, hygiene, and animal welfare standards. It is absolutely necessary to denounce such situations. However, what happens when the farm or the abattoir comes dangerously close to the current utopian models that conform to the current "hygienic" norms (health standards, quality control, environmental sustainability, workers' rights, animal welfare)?

It may happen that slaughterhouses start organizing tours again, start promoting their transparency and their "being on the cutting edge". It may happen that they equip themselves with "glass walls".[5] And so it happens. For example, in Germany, in 2016 a "transparency campaign" was launched; the multinational meat company "Vion" offers video documentations of the slaughtering process (*Vion-transparency*, n.d.); the first slaughterhouses "with glass walls" were opened. A striking example is the Danish Crown pig slaughterhouse in Denmark, one of the largest in the world with an average of 20,000 "processed" pigs, equipped with a glass viewing gallery from which it is possible to see the entire production area and all the operations that take place there. Despite visiting shifts from Monday to Friday, the waiting list "is about four months for day shifts and about six months for evening shifts" (At the abattoirs, n.d.).

It is true that, in these "glass slaughterhouses", what is shown to the public is carefully selected, the *mise en scène* is the result of scrupulous visual architecture, often sparing the actual act of killing. However, this ostentatious exhibition of transparency, of "we have nothing to hide", today as in the mid-nineteenth century – in the light of considerations on the centrality of historically determined conceptions of hygiene – once again problematizes, and perhaps downsizes, the importance of sight, the equation between vision and truth regarding the violence of non-human animal exploitation so crucial to the moral approach to the animal question.

Returning to where we started, the glass-walled abattoir is a popular idea not only within the animal advocacy movement but also in the public opinion. The journalist and author Michael Pollan, for example, suggests: "maybe all we need to do to redeem industrial animal agriculture [...] is to pass a law requiring that the steel and concrete walls of the CAFO's and slaughterhouses be replaced with ... glass. If there's any new 'right' we need to establish, maybe it's this one: 'the right to look'" (2002). Are we sure that this is "all we need to do"?

Notes

1 English and Zacka, drawing on Pachirat's coinage and challenging of the politics of sight, enucleate the three central premises grounding this kind of activist strategy: "(1) that exposing morally repugnant practices will make us see them, (2) that seeing such practices will stop us from acquiescing to them, and (3) that owning up to such practices is preferable to keeping them concealed" (2022, p. 1025). For a further theoretical and philosophical critique of the politics of sight based on an analysis of Western language and sovereign subjectivity, see Castelló (2024).

2 One seemingly paradoxical outcome of this approach is the so-called "New Carnivore" ideology and related movement in which non-human animal slaughter is overtly performed, positively depicted, and proudly advocated for. This ideology is strictly connected to and popularized by television cooking shows, starring celebrity chefs, and cookbooks. See on this Parry (2010). In particular: "In doing so, they [new carnivores] soothe the anxiety that came to characterize the discourse surrounding meat production in the 1980s and 1990s, as spiralling food scares, well-publicized health risks, and increasing popular awareness of the environmental and ethical problems associated

with industrial animal agriculture all combined to undermine the traditional prestige of animal flesh in western societies. They also counter the popular criticism that urban meat-eaters are fundamentally disconnected from the realities of meat production, wilfully ignorant as to the death and violence that made possible the meat on their plates" (pp. 381–382).

3 Other central knowledges of the *dispositif*, which are beyond the scope of the present discussion, are nutrition science, based on the chemical discoveries of the various components of food and in particular Von Liebig's isolation of proteins, on the one hand, and Pasteur's bacteriology, on the other hand, which stabilizes the thrusts for slaughterhouse reform already initiated by the hygienist movement under the miasmatic theory.

4 For another classic text on the cultural history of smell, see Classen et al. (1994).

5 For an analysis starting from the same issue of slaughterhouses with glass walls and transparency campaigns and oriented to problems of representation, see Zenker (2019).

References

Aloe, F. & Stefanoni, C. (2021). Anatomia della nazione. Dalla formula trinitaria alle forme della produzione di popolazione. *Consecutio Rerum*, *10*(1), 357–382.

Arcari, P. (2017). Perverse visibilities? Foregrounding non-human animals in 'ethical' and 'sustainable' meat consumption. *The Brock Review*, *13*(1), 24–52. https://doi.org/10.26522/br.v13i1.1317.

At the abattoirs. (n.d.). Danish Crown. Retrieved January 14, 2024, from https://www.danishcrown.com/global/contact/visit-danish-crown/.

Atkins, P. (2012a). Introduction. In P. Atkins (Ed.), *Animal cities: Beastly urban histories* (pp. 1–17). Ashgate.

Atkins, P. (2012b). Animal wastes and nuisances in nineteenth-century London. In P. Atkins (Ed.), *Animal cities* (pp. 33–66). Ashgate.

Atkins, P. (2012c). The urban blood and guts economy. In P. Atkins (Ed.), *Animal cities* (pp. 77–106). Ashgate.

Ball, M., & Friedrich, B. (2016). The animal activist's handbook: Maximizing our positive impact in today's world. In *The animal ethics reader* (pp. 741–749). Routledge.

Brantz, D. (2002). Stunning bodies: Animal slaughter, Judaism, and the meaning of humanity in imperial Germany. *Central European History*, *35*(2), 167–193. https://doi.org/10.1163/15691610260420656.

Calarco, M. (2015). *Thinking through animals: Identity, difference, indistinction*. Stanford University Press.

Castelló, P. P. (2024). Why seeing is not believing and why believing is seeing: On the politics of sight. *American Political Science Review*, 1–11. https://doi.org/10.1017/S0003055423001491.

Classen, C., Howes, D., & Synnott, A. (1994). *Aroma: The cultural history of smell*. Routledge.

Coleman, W. (1982). *Death is a social disease: Public health and political economy in early industrial France*. University of Wisconsin Press.

Corbin, A. (1986). *The foul and the fragrant: Odor and the French social imagination* (M. L. Kochan, R. Porter & C. Predergast, Trans.). Berg Publishers. (Original work published 1982).

Elias, N. (1978). *The civilizing process: The history of manners* (E. Jephcott, Trans.). Blackwell. (Original work published 1939).

English, J., & Zacka, B. (2022). The politics of sight: Revisiting Timothy Pachirat's every twelve seconds. *American Political Science Review*, *116*(3), 1025–1037. https://doi.org/10.1017/S0003055421001192.

Foucault, M. (1980). The confession of the flesh. In C. Gordon, & C. Gordon (Eds.), *Power/knowledge: Selected interviews and other writings, 1972-1977* (pp. 194–228). Pantheon Books. (Original work published 1977).

Foucault, M. (1994). *The order of things: An archeology of the human sciences* (A. Sheridan, Trans). Vintage Books. (Original work published 1966).

Horowitz, R. (2008). The politics of meat shopping in antebellum New York City. In P. Y. Lee (Ed.), *Meat, modernity, and the rise of the slaughterhouse* (pp. 178–197). University of New Hampshire Press.

La Berge, A. E. F. (1992). *Mission and method: The early nineteenth-century French public health movement*. Cambridge University Press.

MacLachlan, I. (2007). A bloody offal nuisance: The persistence of private slaughter-houses in nineteenth-century London. *Urban History*, *34*(2), 227–254. https://doi.org/10.1017/S0963926807004622.

MacLachlan, I. (2008). Humanitarian reform, slaughter technology, and butcher resistance in nineteenth-century Britain. In P. Y. Lee (Ed.), *Meat, modernity, and the rise of the slaughterhouse* (pp. 107–126). University of New Hampshire Press.

Maehle, A. -H. (2002). Cruelty and kindness to the 'brute creation': Stability and change in the ethics of the man-animal relationship, 1600–1850. In A. Manning & J. Serpell (Eds.), *Animals and human society: Changing perspectives* (pp. 113–137). Routledge.

Marx, K., & Engels, F. (1976). The German ideology. In *MECW* (vol. 5 1845–1847) (C. Dutt, W. Lough & C. P. Magill, Trans.). International Publisher. (Original work published 1932).

Otter, C. (2008). Civilizing slaughter: The development of the British public abattoir, 1850–1910. In P. Y. Lee (Ed.), *Meat, modernity, and the rise of the slaughterhouse* (pp. 89–106). University of New Hampshire Press.

Our World in Data. (2023). Meat and dairy production [Infographic]. https://ourworldindata.org/meat-production#milk-production-across-the-world.

Pachirat, T. (2011). *Every twelve seconds: Industrialized slaughter and the politics of sight*. Yale University Press.

Parry, J. (2010). Gender and slaughter in popular gastronomy. *Feminism & Psychology*, *20*(3), 381–396. https://doi.org/10.1177/0959353510368129.

Pollan, M. (2002, November 10). An animal's place. *New York Times Sunday Magazine*. https://www.nytimes.com/2002/11/10/magazine/an-animal-s-place.html.

Ritvo, H. (2002). Animals in nineteenth-century Britain: Complicated attitudes and competing categories. In A. Manning, & J. Serpell (Eds.), *Animals and human society* (pp. 106–126). Routledge.

Salt, H. (1980). *Animal rights: Considered in relation to social progress*. Clarcks Summit: Society for Animal Rights.

Shukin, N. (2009). *Animal capital: Rendering life in biopolitical times*. University of Minnesota Press.

Singer, P. (2002). *Animal liberation*. Ecco Press.

Tester, K. (1991). *Animals and society: The humanity of animal rights*. Routledge.

Twine, R. (2012). Revealing the "animal-industrial complex": A concept & method for critical animal studies? *Journal of Critical Animal Studies*, *10*(1), 12–39. https://www.criticalanimalstudies.org/wp-content/uploads/2012/10/JCAS±Volume±10±Issue±1±2012±FINAL.pdf.

Vion-transparency. (n.d.). Vion. Retrieved January 14, 2024, from https://www.vion-transparency.com/.

Wadiwel, D. J. (2016). Do fish resist? *Cultural Studies Review*, *22*(1), 196–242. https://doi.org/10.5130/csr.v22i1.4363.

Zenker, F. (2019, May 23). *Transparency without care: Austria's first "Slaughterhouse with glass walls" from a care-ethical perspective* [Conference session]. Rethinking revolution: Nonhuman animals, anstipeciesism, and power. 6th European Association for Critical Animal Studies (EACAS), Universitat Pompeu Fabra, Barcelona, Spain.

6

BEEF, BIBLE, BULLETS

Suicidal Cows and the Ecological Imaginings of Brazil

Jessica Carey-Webb

Edgar Wilson is a man who does not pride himself in his work. Day in and day out he murders cattle at a slaughterhouse where the meat is later turned into hamburgers at a nearby processing plant. He prays for the salvation of the souls of each cow that he stuns and kills. Looking deep into their eyes, he recognizes his own reflection and assumes that it is better for him to do this work than another, less sympathetic murderer. Despite his best efforts, one day the cattle rebel and commit a mass suicide. In this nondescript Brazilian countryside, dystopian dramas unfold within the walls and surrounding fields of the slaughterhouse.

In the global context of inequality and climate crisis, works of Latin American fiction have increasingly addressed the implications of environmental destruction, humans' relationship to the environment, and extractivism in the region. Fiction can help us to examine the environmental toll and complex relationship between nonhuman animals and humans. Brazilian author Ana Paula Maia captures the tensions and cultural impact around meatpacking in her novel *De gados e homens* (2013) or *Of Cattle and Men* (English language translation published in 2023).[1] The book tells the story of Edgar Wilson, a slaughterhouse worker who empathizes with the cattle he is tasked with killing and climaxes when he witnesses the cattle flinging themselves off a cliff.

Of Cattle and Men culminates in this suicide of cattle destined for slaughter. The suicide demonstrates the cattle's cognizance and active resistance to consumer-capitalism. Throughout the book, the lines between cattle and man are continually blurred and complicated. Men are treated as nonhuman animals and, at times, nonhuman animals take on the characteristics of men – calling into question the very idea of humanity. The book focuses on the experiences of Edgar Wilson who interacts with other slaughterhouse workers including a boss, Seu Milo, a new arrival, the half-Indigenous former reindeer slaughterer Bronco Gil (a fixture in

DOI: 10.4324/9781003441908-8

other works by Maia), and another worker, Helmuth. There is also a group of desperate peasants who show up at the plant begging for meat and rejoice after the cows commit suicide, as they gain access to this potential source of nourishment. Throughout the short, 100-page novel, the slaughterhouse is a backdrop akin to a post-apocalyptic horror film. In contrast with colonial chroniclers who described Latin American environments as lush and verdant, the ecology in more recent fiction responds to the climate crisis by depicting drastic wastelands, areas devastated by human encroachment, pesticides, and large-scale agribusiness (Lehnen, 2020, p. 24). The plant and surrounding areas are barren wastelands, demonstrating the toll that the industry has wrought not only on its workers and the nonhuman animals they slaughter, but also on the environment.

Cattle ranching has a massive economic and cultural foothold in Brazil. The desolate surroundings of the slaughterhouse depicted in the novel are emblematic of many places in the Brazilian countryside, where cattle ranching accounts for more than 8.5% of the national gross domestic product (Malafaia et al., 2021). Ranchers fashion themselves as masculine cowboys in the style of the American West, participating in rodeos, wearing cowboy hats, and most alarmingly forming a powerful governmental lobby that supported the ascendance of far-right former President Jair Bolsonaro (in power between 2018 and 2022) (Hoelle, 2015). The agribusiness sector, Evangelical churches, and pro-arms lobby combined to win Bolsonaro the presidency and continue to support far-right policies.

This so-called "beef caucus," or "beef, bible, and bullets caucus" (BBB) is made up of large-scale cattle ranchers and agriculturalists who now populate much of the Brazilian legislature (Maisonnave, 2023). A coalition that formed in 2012, in direct opposition to a legislative discussion about the Brazilian Forest Code, the BBB caucus has expanded beyond environmentally detrimental policies to supporting hardline stances against women, the LGBTQIA+ community, abortion, and gun control. While Brazilian political leadership has shifted back to former center-left-wing and more environmentally friendly President Lula da Silva, the cattle lobby, at the heart of the BBB caucus, still holds a massive sway on the political scene. They aim to legalize more deforestation for cattle ranching, arguing that existing land demarcation for conservation and Indigenous territory is detrimental to the export economy. The cattle industry and BBB politics stems from the colonial history of Brazil and is representative of the pervasive effect of violence and harm toward both human and nonhuman animals. These three pillars are also representative of a masculine hegemony, which is apparent throughout the novel.

With the cattle industry at the heart of the political and cultural landscape, I use *Of Cattle and Men* as an entry point to examine global meat production and its environmental and societal implications, as well as the relationship between nature and culture, "man" and nonhuman animal. Indeed, I loosely follow an organization based on the BBB coalition. In the sections to follow I examine how beef, or large-scale cattle ranching has shaped Brazil, as well as how the novel uses religious imagery (bible) and differing types of violence (bullets) to critique the capitalistic and environmentally destructive cattle ranching industry. While these categories correspond to themes within the novel, they overlap and intersect, as religion is used to justify the violence associated with domination over humans and nonhuman animals. As such, I argue that Edgar Wilson, the protagonist of the novel, demonstrates a level of caring toward nonhuman animals that transforms their mass suicide into an act of resistance that shirks boundaries and opens new possibilities. In this act, the cattle become martyrs and a source of empathetic connection that pushes the reader to question their own consumption. There is a sense of freedom in the suicide, instead of slaughter even at the hands of a sympathetic killer, they choose to jump to their death, rendering their meat almost useless.

The novel works between the categories of human and nonhuman animal – cattle and men – engaging politically to question assumed dichotomies. Indeed, the focus on men rather than women or children is also notable and points to the masculinized politics of the cattle industry. Considering the nonhuman is one way of rethinking the hierarchies of modernity through understanding and respecting not only ecologies, plants, and nonhuman animals as potential soul-bearing beings, but also in recognizing the diversity of humanity itself. Furthermore, the novel exposes the toll of violence toward nonhuman animals, devalued humans, and the environment that are inherent in the cattle industry.

Cattle and Environmental Domination: Beef

Brazil has one of the world's largest cattle industries and is the foremost global exporter of beef (Cheema, 2023). It is also home to leading international meatpacking companies including JBS, Marfrig, and Minerva.[2] JBS alone is one of the world's largest emitters of greenhouse gasses (Gelles, 2023). The Institute for Agriculture and Trade Policy estimates that JBS produces around half of the comparable annual fossil fuel emissions of ExxonMobil, Shell, or BP (IATC, 2022).[3] The Brazilian cattle industry is responsible for deforestation, an outsized carbon footprint, and terrible labor practices that affect worker health and safety. These practices are visible throughout *Of Cattle and Men* as workers struggle with the violence of their job and the countryside suffers from pollution.

The environmental and societal harm wrought by the beef industry and described in the novel centers around varying types of violence. Rob Nixon argues that "progress" often means environmental degradation with devastating consequences suffered not only immediately but also over time especially by the poorest and least powerful, what he deems a "slow violence" (2011). The violence in *Of Cattle and Men* is both direct and bloody in the slaughter and suicide of the cows but is also a slow violence demonstrated by the environmental destruction of the cattle industry, and the psychological toll on the slaughterhouse workers. This violence deeply affects Edgar Wilson as he grapples with his role throughout the novel: "He sighs, sorrowful. It's his job, the only one keeping him alive. He looks back. The ruminants that are grazing quietly, standing in groups or alone – soon he'll come face to face with all of them: Edgar, the murderous beast himself" (p. 15). Due to the violent nature of his work, Edgar transforms into a beast, yet he too is trapped, as this murderous job is the "only thing keeping him alive" – his way of making a living. The justification of killing cattle, as "beasts" thus vanishes, as Edgar, too, sees himself as less than human. The cattle, grazing in the pasture, offer a peaceful vision that is upset by the violence that unfolds within the walls of the slaughterhouse and at Edgar's hands. The cows that Edgar kills form part of a larger culture of domestication of nonhuman animals for economic expansion.

Violence, cattle ranching, and the accompanying meat culture, as well as religious conservatism, can be traced to the colonial history of Brazil. Since colonization, domesticating nonhuman animals was seen as an essential part of civilization and progress, a process that David Nibert deems "domesecration." Domesecration is a term that captures the violence involved in the domestication of nonhuman animals, who are in fact desecrated in the process. In the colonization of the New World, domesecrated nonhuman animals were a key aspect of settler-colonialism and the so-called civilizing mission, crucial to the growth and development of empires. The colonial project was "built on white, 'humanist,' Christian supremacy and fueled by a nonhuman animal, viral, and extractivist economy" (Gerspacher, 2022, p. 70). Indeed, the BBB caucus mirrors the three "C's of colonialism": civilization, Christianity, and commerce. Nonhuman animals, while being dominated themselves, were also used in colonization to dominate nonwhite humans and further extractive industries.

Cattle ranching, like fossil fuels, is an extractive industry because of large-scale farming practices that change soil composition, increase emissions, dump pollutants into the surrounding areas, as well as the deforestation involved in farming cattle and soy for their feed. Extractivism worked hand in hand with the colonial project by creating disposable peripheries with the idea that there will always be a new place to exploit once resources are depleted at current sites of extraction (Klein, 2014, p. 170). Extractive industries create slow violence as a structural and systemic means of providing wealth based on nonrenewable products. For example, Maia describes a bloody river near the slaughterhouse, which is a visible marker of the violence and desecration involved in the nonhuman animal slaughter industry: "This river is where all the slaughterhouses in the area dump thousands of liters of blood and cattle viscera. The river flows into the sea, and so does the blood of these beasts" (p. 4). Not only does the slaughterhouse cause pollution in the areas surrounding it, but the remnants flow to the sea, thus affecting the global environment.

The violence generated toward domesticated nonhuman animals is akin to the violence experienced by "devalued humans," like Edgar and his coworkers. As Nibert explains, "These include invasion, conquest, extermination, displacement, repression, coerced and enslaved servitude, gender subordination and sexual exploitation, and hunger. Accompanying such violence have been deadly zoonotic diseases that have contributed to the destruction of entire cities, societies, and civilizations" (2013, p. 5). The same structures that are used to domesecrate nonhuman animals and exploit the environment are based in colonial, racial, and extractive histories. Expansion was enabled by domesecrating nonhuman animals, as increased farmland was needed to fuel expanding cities and empires.

Domination over nature, nonhuman animals, and "devalued" humans is a central theme in *Of Cattle and Men*. Western thought is based on humanism, which centers the separation between nature and culture, or the so-called civilized and barbarian. Donna Haraway explains that "The discursive tie between the colonized, the enslaved, the noncitizen, and the animal—all reduced to type, all Others to rational man, and all essential to his bright constitution—is at the heart of racism and flourishes, lethally, in the entrails of humanism" (2016, p. 15). Western culture uses categories and dualisms to maintain structures of domination, creating hierarchies of existence based on othering those outside of the category of white, European, and male. To upset this order, as the novel progresses, humanistic categories blur and disappear, "Two enclosures, one for cattle and one for men, standing side by side. Sometimes the smell is similar. Only the voices on one side and the mooing on the other distinguish the men from the ruminants" (p. 9). Within the confines of the slaughterhouse, the only distinction between human and nonhuman animal, separated by enclosures, is who wields power, in this case the men over the cattle, although these same men have very little power outside of the slaughterhouse. The novel plays with the animalism of men and the humanity of nonhuman animals, which culminates in the cattle's suicidal act.

The novel points out the overlap not only between man and cattle and nature and culture, but also the consumer and the product. The book can be read as a sustained critique on consumer-capitalism. The ambiguously racialized workers in *Of Cattle and Men*, through their participation in the labor economy and as men, hold a slightly higher standing than the peasants that populate the countryside and the nonhuman animals that they slaughter. However, slaughterhouses are notorious for their abysmal and abusive labor practices. As Lígia Bezzera explains, Maia's writing focuses on the consequences of consumer society for exploited workers. The characters in Maia's novels "do the dirty work of others" such as laboring in slaughterhouses, coal mines, or in garbage collecting. These workers often find themselves exploited by their bosses or trapped in a system that they cannot escape while others reap the benefits of this "dirty work." In exposing the inner workings of these undesirable jobs that no one else wants to do, the novel questions practices of consumption, and challenges the capitalistic structure that separates man from nature and the consumer from product.

To that end, hamburgers are woven throughout the novel as an anti-capitalist critique. The hamburger is symbolic of Western cultural domination, particularly of the United States with its mass export of McDonalds around the world. When the slaughterhouse workers receive a box of hamburgers to enjoy after work, there is a notable separation between the origin of the food and the food itself, much like a separation between the worker and the product: "Together with his coworkers, they eat the whole box in astonishment. Round and seasoned like that, they don't even look like they had ever been a cow. Not one glimpse of the unbridled horror behind something so tender and delicious" (p. 10). These hamburgers appear unconnected to the "unbridled horror" of slaughter, their round shape and seasoning rendering them something entirely changed from the cows whose souls Edgar is intent on saving.

The novel critiques the consumer-capitalism structure, the environmental and psychological toll of the beef industry, and, indirectly, the colonial legacy of these structures of inequality. Edgar is at once sensitive, empathetic, and highly critical of his job and place in the social hierarchy. As a human with an outsized connection with nonhuman animals, his care and attention to their needs before their slaughter presents a question of potential vindication: can a murderer who sympathizes with his victims in some ways be justified? Or, if the cows will be slaughtered anyway, isn't it better to have them killed at the hands of a sympathetic executioner who will bless their souls? In this case, Edgar's salvation lies in the cow's actions, as his ritualistic slaughter seems to embolden them toward an act that upsets the system, a theme examined in the next section.

Blessing the Beast: Bible

Edgar is different from most of the other slaughterhouse workers. He develops a relationship with each cow before he ends their lives. He gently approaches them and eases their discomfort in a series of rituals. He even goes as far as making the sign of the cross on the forehead of the nonhuman animals before he stuns them. In contrast to his colleagues, he attempts to kill the cattle as quickly as possible, avoiding pain when he can. As the novel asserts: "He's not proud of what he does but if someone has to do it, then let it be him, who has pity on those irrational beasts" (p. 4). Edgar's approach is intentional and ritualistic, he can empathize and "take pity on" the cattle he kills. However, at this initial stage in the novel, they are still "irrational beasts" that, presumably, are deserving of this fate, or at the very least lack an awareness of their situation. Edgar's mission of blessing each cow's soul prior to slaughter can be seen as an allegory for the process of colonization in the Americas, when priests and missionaries sought to "save" Indigenous peoples' souls while forcing them into slavery, stealing their land, and drastically changing their way of life. Again, we can see a structure of dominance with men, particularly white men, at the top.

In his killings, Edgar follows a ritual, echoing religious ceremonies by making the sign of a cross. He establishes a mutual trust with the cow, communing with them just before he takes their lives. While he dislikes this ritual, he continues to perform it. Due to this recognition and increasingly porous borders between species, Edgar is transformed into a kind of missionary tasked with saving the cattle's soul's pre-slaughter: "He doesn't care who eats the cows he slaughters; he cares about ordaining the souls of each ruminant that crosses his path. He believes these animals have a soul, and that he must tend to each one when they die. One soul out of five hundred" (p. 50). Religious undertones pepper Maia's prose, creating a more stringent dichotomy than that between man and nonhuman animal – a strong line between good and bad. As Lidiana Barros and Fernanda Abreu note, religion is the backbone of Edgar's guiding principles (Barros & Abreu, 2022, p. 43). While he realizes that his violent acts will not allow him divine forgiveness, he uses his connection and blessing of the cattle to feel less alone. This is a job that, again, no one *wants* to do. Edgar performs undesirable work, that on a surface level, humans benefit from (by eating hamburgers, for example). Although these benefits, as evidenced by the outsize environmental impact and poor working conditions, are finite, Edgar seems to see his job as a personal sacrifice and, perhaps, by his gentle approach, a means of atonement.

Edgar works with several other characters who set a contrast to his interactions with the cattle. Within the first few pages of the novel, Edgar kills his coworker, Zeca, precisely because of his sadism toward the cows. After repeated warnings to not stun the cattle in a way that would inflict more pain than necessary, Zeca continues to torment the cows and because of this, Edgar kills him much like he would one of the ruminants: "With a mallet, the tool of his trade, he strikes the boy squarely on the forehead, leaving him on the ground in violent spasms and low moans. Edgar Wilson makes the sign of the cross before lifting Zeca's dead body and wrapping it in a blanket. Not one drop of blood was spilled. His work is clean" (p. 10). While Edgar performs the "dirty work" of others, this murder of an actual human is, instead, clean. Edgar does not view his cattle slaughter as a justifiable act, yet he does view Zeca's murder as such, pointing to his sympathy and caring toward the nonhuman animals, and his unique moral compass. When Edgar's boss asks about Zeca, Edgar informs him honestly and directly of his murder, and after no one shows up looking for him, the matter is quickly forgotten. In adding this story toward the beginning of the book, Maia once again sets up an ethical and moral conundrum through blurring the lines between cattle and man. Ultimately, in killing Zeca, Edgar equalizes the playing field between man and nonhuman animal. If one is causing outsize damage to the other, then they must suffer the consequences.

As the novel progresses, Edgar's connection with the nonhuman animals he kills grows: "Edgar feels so connected with the ruminants, with their unfathomable gaze and the vibration of their bloodstream, that sometimes he loses his conscience when questioning who is the man and who is the ruminant" (p. 68). Seemingly it would be his conscience that pushes him to spare the nonhuman animals any outsize pain before their slaughter. The warmth and vibration of the cattle's bloodstream, akin to human blood and warmth, pulls them closer to humanity, and, as such, Edgar. Yet, as he grows more connected with the cows by the prolonged gazes that he shares with them, he loses his sense of personhood, connecting more with the nonhuman animals than other humans like Zeca. As Bezerra points out, "The nonhuman animalization of humans and humanization of animals that we see in Maia's novels suggests their shared condition as victims of the same oppressive system in which the many are exploited in order to advance the goals of the few" (2022, p. 124). Consciousness, or the conscience mentioned here is a fundamental part of what is regarded as a soul. In the colonial project, saving the souls of Indigenous peoples was used as a justification for expansion and in many cases enslavement, forming one of the three c's – Christianity– or in the BBB designation, the bible. In this novel, human and nonhuman animal recognize themselves in each other as soul or conscience bearing beings.

After the cattle's suicide (explored in detail in the next section), the characters are left grappling with what this act means. As such, they, like Edgar, turn to religion: "The cattle's collective suicide would remain unexplained. Perhaps it was Divine Providence answering the prayers of the local residents who begged for food, especially meat. Just as the Israelites were met with a shower of quail, people of other deserts received a shower of cattle: meat from the heavens; death that gives life" (p. 95). The slaughterhouse workers have better circumstances than the peasants who periodically appear at the gates of the slaughterhouse to beg for meat. While the slaughterhouse workers' employment is far from glamorous, they are employed, and as such are able to regularly buy the very hamburgers they have a hand in making, the catch-22 of consumer-capitalism. When the cattle commit suicide, these poor country-folk flock to the dead cattle to get what they can before it spoils ("The cows jumped off the cliff. Our prayers have been answered" (p. 88)). As such, the suicide also upsets the capitalist order. In their suicidal rebellion, the cattle have sacrificed themselves to feed the hungry.

Domesecrated nonhuman animals, a key aspect of the colonial project, can be understood here as an allegory for economic expansion, and domination by white European powers, factors often justified by a divine mission to evangelize and "save the souls" of Indigenous peoples. The religious undertones that pepper the novel again point to the overlap between categorical distinctions and the links between the animal-industrial complex to colonialism and consumer-capitalism.

Suicide and Violence: Bullets

The global cattle industry is fueled by violence. While the "bullets" of the BBB caucus represent the gun lobby, in this section I use "bullets" to analyze the types of violence that permeate the narrative. The novel draws on both immediate, visceral violence (slaughtering of the animals, murder of a coworker, the collective suicide) and slow violence exemplified by the environmental impact of the slaughterhouse. The novel also acknowledges a gender dimension, or the role of men in perpetuating these varying types of violence. Men, particularly those in positions of power within Western industrial sectors, are the most resolute supporters and beneficiaries of extractive practices and climate change denial, with the term "industrial" emphasizing their role in wealth creation. The animal-industrial complex can therefore be seen as part of this societal structure where men are the primary perpetrators of slow violence and social inequalities (Hultman & Pulé,

2020, p. 1). This spectrum of violence is demonstrative of the pervasive effects of violence against nonhuman animals, which is a gendered enterprise, a product of a consumer-capitalist economic structure.

Visceral violence is most acute with the climactic suicide of the cattle. After noticing that something is "off" with the cattle, Edgar invites two other slaughterhouse workers, Bronco Gil, and Helmuth, to stalk the cattle and watch their movements unfold:

> The first cow jumps and then the second. Bronco Gil tries to put a stop to it, but is blocked by Edgar and Helmuth, who have decided to just watch the horrific spectacle. And so the cattle keep jumping, one after the other, until all of them have leapt over the edge, emitting long moos.
>
> *(p. 86)*

In this case seeing is believing, as the workers, aside from Edgar, were previously incredulous that cattle could enact free will and kill themselves. As they witness the spectacle, Edgar does not want to interfere, but rather watches in awe. Their leap, punctuated by moos, becomes at once humorous and freeing. The cows' jump symbolizes an end or at the very least a disruption to the consumer-capitalist structure (Bezerra, 2022, p. 128).

When the men confront the slaughterhouse boss, Seu Milo, with news of the cattle's suicide, he states blankly that "Cows don't kill themselves, we're the ones that kill them" (p. 88). Despite his protests, Seu Milo seems to respect the cattle because of their choice, even though he loses money with their jump. And, when he witnesses the dead cow's bodies, he transforms. As the novel ends: "Milo looks out over the cliff. The horizon is aglow, and the sun is partially risen behind the mountains, peeking out in golden rays. All the weight on Milo's face vanishes. He takes a deep breath and, for a moment, feels peace. A small peace, but sensitive to his spirit" (p. 89). As Seu Milo looks over the cliff where the dead cattle are piled, he gains a sense of calm and harmony. This peace fuels his spirit, again linking to the conscience and the question of what beings carry a soul. Despite losing money, he gains peace. This moment is a subtle rebellion against consumer-capitalism and the environmental toll of cattle ranching as Seu Milo communes with nature, feeling the sun on his face. It also seems to symbolize the dawn of a new day where perhaps the cattle's act will restructure the consumptive practices that have contributed to this dystopia.

Maia was influenced by a *Daily Mail* article from 2009, headlined "Police baffled as dozens of 'suicidal' cows throw themselves off cliff in the Alps" about the mysterious death of cattle in Switzerland (Mail Foreign Service).[4] Dozens of cows threw themselves off a cliff in a matter of three days, with no explanation. She said that the picture of the cows, their brown hides piled on top of one another at the bottom of a cliff inspired her to write the book. She later did research on slaughterhouses, their impact on surrounding areas, and even visited several while working on the novel. Despite attention-grabbing headlines like the one that inspired Maia, can nonhuman animals commit suicide? And if so, what are the implications for thinking about human-nonhuman animal relationships?

There is a debate among scientists and scholars about whether nonhuman animals are capable of committing suicide, an act that demonstrates a certain amount of free will, one of the defining factors used to differentiate humans and nonhuman animals. David Peña Guzmán posits that, because humans are taught their superiority over nonhuman animals from an early age, they have long assumed nonhuman animals to be incapable of committing suicide (2017, p. 11). However, the cow's suicide within the novel is a cognitive choice that challenges the supposed superiority of nonhuman animals to humans. If these cows know they are destined to be killed, why not take charge and kill themselves, subsequently (although indirectly) rendering their meat, themselves as a product, relatively useless.

The cow's suicide is not so much a sacrifice as an act of rebellion demonstrative of cognizance.[5] Throughout history, nonhuman animals have exhibited social and political resistance against human domination, persisting with agency despite widespread exploitation in a global capitalist system, yet their centuries-long struggle for liberation and justice often remains unacknowledged in recorded history (Colling, 2020, xiv). Within the novel, while the men and cattle are killed to continue to produce meat, and the peasants of the countryside benefit from the cow's self-destruction, I read this act as an upset in the social and capital order designed to liberate the cattle, and subsequently, the men. As Eva Meijer explains, animal rights theorists have challenged the perception of nonhuman animals as mere objects within human legal, political, and cultural systems, advocating since the 1970s for their recognition as sentient beings deserving of inclusion in moral communities, prompting recent work to consider them as political groups and members of interspecies communities, shifting the focus from how they should be treated to understanding their desires for life, relationships, and coexistence on the planet (2019, p. 4). Much of this research is enmeshed in recognizing the pointed resistance to human domination of many nonhuman animals. In *Of Cattle and Men*, the suicide frees the workers from their oppressive role in continual killing, at least for a short time. It also showcases the autonomy of the nonhuman animals themselves. These cattle refuse to be domesecrated, choosing to instead fling themselves to their death, and, in turn, causing reflection and potential transformation on the part of the men involved in this industry.

Conclusion

Although this chapter is structured around the themes and political implications of BBB, it is easy to note that none of these categories are binding, but rather they overlap and at times are indistinguishable, much like the distinctions between humans and nonhuman animals within the novel. Coursing through each category is violence: toward men, toward nonhuman animals, toward the environment. Indeed, the environmental and societal harm toward domesecrated animals is enmeshed with violence against devalued humans, and reveals the gendered markers that are used to justify consumption and extraction. Beyond exposing the violence of this industry, *Of Cattle and Men* also reveals a sensitivity and connection between species that transcends domesecration.

The novel offers a suggestion toward how human and nonhuman animal relationships can transform. If we can recognize the free will in the suicidal act of the cattle, or we can connect with their eyes, a so-called "window to the soul," perhaps there can be a reordering of the power dynamic. In recognizing that the round shape of a seasoned hamburger came from a soul-bearing being with much in common with humans, perhaps we can consume fewer steaks on the "other side of town." The book offers a view into the interactions between humans and nonhuman animals, the porous nature of humanity, and the slow violence of the cattle industry.

Notes

1 In this chapter, I use English language translation for relevant quotes.

2 The United States has heavily invested in the Brazilian meat industry: "The big three Brazilian meat companies—JBS, Marfrig, and Minerva—were all made possible by decades of business dealings and mergers facilitated by a history of US military and economic intervention in Latin America in the 1960s and '70s, as well as by loans from the World Bank and other neoliberal trade mechanisms" (Gerspacher, p. 71).

3 The Brazilian cattle industry's shady practices were also at the heart of the *lava jato* scandal that ousted former President Dilma Rousseff and led to Bolsonaro's election.

4 I had the opportunity to speak with the author about the novel, her process, and inspiration in April 2023.

5 Bezerra reads the cattle's suicide as a "sacrificial ritual transmuted into transgression": "The literal slaughtering of the cows and the metaphorical slaughtering of the men would represent the hecatomb of consumer capitalism: the sacrifice of the many so that the few can prosper. On the other hand, a mysterious mass suicide committed by the cattle can be interpreted as a sacrificial ritual transmuted into transgression" (p. 126).

References

Barros, L. O., & Abreu, F. M. (2022). De gados e homens: a carne processada na obra de Ana Paula Maia. *Terra Roxa e outras terras: Revista de Estudos Literários*, *42*(1), June.

Bezerra, L. (2022). *Everyday consumption in twenty-first century Brazilian fiction*. Purdue University Press.

Cheema, R. (2023, February 3). 15 countries that export the most beef in the world. *Yahoo Finance*. https://finance.yahoo.com/news/15-countries-export-most-beef-181025270.html?guccounter=1&guce_referrer=aHR0cHM6Ly93d3cuZ29vZ2xlLmNvbS8&guce_referrer_sig=AQAAAD-vKcUKGp5185L_Q9GOAPo18zq-avQw4KAD-JoKv_ZAAHd83MJEtmgb8Q38JVKQ4UUjHKVAB0aoW7tbzCjaJiL4y-6-2ipXjX-lMJBEylJymU5fZc4Dtg-vm3v_XOm1iGtmYakYbOE-EGeEsfL78Bj9YbSDU_JVUSs87n-FJwASPF.

Colling, S. (2020). *Animal resistance in the global capitalist era*. Michigan State University Press.

Gelles, D. (2023, September 28). Grilling the world's biggest meat producer. *The New York Times*.

Gerspacher, A. (2022). Zoonotic undemocracy. *October Magazine, Ltd. And Massachusetts Institute of Technology*, *181*, 61–92.
Haraway, D. (2016). *Manifestly Haraway*. University of Minnesota Press.
Hoelle, J. (2015). *Rainforest cowboys: The rise of ranching and cattle culture in Western Amazonia*. University of Texas Press.
Hultman, M., & Pulé, P. (2020). "Ecological masculinities: A response to the societal crises of our time. *Population*, *23*(2), 1–13.
Institute for Agriculture and Trade Policy. (n.d.). *Media Brief: JBS increases emissions 51 percent*. https://www.iatp.org/media-brief-jbs-increases-emissions-51-percent.
Klein, N. (2014). *Beyond extractivism*. Simon & Schuster.
Lehnen, L. (2020). Ecocriticism in Brazil: The wastelands of Ana Maria Maia's fictions. *Romance Quarterly*, *67*(1), 22–35.
Maia, A. P. (2013). *De gados e homens*. Editora Record.
Maia, A. P. (2023). *Of cattle and men*. Translated by Z. Perry. Charco Press.
Mail Foreign Service. (2009, August 28). Police baffled as dozens of 'suicidal' cows throw themselves off cliff in the Alps. *Daily Mail*. https://www.dailymail.co.uk/news/article-1209638/Scientists-baffled-suicidal-cows-throw-cliff-Switzerland.html.
Maisonnave, F. (2023, June 1). Brazil's congress weakens pro-environment ministries in a rejection of Lula. *AP News*. https://apnews.com/article/brazil-environment-amazon-climate-change-lula-agriculture-lobby-bfc3ce75122fe97e9266e8e27952e39e.
Malafaia, G. C., Mores, G. V., Casagranda, Y. G., Barcellos, J. O. J., & Costa, F. P. (2021). The Brazilian beef cattle supply chain in the next decades. *Livestock Science*, *253*, 104704. https://doi.org/10.1016/j.livsci.2021.104704.
Meijer, E. (2019). *When animals speak: Toward an interspecies democracy*. New York University Press.
Nibert, D. (2013). *Nonhuman animal oppression and human violence: Domesecration, capitalism, and global conflict*. Columbia University Press. http://ebookcentral.proquest.com/lib/unm/detail.action?docID=1188374.
Nixon, R. (2011). *Slow violence and the environmentalism of the poor*. Harvard University Press.
Peña-Guzmán, D. M. (2017). Can nonhuman animals commit suicide? *Nonhuman Animal Sentience*, *20*(1). http://doi.org/10.51291/2377-7478.1201.

7

RE-EXAMINING THE MEATPACKING-METHAMPHETAMINE HYPOTHESIS

Cindy Brooks Dollar and Josh Hendrix

We, the authors, met when we were graduate students at North Carolina State University, a university originally established as an agricultural college, and which still boasts agriculture and engineering as premier areas of study. Although not a part of our curriculum or class work, we separately, and without knowledge of the others' interest, read a book titled *Fast Food Nation*. One night following our Social Control class, we had a conversation and realized our shared fascination with some of the topics presented in the book. To be sure, millions of others read it. The year of its publication, Eric Schlosser's *Fast Food Nation* became a New York Times bestseller. It garnered so much public attention that a video documentary soon followed.

The book argues that the fast-food industry is reflective of American society's values and has fundamentally changed the world economy, global landscape, public health concerns, and popular culture. Providing a critique of the U.S. fast-food industry, Schlosser claims that the proliferation of fast food has resulted in significant environmental harm, the loss of small-scale agricultural farms, a sharp rise in animal deaths because of industrialized slaughtering, and growing health concerns for humans whose diet is largely comprised of fast food. Although some of these claims were presented in previous investigations of the American meatpacking industry (e.g., Sinclair, 1905), Schlosser's account was noteworthy in its contemporary depiction of American life.

Being interested in the purpose, prevalence, and patterns of substance use, Schlosser's work caught our attention through a rather inconspicuous footnote in which he claimed that slaughterhouses were a breeding ground for illicit methamphetamine use. Specifically, Schlosser's writes,

> The unrelenting pressure of trying to keep up with the line has encouraged widespread methamphetamine use among meatpackers. Workers taking "crank" feel

DOI: 10.4324/9781003441908-9

> charged and self-confident, ready for anything. Supervisors have been known to sell crank to their workers or to supply it free in return for certain favors, such as working a second shift.
>
> *(Schlosser, 2001, p. 179)*

Prior to pursuing her Ph.D., the first author, Cindy, had been employed with a plaintiff civil attorney firm. As a part of that work, she had assisted in investigating concerns about the marketing and access of pseudoephedrine, an ingredient that became widely associated with methamphetamine production. Through that work, industry accounts from the mid-to-late 1990s were uncovered that identified knowledge of the link between over-the-counter pseudoephedrine products and problematic methamphetamine manufacturing. Cindy also reviewed at least one document that hinted at a potential connection between methamphetamine production and areas with abundant farming and meatpacking plants, likely because it reminded her of the area in which she was raised – a rural town where economic gloom and illicit drug use was common. Despite this curiosity, she didn't think much more about it until reading *Fast Food Nation*. Josh, the second author, spent the first few years of graduate school studying precarious and often undesirable work arrangements that countless Americans take on to make ends meet. His work primarily focused on nonstandard work arrangements, like shift work, and their effects on individuals and their family lives. Josh was also interested in the various push-pull factors that explain why people begin – and *keep* working jobs they may not want. Employment at a meatpacking plant is especially taxing as it demands workers to be in the constant presence of, and sometimes active participants in, killing animals. As a work and family researcher, the potential connection between methamphetamine use and meatpacking was an intriguing idea to him because it posited that illicit drug use, while in many ways detrimental to the health of workers, could actually act as a facilitator of their employment. He wondered: Do meatpacking workers use methamphetamines as a way of overcoming the speedy and unsavory nature of animal slaughter?

Our mutual curiosities and past experiences urged us to further examine Schlosser's claim, although we did not fully commit to doing so until after we completed our doctoral degrees. As we dove into the literatures on slaughterhouse labor, we read of slaughterhouse workers' relatively high frequency of substance use. For example, Broadway (1990) concluded that the 1970s growth of the beef-packing industry in southwest Kansas was met with increases for alcohol-related problems. Over a decade later, speaking beyond the confines of a single state site, Macnair (2002) asserted that slaughterhouse workers involved in "active participation of causing trauma" suffered from frequent drug use due largely to "psychological consequences" related to killing. We did not, however, find scholarship investigating clear associations between the meatpacking industry and the specific use of methamphetamines. Given Schlosser's claim and other nationally publicized reports that such associations were apparent (Reding, 2010; Saulny, 2008), we thought such an investigation was timely. So, Josh located publicly available agricultural and drug-related data gathered throughout the United States and we began our first investigation.

Our first paper on this matter was published in *Organization & Environment*, a journal then dedicated to highlighting organizational management's link to social issues. The paper offered a macro-level examination of the meatpacking-methamphetamine hypothesis. By compiling data from five sources, including the Federal Bureau of Investigation, U.S. Census Bureau and Departments of Health & Human Services, Justice, and Agriculture, we examined whether state-level variations in commercial meat production were associated with problematic methamphetamine use over a time span of 11 years (2001–2012). Soon after its publication, SAGE Publishing presented a special feature on the article. In this chapter, we revisit our original study. Enhancing our original 11-year longitudinal study with an additional 7 years of data, this chapter investigates the touted meatpacking-methamphetamine link across nearly two decades, from 2001 to 2019.

Since our first publication on this topic, animal meat eating has continued to increase (Ritchie, et al., 2023), and methamphetamine use, described now as "chronic," contributes to rising overdose deaths (ONDCP, 2022). Although overdose deaths are seen across all races and ethnicities in the working age U.S. population (Han, et al., 2021a), it is especially significant among certain marginalized populations (Han et al., 2021b), including racial-ethnic minorities and immigrants of color who are often employed in the most dangerous occupations of slaughterhouses (Pachirat, 2013). These increasing trends have occurred alongside deregulation of meat and poultry plants. In fact, in 2017, the maximum slaughter line speed was revoked, allowing meat processing centers to now process meat even faster than before, despite evidence that doing so adds additional risks to worker safety (Human Rights Watch, 2019). These factors coupled with the continued dearth of knowledge available about the links between the animal slaughter and illicit stimulant use, like methamphetamines, make it worthy of re-examination.

The Methamphetamine-Meatpacking Hypothesis

We have previously proposed how the methamphetamine-meatpacking hypothesis may operate at the macro-areal level (Hendrix & Dollar, 2017), so herein we provide only a summary of our earlier arguments. The logic of our argument builds on the legacy of work about employment's influence on individual-level behavior. Research consistently concludes that job (in)stability is related to various behavioral outcomes, including violence and illicit substance use (Dollar & Hendrix, 2015; Dollar & Ray, 2013; Kandel & Yamaguchi, 1987; Wright & Cullen, 2004). Some scholarship similarly concludes that labor market conditions and the type and quality of work influence these behaviors (Conway et al., 1981; Fennell et al., 1981; Gordon et al., 1986; Guinn, 1983; House et al., 1986; Laub & Sampson, 2003; Tagliacozzo & Vaughn, 1982).

Meat processing is a dangerous job. The meatpacking industry thrives on physically and emotionally demanding work. Given the nature of the jobs, workers in slaughterhouses are exposed to a wide range of potential injuries. For example, slaughterhouse workers are at high risk for developing musculoskeletal disorders and disabilities that develop from forceful, repetitive body movements (Fitzgerald, 2010; Frost et al., 1998; Gustafsson, 1997; Hansen & Bernard, 1982; Leclerc et al., 2004; Sundstrup et al., 2013; Viikari-Juntara, 1983). Bacterial and fungal contaminants, which spread as animal skin and hair are removed and organs are cleaned, are also a hazard. Skin diseases and various issues of respiratory distress and inflammation are well-documented consequences as a result (e.g., Aziz et al., 1996; Gabal & el Geweily, 1990; Gilbert et al., 2012). Other job-related injuries are sustained due to working quarters being covered with blood, grease, and chemicals alongside sharp instruments and machines that are needed to kill, cut, debone, and package meat. A 2017 Monthly Labor Review finds that animal slaughter workers have a higher rate of injury and illness than all other manufacturing and private industry workers (also see Stull & Broadway, 2004). Such health risks negatively affect the level of job stability available for those in the meat production industry as these unsafe and unsanitary work conditions may explain the high worker turnover (Cudworth, 2011; Ehman et al., 2012).

Rapid speed and hazardous workspaces also bring about deleterious emotional tasks, which increase work stress and decrease job satisfaction. The excessive pace is one of the most common complaints among industrial slaughterhouse employees (Dillard, 2008), but emotional challenges are also noticeable in research investigating the psychological tensions of workers who participate in killing and dismembering animals (Smith, 2002; Pachirat, 2013). Eisnitz's (1997) study of American slaughterhouses reveals that slaughterhouse butchers express remorse about their job duties. These findings are bolstered by scholarship suggesting slaughterhouses employees cope with work-related anguish by finding ways to emotionally disconnect from their work (Richards et al., 2013; Vialles, 1994), which includes substance consumption (Macnair, 2002).

Low wages further detract from the worker satisfaction in this industry. According to long-standing data from the Bureau of Labor Statistics (BLS), the median hourly wage of a slaughterhouse worker falls well below the median hourly wage nationwide. What is more, these jobs are often concentrated in rural areas where other work options are scarce. For instance, in the Southeastern part of North Carolina, slaughterers, meat packers, and meat processors are among three the top five most prevalent jobs (Walljasper, 2018). The Human Rights Watch (2004) has noted these depressed wages, which arguably results from the severe power imbalance among industry leaders and workers. The meat processing industry is owned by a handful of large corporations. According to a 2021 White House Document, four conglomerates (Tyson, JBS, Marfrig, and Seaboard) control 55% to 85% of the market for pork, beef, and poultry. Slaughterhouse workers, however, tend to come from vulnerable populations. Many are politically disenfranchised, economically underprivileged, and often live in communities that are relatively socially isolated from mainstream (i.e., dominant) institutional supports, leaving them with limited collective bargaining power. In short, there are multi-level disadvantages that accumulate and exacerbate poor working conditions.

Some persons may perceive the use of methamphetamines as a calculated strategy to cope with such circumstances. Indeed, methamphetamine users often report the drug's ability to "numb the mind," move quicker and more effectively, and perform tasks longer without breaks (Boeri et al., 2009; Joe, 1996; Kurtz, 2005; Lende et al., 2007; Sherman et al., 2008; Von Mayrhauser et al., 2001). It is possible, then, that methamphetamines allow workers to escape feelings of discomfort, tension, or guilt associated with being involved in an industry that harms and kills millions of animals per day. Given methamphetamine's ability to deliver confidence and fearlessness (Anglin et al., 2000; Bungay et al., 2006), the consumption of these substances may provoke feelings of empowerment associated with the domination of animals.

All these potential issues extend beyond slaughterhouse workers and into the communities in which slaughterhouses operate. Although Schlosser (2001) focused on methamphetamine use among slaughterhouse workers, we broaden the meat processing-methamphetamine hypothesis to provide a more aggregated, areal-level analysis. Ample evidence indicates the ways in which concentrated industry may influence criminogenic place-based factors. Social disorganization, in its original and systemic forms, posits that neighborhood-level socio-political isolation and economic segregation fosters the emergence of cultural (mal)adaptations, resulting in higher rates of antisocial behavior (Bursik & Grasmick, 1993; Shaw & McKay, 1942). Empirical analyses reveal that areas with a large percent of low-wage, unskilled workers experience more significant problems because of the lack of strong socio-political networks extending within and beyond the local community (e.g., Kasarda & Janowitz, 1974; Kubrin & Weitzer, 2003; Sampson et al., 2002). These discoveries are well-known in "mainstream" criminology but reflect more critical criminological arguments about the political economy's influence on systemic harms, whether presently defined as crimes or not.

Aulette and Michalowski (1993) demonstrate how several fatalities at a chicken processing plant in North Carolina were caused by multifaceted forces that reflect global, national, organizational, community, and individual level factors. In this way, they show that what happens in a single workplace or among only a subset of workers reflects macro-social circumstances. Similarly, while it is common to credit green criminology with arguments about the importance of co-considering harm to people, nonhuman animals, and the "natural" environment, conventional criminologists would likely agree. Shaw and McKay's (1942) social (dis)organization model highlighted relationships between human life and ecological relations. As such, it is not only possible, but arguably necessary, to merge components of the so-called conventional criminology with elements of critical criminology. In doing so we may be better equipped to appreciate how political agencies and economic-driven organizations combine, whether through direct action or implicit inaction, to influence the livelihood and health of all worldly species.

In summary, we hypothesize that deleterious effects of the slaughterhouse industry are not exclusive to its workers as these effects can effectively spillover to the community at large, thus increasing the potential for problematic methamphetamine use among community residents. As pointed out earlier, many slaughterhouses operate in rural areas where work options are scarce, and the industry often disproportionately employ immigrants, refugees, and people of color who have difficulties finding safe, stable, well-paying jobs (e.g., BLS Reports, 2019; Borjas et al., 2006; Penner, 2008; Reid & Rubin, 2005). Being anxious about employment and opportunities for economic and social growth may hinder an ability to build and maintain social solidity and mutual support, especially if areas are marked with deep-seated ethnic and racial tensions. Extant studies, indeed, show that the slaughterhouse industry is associated with areal-level violent crime, which seems to be partially explained by a job-seeking population influx and subsequent loss in social cohesion (e.g., Fitzgerald et al., 2009). When populations lack cohesion and sense social isolation, they may increase activities meant to alleviate the negative feelings associated with systemic- *and* personal-level stressors (Agnew, 2006). Substance use is one way that such strains may be relieved.

Data Description

Our first examination of the meatpacking-methamphetamine hypothesis examined 11 years of state-level data, from 2001 through 2012. In this chapter, we re-examine our earlier work after obtaining data from the same sources for 7 additional years, providing us with a total of 18 years of observable data.[1] Although our earlier measures for methamphetamine use, population size, crime rate, and number of law enforcement officers, are still accessible, more recent USDA data are limited to chicken and red meat processing centers; therefore, our analysis is limited to these meats.[2] We are not especially concerned about this data limitation because chicken and beef are the most consumed meats in the United States and have been since 2008. In fact, while beef consumption has steadily declined since the 1970s, chicken consumption has increased, overtaking beef as the preferred meat about 10 years ago. To our knowledge, this examination thus reports the largest state-level longitudinal analysis of our meatpacking-methamphetamine hypothesis.

We conducted a preliminary examination of the 2013–2019 data to narrow our attention on the most recently acquired data. We find that the amount of commercially produced red meat has increased each year. Although all states report information on the amount of commercialized red meat processed, fewer states appear to be involved in the commercial processing of chicken. Still, among those states with reported data, the amount of chicken processed has also increased each year. Congruent with other scholars (Han et al., 2021a, b), we also find that the rate of hospitalizations for methamphetamine-related injuries continued to climb annually. In fact, the average number of hospital admissions for methamphetamine-related injury increased by nearly 65% between 2013 and 2019. Importantly, 2013's average was also higher than 2012, showing that the rise has been steady for several years.[3]

Bivariate correlations between commercialized meatpacking and problematic methamphetamine remain positive and statistically significant between 2013 and 2019, but as was true with our original study, the relationship is relatively weak (r ~ .23).

To assess the potential meatpacking-methamphetamine relationship more completely, we appended the recently acquired data to our original dataset, leaving us with a single dataset encompassing 18 years of data (2001–2019). We employed regression techniques that allow us to account for other potentially relevant factors, including state-specific population size, crime rate, and size of law enforcement. Decomposition models, also known as hybrid regression models, can produce within and across state estimates. Importantly, these models identify observed time-invariant parameters, so unlike fixed effect models, which estimate longitudinal effects but leave across-unit effects as fixed, decomposition models provide separate estimates for within-unit and across-unit variation in a single model (Allison, 2005).

We examined the effect of meatpacking on problematic methamphetamine use in a few ways. First, because the data were originally disaggregated by type of meat, we estimated the separate effect of red meat and poultry processing on problematic methamphetamine use. However, given changes in meat preferences over time (Daniel et al., 2011; Haley, 2000; USDA/NASS Agricultural Statistics, 2000), we also created a combined score to form a single meat production variable. Across every model, we find a positive relationship between commercialized meat processing and methamphetamine hospitalizations, net of other factors in the model. Table 7.1 shows the results of our full model. As illustrated, the effect

TABLE 7.1 Decomposition Models Predicting Methamphetamine Use, 2001–2019

	Total (Combined) Meat Processing
Longitudinal	
Meat processing	.0003 [0.0004]*
Population size	.0005 [0.0009]*
Size of police	281.19 [413.40]
Crime rate	–115.28 [.29]*
State comparison	
Meat processing	.0010 [.0002]*
Population size	.0004 [.00008]*
Size of police	–.0450 [.0174]*
Crime rate	–.1901 [.5476]*
Time invariant	
West	22740.64 [3779.38]*
Midwest	4537.05 [2710.51]**
Northeast	712.80 [2764.99]
R2 (overall)	.69

Note: Unstandardized regression coefficients [standard error in brackets].

*$p < .05$.

**$p < .10$.

size of commercialized meat processing on problematic methamphetamine use is small, but when all variables are included in the model (i.e., total amount of meat commercially processed, methamphetamine use, state population size, crime rate, police size), a notable 69% of variance is explained.[4]

Conclusion

Links between the slaughterhouse industry and methamphetamine consumption have fueled ample political and public attention (e.g., Bloom, 1997; Schlosser, 2001). Although varied speculations exist, assertions about the relationship between these two factors were promoted by media outlets for decades (e.g., Garrett 2013; Reding, 2010; Saulny, 2008). In fact, the widespread yet diverse media coverage partially prompted our first empirical investigation of the matter.

In this chapter, we revisit our earlier investigations by re-evaluating the association between meat production and methamphetamine use with recently obtained data. Examining 18 years of state-level data on commercialized animal slaughter and problematic methamphetamine use, we find a consistent positive relationship between the two even after controlling for relevant legal and population factors. Indeed, the models explain a considerable amount of variance in methamphetamine use.

While we feel confident that the models illustrate a clear connection between commercialized meatpacking and problematic methamphetamine use, our models are unable to reveal precisely *how* these factors are related. As discussed at the beginning of this chapter, Schlosser (2001) articulated the relationship as stemming from poor working conditions. The physical and emotional demands of commercialized animal slaughter were said to result in employees self-medicating with methamphetamines to relieve pain and discomfort, a practice encouraged by employers (Schlosser, 2001, p. 179). Our data preclude any direct test of this hypothesis. We nonetheless find our work informative. In fact, we humbly submit that our investigations provide a creative way to systematically assess the relationship of animal slaughter and methamphetamine use. Our analysis has limitations to be sure, and our inability to assess the mechanisms by which meatpacking and methamphetamine are related is a consequence of some of those limitations. Still, our findings indicate that this area of inquiry is worthy of additional investigation. To this end, we close this chapter by briefly proposing a few ideas that scholars may wish to explore in future studies.

Ethnographic accounts of slaughterhouse work exist (e.g., Pachirat, 2013; Wentworth, 2016), but an ethnography that focused more strictly on examining if and how slaughterhouse culture reflexively shapes substance use among those who work in the industry would be instructive. Such projects may reveal not only the presence and extensiveness of drug use but also if certain types of substances are more prevalent than others and why. Ethnographic inquiries could also reveal if patterns of substance use vary by job task, which would be edifying given some of the inconsistent findings about employment's association with illegal substance use.

Other research methods are also suitable for this area of study. Because meatpacking processing plants open and close their doors over time, there is an opportunity to investigate if the entry or exit of slaughterhouses is followed by changes in methamphetamine use within the same (and/or neighboring) area. We can imagine this work being conducted with a variety of methodological techniques. Analyses utilizing geographic information system (GIS), quantitative data analysis of localized secondary data, or primary data collected via interviews or surveys of those who have recently entered or exited jobs in a slaughterhouse are all appropriate strategies. Finally, a comparative analysis examining slaughterhouse workers and workers in other industries that are deemed "dirty," difficult, or degrading may help us better detect if and how certain forms of labor participation affect specific types of substance use.

Even though research in Critical Animal Studies (CAS) and the animal-industrial complex (A-IC) focus heavily on the slaughter industry, these areas of inquiry encompass investigations of a wide variety of activities that systematically exploit animals (Nibert, 2011; Noske, 1989). Activists have previously linked CAS and A-IC to human's commodification and exploitation of nonhuman animals (e.g., Stallwood, 2014), which inevitably reflect the speciesist ideology that humans are superior to other animals. But speciesism is just one of many "othering" practices pervading our modern systems and interactions. In other words, while speciesism certainly emboldens harm against nonhuman animals, the logic of domination (Warren, 1990) that underlie speciesism is apparent in countless contexts. When we recognize that the same mechanisms that promote speciesism promote other forms of domination, the co-mingling of harming nonhuman animals and harming human animals is revealed.

Prior scholarship has, for example, concluded positive relationships between animal slaughter and participation in violence against humans (Fitzgerald et al., 2009; Jacques, 2015). Substance use has also been positively associated with factory food labor participation (Broadway, 1990; Ehman et al., 2012; Macnair, 2002).[5] Taken together, this research suggests that violence against humans and violence against nonhumans is inseparable. Importantly, this is so even when violence against nonhuman animals is conducted through legal (i.e., "legitimate") means of employment. When viewed from this lens, questions about how lawful, institutionalized activities may produce damaging results are raised, which encourages queries about the presence and maintenance of a seemingly ubiquitous culture of harm.

Describing the processes by which human animals become justifiably caged, sociologists and criminologists have long noted how "othering" practices catalyze a culture of harm (e.g., Goffman, 1961; Sykes, 1958). Using this work as an underpinning, we can begin to ask questions about the extensiveness of "othering" practices. Research on sexism, colorism, xenophobia, racism, ethnocentrism, homophobia, and other discriminatory activities describe othering as a symbolic boundary that separate people or objects (Lamont & Fournier, 1992) for the purpose of perpetuating inequality through devaluation and targeted exploitation (Schwalbe et al., 2000; Small et al., 2010). This research nearly exclusively focuses on within-species "othering" of humans. But how might we, as humans, "other" nonhumans, and (how) are these othering processes similar across species?

Over 20 years ago bell hooks wrote,

> Cultures of domination rely on the cultivation of fear as a way to ensure obedience...Fear is the primary force upholding systems of domination. It promotes the desire for separation...When we choose to love, we choose to move against fear – against alienation and separation. The choice to love is the choice to connect – to find ourselves in the other.
>
> *(2001, p. 93)*

hooks reminds us that practices of othering can be detrimental and encourages us to realize our relation to one another. In doing so, we not only gain the potential to assess how symbolic, structural, and cultural violence is perpetuated, but we also accept the power of agency and perhaps take on the responsibility to choose a different way to live.

Notes

1 Throughout 2020 and 2021, the amount of missing data on methamphetamine and meat processing measures was substantial, so our current analysis encompasses 2001 through 2019.
2 Red meat production includes beef, lamb and mutton, venison, and goat, but beef is the most common meat produced in these factories.
3 Missing data is rare, but Oregon does not report methamphetamine hospitalizations from 2013 to 2019, and South Carolina has no reported data for 2019.
4 The variance explained increases greatly when the control variables are included in the model, indicating that these variables are important covariates.
5 Here, we posit problematic human drug consumption as a form of internalized violence against oneself.

References

Agnew, R. (2006). *Pressured into crime: An overview of general strain theory*. Oxford University Press.
Allison, P. D. (2005). *Fixed effects regression methods for longitudinal data using SAS*. The SAS Institute.
Anglin, M. D., Burke, C., Perrochet, B., Stamper, E., & Dawud-Noursi, S. (2000). History of the methamphetamine problem. *Journal of Psychoactive Drugs*, *32*, 137–140.
Aulette, J., & Michalowski, R. (1993). A fire in hamlet: A case study of state-corporate crime. In Kenneth Tunnell (Ed.), *Political crime in contemporary America*. Garland.
Aziz, M. A., Bahamdan, K., & Moneim, M. A. (1996). Prevalence and risk factors for warts among slaughterhouse workers. *East African Medical Journal*, *73*, 194–197.
Bloom, L. (1997). *PDFA initiatives*. Paper presented at the National Methamphetamine Drug Conference: Workgroup 1, Prevention: Public Information Initiatives at Home and

Work, hosted by the Office of National Drug Control Policy, Executive Office of the President of the United States.
BLS Reports. (2019). Labor Force Characteristics by Race and Ethnicity, 2018. *US Bureau of Labor Statistics*. https://www.bls.gov/opub/reports/race-and-ethnicity/2018/home.htm.
Boeri, M. W., Harbry, L., & Gibson, D. (2009). A qualitative exploration of trajectories among suburban users of methamphetamine. *Journal of Ethnographic and Qualitative Research*, *3*, 139–151.
Borjas, G. J., Grogger, J., & Hanson, G. (2006). *Immigration and African-American employment opportunities: The response of wages, employment, and incarceration to labor supply shocks*. National Bureau of Economic Research, Working Paper 12518.
Broadway, M. J. (1990). Meatpacking and its social and economic consequences for Garden City, Kansas in the 1980s. *Urban Anthropology*, *19*, 321–344.
Bungay, V., Malchy, L., Buxton, J. A., Johnson, J., Macpherson, D., & Rosenfeld, T. (2006). Life with jib: A snapshot of street youth's use of crystal methamphetamine. *Addiction Research and Theory*, *14*, 235–251.
Bursik, R. J., & Grasmick, H. G. (1993). *Neighborhoods and crime: The dimensions of effective community control*. Lexington Books.
Conway, T. L., Vickers, R. R., Ward, H. W., & Rahe, R. H. (1981). Occupational stress and variation in cigarette, coffee, and alcohol consumption. *Journal of Health and Social Behavior*, *22*, 155–165.
Cudworth, E. (2011). Climate change, industrial animal agriculture and complex inequalities. *International Journal of Science in Society*, *2*, 323–335.
Daniel, C. R., Cross, A. J., Koebnick, C., & Sinha, R. (2011). Trends in meat consumption in the United States. *Public Health Nutrition*, *14*, 575–583.
Dillard, J. (2008). A slaughterhouse nightmare: Psychological harm suffered by slaughterhouse employees and the possibility of redress through legal reform. *Georgetown Journal on Poverty Law & Policy*, *15*, 391–408.
Dollar, C. B., & Hendrix, J. A. (2015). The importance of romantic and work relations on nonmedical prescription drug use among adults. *Sociological Spectrum*, *35*, 465–481.
Dollar, C. B., & Ray, B. (2013). Adult nonmedical prescription drug use: An examination of bond theory. *Deviant Behavior*, *34*, 932–949.
Ehman, A., Yildiz, A. S., Bez, Y., & Kingir, S. (2012). Psychological symptom profile of butchers working in slaughterhouse and retail meat packing business: A comparative study. *Kafkas Üniversitesi Veteriner Fakültesi Dergisi*, *18*, 319–322.
Eisnitz, G. A. (1997). *The shocking story of greed, neglect, and inhumane treatment inside the U.S. meat industry*. Prometheus Books.
Fennell, M. L., Rodin, M. B., & Kanto, G. K. (1981). Problems in the work setting, drinking, and reasons for drinking. *Social Forces*, *60*, 114–132.
Fitzgerald, A. J. (2010). A social history of the slaughterhouse: From inception to contemporary implications. *Research in Human Ecology*, *17*, 58–69.
Fitzgerald, A. J., Kalof, L., & Dietz, T. (2009). Slaughterhouses and increased crime rates: An empirical analysis of the spillover from "The Jungle" into the surrounding community. *Organization and Environment*, *22*, 158–184.
Frost, P., Andersen, J. H., & Nielsen, V. K. (1998). Occurrence of carpal tunnel syndrome among slaughterhouse workers. *Scandinavian Journal of Work, Environment, & Health*, *24*, 285–292.
Gabal, M. S., & el Geweily, M. (1990). Dermatologic hazards among slaughterhouse workers. *The Journal of the Egyptian Public Health Association*, *65*, 191–206.
Garrett, R. (2013). How a Common Fertilizer Became a Meth Magnet. *Modern Farmer.* https://modernfarmer.com/2013/05/how-a-common-fertilizer-became-a-meth-magnet/.
Gilbert, M. J., Bos, M. E., Duim, B., Urlings, B. A., Heres, L., Wagenaar, J. A., & Heederik, D. J. (2012). Livestock-associated MRSA ST398 carriage in pig slaughterhouse workers related to quantitative environmental exposure. *Occupational and Environmental Medicine*, *69*, 472–478.

Goffman, E. (1961). *Asylums: Essays on the social situation of mental patients and other inmates*. Aldine Transaction.
Gordon, N. P., Cleary, P. D., Parker, C. E., & Czeisler, C. A. (1986). The prevalence and health impact of shiftwork. *American Journal of Public Health*, *76*, 1225–1228.
Guinn, B. (1983). Job satisfaction, counter-productive behavior and circumstantial drug use among long-distance truckers. *Journal of Psychoactive Drugs*, *15*, 185–188.
Gustafsson, B. (1997). The health and safety of workers in a confined animal system. *Livestock Production Science*, *49*(2), 191–202.
Haley, M. M. (2000). Changing Consumer Demand for Meat: The U.S. Example, 1970–2000. Economic Research Service, USDA. Changing Structure of Global Food Consumption and Trade, WRS-01-I, 41-48. Available at https://www.ers.usda.gov/webdocs/outlooks/40303/14976_wrs011g_1_.pdf?v=1365.
Han, B., Compton, W. M., & Jones, C. M., Einstein, E. B., & Volkow, N. D. (2021a). Methamphetamine use, methamphetamine use disorder, and associated overdose deaths among US adults. *JAMA Psychiatry*, 78(12), 1329–1342.
Han, B., Cotto, J., Etz, K., Einstein, E. B., Compton, W. M., & Volkow, N. D. (2021b). Methamphetamine overdose deaths in the US by sex, race, and ethnicity. *Research Letter*. https://jamanetwork.com/journals/jamapsychiatry/fullarticle/2774859.
Hansen, N. S., & Bernard, J. (1982). Incidence of disability pensions among slaughterhouse workers in Denmark with special regard to diagnosis of the musculo-skeletal system. *Scandinavian Journal of Public Health*, *10*, 81–85.
Hendrix, J., & Dollar, C. B. (2017). American slaughterhouses and the need for speed: An examination of the meatpacking-methamphetamine hypothesis. *Organization & Environment*, *30*(2), 1–19.
hooks, b. (2001). *All about love: New visions*. HarperCollins Publishers.
House, J. S., Strecher, V., Metzner, H. L., & Robbins, C. A. (1986). Occupational stress and health among men and women in the Tecumseh community health study. *Journal of Health and Social Behavior*, *27*, 62–67.
Human Rights Watch. (2004). Blood, sweat, and fear: Worker's rights in U.S. meat and poultry plants. *A Human Rights Watch Report*. http://www.hrw.org/reports/2005/usa0105/.
Human Rights Watch. (2019). Blood, sweat, and fear: Worker's rights in U.S. meat and poultry plants. *A Human Rights Watch Report*. http://www.hrw.org/reports/2005/usa0105/.
Jacques, J. R. (2015). The slaughterhouse, social disorganization, and violent crime in rural communities. *Society & Animals*, *23*(6), 594–612.
Joe, K. A. (1996). The lives and times of Asian-Pacific American women drug users: An ethnographic study of their methamphetamine use. *Journal of Drug Issues*, *26*(1), 199–218.
Kandel, D. B., & Yamaguchi, K. (1987). Job mobility and drug use: An event history analysis. *American Journal of Sociology*, *92*, 836–878.
Kasarda, J. D., & Janowitz, M. (1974). Community attachment in mass society. *American Sociological Review*, *39*, 328–339.
Kubrin, C. S., & Weitzer, R. (2003). New directions in social disorganization theory. *Journal of Research in Crime and Delinquency*, *40*, 374–402.
Kurtz, S. (2005). Post-circuit blues: Motivations and consequences of crystal meth use among gay men in Miami. *AIDS and Behavior*, *9*, 63–72.
Lamont, M., & Fournier, M. (1992). *Review of 'cultivating differences: Symbolic boundaries and the making of inequality*. University of Chicago Press.
Laub, J. H., & Sampson, R. J. (2003). *Shared beginnings, divergent lives: Delinquent boys to age 70*. Harvard University Press.
Leclerc, A., Chastang, J., Niedhammer, I., Landre, M., & Roquelaure, Y. (2004). Incidence of shoulder pain in repetitive work. *Occupational and Environmental Medicine*, *61*, 39–44.
Lende, D. H., Leonard, T., Sterk, C. E., & Elifsonc, K. (2007). Functional methamphetamine use: The insider's perspective. *Addiction Research & Theory*, *15*, 465–477.

Macnair, R. M. (2002). *Perpetration-induced traumatic stress: The psychological consequences of killing*. Greenwood Publishing Group.
Nibert, D. (2011). Origins and consequences of the animal industrial complex. In S. B. Best, et al. (Eds.), *The global industrial complex: Systems of domination* (pp. 197–209). Rowman & Littlefield.
Noske, B. (1989). *Humans and other animals: Beyond the boundaries of anthropology*. Pluto Press.
ONDCP. 2022. Plan to address methamphetamine supply, use, and consequences. Executive Office of the President. Washington, D.C. Available at https://www.whitehouse.gov/wp-content/uploads/2022/05/ONDCP-Plan-to-Address-Methamphetamine-Supply-Use-and-Consequences.pdf
Pachirat, T. (2013). *Every twelve seconds: Industrialized slaughter and the politics of sight*. Yale University Press.
Penner, A. M. (2008). Race and gender differences in wages: The role of occupational sorting at the point of hire. *The Sociological Quarterly*, *49*(3), 597–614.
Reding, N. (2010). *Methland: The death and life of an American small town*. Bloomsbury Press.
Reid, L. W., & Rubin, B. A. (2005). Integrating economic dualism and labor market segmentation: The effects of race, gender, and structural location on earnings, 1974–2000. *The Sociological Quarterly*, *44*(3), 405–432.
Richards, E., Signal, T., & Taylor, N. (2013). A different cut? Comparing attitudes towards animals and propensity for aggression within two primary industry cohorts-farmers and meatworkers. *Society & Animals*, *21*, 395–413.
Ritchie, H., Rosado, P., & Roser, M. (2023). Meat and dairy production. *Our World in Data*. Available at https://ourworldindata.org/meat-production.
Sampson, R. J., Morenoff, J. D., & Gannon-Rowley, T. (2002). Assessing "neighborhood effects": Social processes and new directions in research. *Annual Review of Sociology*, *28*, 443–478.
Saulny, S. (2008, May 13). Hundreds are arrested in U.S. sweep of meat plant. *New York Times*.
Schlosser, E. (2001). *Fast food nation: The dark side of the all-American meal*. Houghton Mifflin Company.
Schwalbe, M., Holden, D., Schrock, D., Godwin, S., Thompson, S., & Wolkomir, M. (2000). Generic processes in the reproduction of inequality: An interactionist analysis. *Social Forces* 79(2), 419–452.
Shaw, C. R., & McKay, H. (1942). *Juvenile delinquency and urban areas*. University of Chicago Press.
Sherman, S. G., German, D., Sirirojn, B., Thompson, N., Aramrattana, A., & Celentano, D. D. (2008). Initiation of methamphetamine use among young Thai drug users: A qualitative study. *Journal of Adolescent Health*, *42*, 36–42.
Sinclair, U. (1946/1905). *The jungle*. The Viking Press.
Small, M. L., Harding, D. J., & Lamont, M. (2010). Reconsidering culture and poverty. *ANNALS of the American Academy of Political and Social Science*, *629*(1), 6–27.
Smith, M. (2002). The 'ethical' space of the abattoir: On the (in)human(e) slaughter of other animals. *Human Ecology Review*, *9*, 49–58.
Stallwood, K. (2014). Animal rights: Moral crusade or social movement?. In J. *Sorenson (Ed.), Critical animal studies: Thinking the unthinkable* (pp. 298–317). Canadian Scholars' Press.
Stull, D. D., & Broadway, M. J. (2004). *Slaughterhouse blues: The meat and poultry industry in North America* (2nd ed.). Wadsworth/Cengage Learning.
Sykes, G. (1958). *The society of captives: A study of a maximum security prison*. Princeton University Press.
Tagliacozzo, R., & Vaughn, S. (1982). Stress and smoking in hospital nurses. *American Journal of Public Health*, *72*, 441–48.

USDA/NASS. *2000 agricultural statistics annual*. (2000). Retrieved October 19, 2015, from http://www.nass.usda.gov/Publications/Ag_Statistics/2000/index.asp (Not accessible as of [05/07/2024]).
Vialles, N. (1994). *Animal to edible*. Cambridge University Press.
Viikari-Juntara, E. (1983). Neck and upper limb disorders among slaughterhouse workers: An epidemiologic and clinical study. *Scandinavian Journal of Work, Environment & Health*, *9*, 283–290.
Von Mayrhauser, C., Brecht, M. L., & Anglin, M. D. (2001). Use ecology and drug use motivations of methamphetamine users admitted to substance abuse treatment facilities in Los Angeles. *Journal of Addictive Diseases*, *21*, 45–60.
Walljasper, C. (2018). *Slaughterhouses provide low, yet stable floor for employment in rural America*.
Warren, K. J. (1990). The power and the promise of ecological feminism. *Environmental Ethics*, *12*(2), 125–146.
Wentworth, K. N. (2016). *Performing the slaughterhouse: Making meaning and words in daily practice*. Dissertation. University of California, San Diego.
White House Briefing Room. (2021). *Recent Data Show Dominant Meat Processing Companies Are Taking Advantage of Market Power to Raise Prices and Grow Profit Margins*.
Wright, J. P., & Cullen, F. (2004). Employment, peers, and life-course transitions. *Justice Quarterly*, *21*, 183–205.

PART III

Cultural Hegemony of Animal Cruelty Commerce

8

SELFIE SAFARIS

The Violence of Contemporary Camera Hunting and Trophy Shot Selfies

Corina Medley

Walton Ford – an artist whose paintings of animals are steeped in historical analyses that interrogate "how humans interact with non-domestic animals, how animals live in the human imagination, how humans use animals as metaphors for their own insecurities (2021)" – invites us to think beyond representations of animals and consider the realities in which they are produced. Particularly known for his work on and around Audubon and his paintings, Ford explains how he adapted an Audubon painting by incorporating the violent realities behind Audubon's work into his own rendition of it:

> I would take a particular plate from the Audubon portfolio and amplify it in some weird way. Audubon painted a sparrow hawk with a sparrow. But I made a painting with a sparrow hawk sitting on top of a huge pile of sparrows because this was how Audubon went about shooting birds. He was insatiable. He shot birds off the deck of ships. People say, 'Oh, he shot birds because he was studying them.' No! He shot them without collecting them. He talks about raking them up into big piles on the beach and counting them, saying, 'We had a great day, we shot 500 birds today.' This idea of him as a conservationist is bullshit; it's something we put on him. He was just thinking like any frontiersman.

Ford named it *Killy*. If you look at it without this context, it is sublime. Viewing it with this background in mind, the meaning of the painting changes – it is no longer a beautiful depiction of unreal horror, it represents real horror, beautifully. And along with that realization, once unseen contradictions and connections between cultural narratives, objects, and actors also become visible. Knowing this also makes it difficult to see Audubon's paintings the same way. Once Audubon himself is conjured through his own paintings, the striking images of birds that he

DOI: 10.4324/9781003441908-11

produced no longer invoke a gaze that venerates animals, a spirit that wants to conserve them. We are left instead with a handsomely produced marker of gratuitous violence, a painting that is little more than an oil-based trophy shot that symbolizes the death of not only the birds that are represented in them, but also countless others that we cannot see.

It would be wrong to think of Audubon as an anomaly. Prominent individuals associated with nascent conservationism in the United States inflicted violence against animals in ways that were antithetical to the ethos of the conservationist movement, often in the name of conservation itself. That violence concurrently stemmed from insecurities around human-animal boundaries and relations, as well as human-human ones tied to social categories such as gender, sexuality, class, race, and nationality (Haraway, 1984; Purdy, 2015). Although it may be invisible, that violence is also embedded in other representations of animals produced at that time, such as scientific illustrations and engravings, museum exhibits, and taxidermy. And as cameras became more accessible, that also includes photographs of them (Brower, 2005, 2010; Dunaway, 2000; Haraway, 1984).

Camera hunting emerged in the late 1800s, and gained popularity, especially among conservationists, in the early 1900s. It was seen as an ethical alternative to hunting with a gun because the goal was not to kill an animal, it was to capture their image with a camera. As a hobby, the skills and sportsmanship associated with gun hunting were transferred onto camera hunting. The major difference between them, for the hunter and the hunted, was the prize: a photograph – not a carcass – becomes their trophy. Given the choice between mortally wounding animals for recreation or immortalizing their lives on film, "hunting with a camera" was considered to be more or less "a harmless sport" (Shiras, 1900, p. 1364). But as with Audubon's paintings, there was a gulf between the narrative of camera hunting, the beautiful trophy photographs produced by camera hunters, and the reality behind their production. Animals did suffer at the hands of camera hunters, and that suffering can also be traced to elite conservationists, and anxieties articulated around and through the notion of a nature/culture divide (Brower, 2010; Haraway, 1984).

Over a century later, backdropped by the neoliberalization of wildlife tourism (Duffy, 2014; Fletcher, 2023, Moorhouse et al., 2017; Vasan, 2018) and the proliferation of selfie culture, shooting animals with a camera is popular worldwide, particularly where those two meet: taking selfies with wildlife. Similar to the trophies produced by camera hunters generations ago, these images – and the fun-loving or caring narratives typically tied to them – are not as innocuous as they may appear: wildlife selfies are not only desirable and marketable, they are harmful (Carder et al., 2018; Carr & Broom, 2018; Lemieux et al., 2018; World Animal Protection, 2017). Although some fin de siècle camera hunters burdened or killed animals to obtain their trophy shot, this type of contemporary camera hunting and the trophies it produces – trophy shot selfies – mostly entails violence.

To grapple with that violence, we must first make it visible. For that, we turn to another discussion on method for another artist that produces brilliant, historically grounded representations of animals, Mark Dion. In a Q&A session for the ICA Boston museum, he provides a glimpse of his approach for producing art (ICA Boston, 2017):

> **Your practice is both research-oriented and process-based. Do your works more often start with an image or an idea? A sense of the process, or of the final product?**
>
> Images are critical, and so is conventional research, but the most important starting point is always place. The work is mostly contextual in nature, and listening to place, to the original site of project or sculpture, is the first step for me. I search the history for paradigmatic characters, events, anomalies, moments which encapsulate an essential aspect of what makes this context or place different, curious, or important.

Following Dion, we must start with the spaces tied to the creation of trophy shot selfies to understand what makes this form of contemporary camera hunting so violent. Reverse engineering his approach, we can locate the violence of wildlife selfies by starting with the product itself – the trophy shot selfie – then tracing it to the spaces where they are produced. Once we connect wildlife selfies to the places and people that make them possible, the harms of them are more easily seen. Ultimately taking both Dion and Ford's advice into account, we should think through, but not limit ourselves to, the image itself to understand what representations of animals mean to us – and what producing them also means for animals.

Approached as a form of contemporary camera hunting driven by selfie culture, this chapter addresses the harms of taking selfies with wildlife in leisure spaces. While relying on an interdisciplinary framework grounded in green criminology and a deviant leisure perspective, the aim is to ultimately reveal and make sense of the violence imbued in the type of trophy ultimately produced by these camera hunters – the trophy shot selfie. Using the trophy shot selfie as a heuristic to locate sites and encounters that produce these representations of animals, two broad types of camera hunting are identified and discussed here. The first is canned camera hunting, which occurs in spaces such as wildlife tourist attractions, social venues, and ersatz sanctuaries. The second is open season wildlife selfies, which occurs in natural environments in which encounters with wildlife take place in their own habitats. Together, these are known here as selfie safaris: camera hunting encounters in leisure spaces that ultimately produce a trophy shot selfie – and violence against animals.[1]

Green criminology is a subfield of criminology concerned with environmental crimes and harms (see, for example: Brisman & South, 2013, 2020), and it offers a particularly useful lens for understanding selfie safaris for two main reasons: a green criminological perspective emphasizes the importance of taking the de jure/de facto divide into account when locating harm, and it also expands – and in many ways, subverts – conventional notions of who or what counts as a subject in criminology, who or what is a worthy – and arguably necessary – object of inquiry.

For the first, while green criminology points us to different relationships between trophy shot selfies and crime proper – that is, formal deviance and sanctions, various laws, and criminal justice systems – it also directs us to harms and injustices situated beyond them. More than merely being a subfield within criminology, a theoretical and empirical catchall, green criminology is a paradigm with a critical foundation that focuses on economic arrangements, power dynamics, and systems of inequality that impact, and are impacted by, environmental crimes and harms (see, for example: Lynch & Long, 2022). Consequently, many critical strands of thought and research clusters within the field of green criminology implicitly or explicitly overlap with zemiology or adopt a zemiological perspective, which is the study of social harms. In regard to environmental harms and crimes – and harm and crime, more broadly speaking – green criminology reminds us that many things that are harmful are not criminalized, and many things that are criminalized are not harmful.

This aspect of the green criminological perspective is especially important for the task of identifying and thinking about the violence of selfie safaris, as it situates harms and crimes against animals within the historical, cultural, and economic context that is broadly understood as late modernity, late capitalism, or what cultural theorist Mark Fisher (2009) calls capitalist realism. Making these connections is not unique to green criminology, scholars in other fields have long recognized the importance of tracing human-animal relations to economic ones. For example, those concerned with critical animal studies in anthropology and sociology have established the unequivocal need for grappling with the "animal-industrial-complex" if we are to reduce violence against animals (see, for example: Noske, 1989; Twine, 2013). And scholars concerned with the animal question in cultural geography have similarly illustrated how the "animal economy" and violence are constitutive (Wolch & Emel, 1998). A preponderance of harms tied to social arrangements like the animal-industrial complex and the animal economy are not illegal – they are tied to the normal functioning of society, and they are typically overlooked in everyday life, if not celebrated. When laws are in place to protect animals against things such as neglect, cruelty, poaching, and trafficking, the law in and of itself is not necessarily a deterrence, as violence persists despite codification and the punishments accompanying it. De jure recognition and protection can benefit animals, but we cannot depend on it or the state alone to spur meaningful social change – particularly when capital, or what feminist science and technology scholar Donna Haraway calls "lively capital" (2008), is at stake.

Even so, more laws do not equal more justice – the state inflicts immeasurable harm in the name of justice, particularly through the criminal justice system itself. Scholars and activists working in the prison abolition movement and fields like abolitionist criminology have shown how the "prison-industrial-complex" and mass incarceration work in tandem via the commodification of crime and criminalization of poverty (see, for example: Davis, 2011, 2014). Where human and non-human abolitionist movements intersect, the state and dominant institutions are implicated in both human and inter-species harms, many of which operate simultaneously (see, for example: Nagel & Nocella, 2013). Work around the "military-animal industrial complex" is one example of this, revealing how anthropocentrism, the ethos of militarism, and the spirit of capitalism join forces to concurrently inflict harm on humans and non-humans alike (Nocella et al., 2013). The state does not act alone or only on its own behalf, many environmental crimes and harms are forms of state-corporate crimes that stem from symbiotic relations between the state and corporations (see, for example: Rothe & Medley, 2019; Ruggiero & South, 2013). The blurred boundary between them is not only evident in laws that allow corporations that cause suffering to animals to flourish, but also ones that use the criminal justice system to protect businesses from those that oppose their violence, such as the criminalization of making systemic horrors known through Ag-Gag laws. And protest as a form of resistance continues to be met with draconian consequences. Protestors not only face police brutality, fines, and incarceration for speaking out about animal abuse and environmental harms, they also risk being labeled as terrorists by the state – that is, formally marked as an enemy of the state for attempting to put an end to environmental harms and crimes, including violence against animals.

In short, the procedural-legal and socio-legal distinction made in green criminology is important here (Brisman, 2008). The former addresses harms against animals that are only recognized by law and the latter is concerned with violence against animals that is not necessarily criminalized. Although selfie safaris are intertwined with important issues such as animal welfare laws and law enforcement efforts aimed at protecting animals, this chapter emphasizes the socio-legal end of the continuum in order to locate and make sense of some of the underlying ideological and structural mechanisms fueling the desire to accumulate and display trophy shot selfies in the first place. Until these broader influences change, harms against animals will continue, whether or not they are criminalized. Much of the violence tied to selfie safaris is illegal, and despite formal attempts at sanctioning things like harassing, capturing, or trafficking animals it remains a problem.

For the second, green criminology provides the space for non-anthropocentric analyses. Although including humans and human relations in theoretical and empirical work pertaining to harm, violence, or crime is necessary, a green criminological paradigm also recognizes that we must not limit our scope to them. Other species of animals besides humans – among other living beings – must also be acknowledged as subjects or actors, or at the very least, entities worthy of analysis. This of course alters not only who or what is framed as a perpetrator of harm or a criminal, but also who or what is considered a victim (see, for example: White, 2018). These shifts inevitably change established criminological notions of what justice is and how it should be done, and they lend themselves to imagining what non-anthropocentric social justice might look like, such as ecological justice and species justice. In other words, green criminology fosters theoretical and empirical work – as well as praxis – that de-centers humans and takes other animals seriously.

In some cases, a non-anthropocentric approach means initiating the animal turn in existing subfields such as victimology (see, for example: Flynn & Hall, 2017), or developing new subfields within (green) criminology, such as animal abuse studies (see, for example: Maher et al., 2017), nonspeciesist criminology (see, for example: Beirne, 1999), and wildlife criminology (see, for example: Nurse & Wyatt, 2020). Much of the violence against animals addressed in these areas are linked to the aforementioned issue – the confluence of symbolic and systemic violence vis-à-vis dominant culture and the economy, through various industries such as the pet industry, industrial or factory farming, entertainment and edutainment industries, and the fashion industry.

Deviant leisure is a criminological theory that complements the critical elements of green criminology discussed above (see, for example: Smith & Raymen, 2018). It relies on a zemiological approach to harm and crime, as it links many of the harms observed in contemporary society to neoliberal, consumer capitalism: most of the harms tied to consumption and leisure industries are not criminalized, or even remotely deviant. If anything, they are markers of hyper-conformity. Environmental crimes and harms are also a domain of the deviant leisure perspective, which has been applied to areas such as ecotourism (White, 2019), luxury canned hunting experiences (Smith, 2019), and plastics (Smith & Brisman, 2021). It is a psychosocial framework that takes both individual-level and structural analyses into account, which directs us to motivations and drives for obtaining a trophy shot selfie alongside the industries that shape them. Finally, the harms of selfie safaris are situated here as acts that disregard or deny animal subjectivity (Wolch, 1998), agency (Haraway, 2003), and autonomy (Rowlands, 2002). It also entails suffering, which is not limited to physical pain and injury, and includes animal fear and anxiety (Rowlands, 2002).

Canned Camera Hunting: The Harms of Trophy Shot Selfies With Captive Wildlife

A critical difference between camera hunting when it was practiced as "a harmless sport" decades ago and forms of camera hunting popular today, is that pictures of wildlife at that time were – fittingly enough – taken in the wild: animals were photographed in their natural habitats and left undisturbed, as camera hunters remained unseen and undetected to capture the shot and secure their trophy. That distinction has many consequences for animals forced to perform for and with camera hunters seeking selfies.

First, the animals are not only disturbed, they are taken out of the wild – their habitat – and held in captivity in order to be photographed. The methods of capturing animals are often cruel, and captivity itself usually involves abuse or neglect, or both, and fatalities are also common (Carder et al., 2018; Carr & Broom, 2018; Glass & Fennell, 2018; Lemieux et al., 2018; Turner, 2018; World Animal Protection, 2017). Furthermore, once captured, their immediate social groups and species are also jeopardized, particularly when protected, threatened, or endangered animals are involved (Carder et al., 2018; World Animal Protection, 2017). Just as there are those in traditional gun hunting who prefer to hunt animals "in the wild," and there are others that engage in canned hunting, this version of camera hunting is akin to the latter – it is essentially canned camera hunting. The second difference is that the camera hunter is not only seen and detected by the animal, but they are also part of the picture itself. So, in addition to the baseline harms that wildlife experience during captivity and in the process of being captured, close and direct contact with tourists also threatens their well-being (Carder et al., 2018; Carr & Broom, 2018; Glass & Fennell, 2018; Turner, 2018; World Animal Protection, 2017). Below, these harms are illustrated through manifestations of canned camera hunting that are understood in relation to anthropocentric and anthropomorphic beliefs and practices – and more specifically, how each perpetuates harm through the avowal and disavowal of animal abuse, respectively.

Anthropocentric Avowal of Harm: The Commodification of Trophy Shot Selfies

Given the ubiquity of places that offer these forms of interaction, contact with wildlife is more accessible than ever. For contemporary camera hunters, that means being able to shoot captive wildlife where they live, or having captive animals come to them. Both forms of canned camera hunting guarantee a trophy shot selfie, as these animals are completely stripped of their agency and subjectivity through the commodification of trophy shot selfies. For those that sell and purchase these encounters, wildlife are merely property (Nurse & Wyatt, 2020), something we attribute value to (van Uhm, 2018). Here we see that harm is perpetuated through an anthropocentric avowal of animal abuse, one that justifies objectifying wildlife as commodities. As they are no longer situated as sentient beings capable of suffering, the well-being of animals within leisure industries is secondary to human pleasure and profit, granted their well-being outside of their economic value is even acknowledged at all. Even so, the harms of camera hunting do not stem from narratives of overt violence, even when the harms are avowed through anthropocentrism, as they are packaged as thrills and excitement – good times – for human consumers.

Turning to the deviant leisure perspective, the violence endemic to these sites and encounters can be understood as form of anthropocentrism shaped by what criminologist Steve Hall calls special liberties, which "is synonymous with success in the hyper-competitive individualized environs of the current neoliberal social order" (Smith & Raymen, 2018 p. 67). While the subjectivity of wildlife is denied, camera hunters are validated as subjects through consumption, their ability to access and purchase leisure experiences with wildlife. Any moral or ethical duty to others – human and non-human – is deemed irrelevant or inconsequential because they can be who they want to be and get to do what they want to do as subjects of consumer capitalism. In short, the ethos of special liberties "allows the individual to abdicate from moral responsibility and perpetuate the economic, social and environmental forms of inequality that leak from commodified leisure" (Smith & Raymen, 2018, p. 72). While this is obviously tied to conspicuous consumption, the Veblenian notion of the leisure class has shifted, it "has mutated into a more democratised form of leisure whereby many forms of leisure that would previously have been the preserve of a wealthy elite are now conceived of as rights or expectations" (Raymen & Smith, 2019, p. 2). A psychosocial factor at play here is the Žižekian concept of the cultural injunction to enjoy – the notion that it is imperative, as subjects of consumer capitalism, to constantly seek out enjoyment and perpetually have fun. Consequently, that means the "capacity for harm is embedded within culturally acceptable, value-normative behaviours" (Smith & Raymen, 2018 p. 67). With incessant consumption, conspicuous displays of leisure activities, and the constant pursuit of enjoyment as dominant expectations, what matters here is that humans are entertained, the terror and misery they are causing animals does not. Below, different forms of commodified canned camera hunting and some of the harms associated with them are briefly illustrated through three broad sites: wildlife tourist attractions, wildlife-for-hire, and ersatz sanctuaries.

Wildlife Tourist Attractions

Wildlife tourist attractions across the world take on many forms, such as zoos, roadside zoos, safaris, marine parks, and aquariums. Places like these – and the sites below – objectify animals by turning them into photo props, photo opportunity models, or a thing to be included in a souvenir photo, like posing in front of the Eiffel Tower when in Paris. Because wildlife are subjects, not objects, additional harms are imposed on them to make them less likely to object to their objectification. Techniques used to make animals comply, be less animal-like, are intended to make the experience safer and more satisfying for the consumer. The specific tactics an attraction will use varies by species, as the specific behavior or body part deemed problematic depends on the animal.

Sometimes suffering stems from the animal not doing what the venders want them to do when it is time for a photo. For example, if a tiger is not looking up when the customer is trying to take a photo, igniting a lighter near their nose leather and mouth is one strategy used to make them to look up – what makes this especially awful is that some of these lions look down because they cower in fear when people raise their arms or hands around them, another obvious sign of abuse.[2] To make sure wildlife stay in place or do not run away, many are chained by their necks or legs, or kept in very small enclosures so that the camera hunter can secure a shot next to the cage. Cruel modes of confinement create stress and anxiety, which is visible through stereotypic behaviors, such as incessant pacing and gnawing. These signal mental deterioration – psychosis – from confinement, which is sometimes referred to as "zoochosis." On the other end of the continuum, some animals are clearly lethargic and unusually calm, which is the result of being drugged. This strategy, for example, has been used to sedate tigers so they will not react negatively – or at all – to camera hunters seeking a trophy shot selfie with them (Cohen, 2013). Abuse is not limited to making them better props, sometimes wildlife suffer to increase the safety of encounter for the customer. For example, attractions that feature slow loris commonly clip their teeth (Kitson & Nekaris, 2017; Osterberg & Nekaris, 2015), this practice also used for sloths (Carder et al., 2018). Other places declaw animals, which permanently disfigures them and can lead to a lifetime of suffering and limping.[3] Taping the mouths of animals such as snakes shut is also practiced at some wildlife tourist attractions.

In addition to sites like these, there are also tourist attractions that offer experiences with wildlife to increase business, such as tourist shops. These are not wildlife tourist attractions proper, but similar techniques are used to make the animals more conducive to being posed or handled for photos. For example, to increase business, a souvenir shop in Florida kept a baby alligator with their mouth taped shut at the shop so customers can take their photo with them (Bartoli, 2023). And Pizza, known by many as the World's Saddest Polar Bear, developed "zoochosis" after being acquired by a mall in China to attract more customers. Pizza was kept in a small glass and concrete enclosure, without privacy or enrichment, so customers could get selfies with him.

Wildlife-for-Hire

While some camera hunters prefer to travel, others rent wildlife from vendors that bring animals to them. Customers that secure trophy shot selfies for themselves and others attending their events not only contribute to suffering tied to captivity, but also the added misery of being transported to and from new and different spaces, many of which are loud and crowded. In recent years these services have been met with some resistance, and they show the range of spaces that trophy shot selfies are captured in different regions through wildlife rentals. For example, in Florida controversy ensued after a high school in Miami High rented a tiger for prom, with video footage showing a stereotypic, stressed out tiger in a small cage surrounded by teenagers.[4] In Australia, a nightclub rented animals for influencers to take photos with to promote the launch of a new club.[5] And in Long Island New York, a business selling sloth selfie encounters in a strip mall began advertising home visits after the owner was forced to close the location because it was not a legitimate zoological park.[6]

Ersatz Sanctuaries

Many tourists visit attractions that blatantly undermine the well-being of non-human animals just so they can be photographed with them. Customers cannot deny that animals are being mistreated or exploited solely for their use when the services that they are paying for include animals with their mouths taped shut, that are tied or chained down, or have missing teeth or claws. But there are others that do care, and they are duplicitously led by those soliciting the attractions to believe that their services are harmless to animals or even benefits them. Animals are not only commodified in these spaces, so is any compassion humans may have toward them. This is an example of what legal scholar Delcianna Winders calls humane washing: "Through 'humane-washing' – using unregulated terms like 'sanctuary' and participating in misleading accreditation programs – captive wildlife facilities are profiting from making consumers feel better" (Winders, 2017, p. 161). For ersatz sanctuaries across the world, animal welfare is nothing more than branding and packaging to generate profit, a narrative and label that tourist attractions and sham accreditors use to make money from the suffering of wildlife.

Anthropomorphic Disavowal of Harm: Trophy Shot Selfie Menageries

While anthropocentrism is more obviously tied to animal abuse, anthropomorphism also fosters violence. Camera hunters that anthropomorphize animals concurrently disavow harm while perpetuating it, as they do not think they are contributing to harm because they do not know what suffering looks like. Not knowing how certain species typically behave, these selfie seekers assume they know how animals are supposed to look and act, based on their own anthropomorphic vantage. As humans project their own personal and cultural meanings onto wildlife, those meanings become imbued on the bodies of individual animals and embedded in our interactions with them. These projections have consequences for animals and absolves the camera hunter of guilt or responsibility of wrongdoing. Ultimately, these camera hunters cannot see beyond themselves, their collection of trophy shot selfies becomes a digital menagerie that overlooks the violence and suffering of animals forced to live in captivity and perform as photo props.

On the cover of historian Erica Fudge's book, *Animal*, is a photo of a monkey named Ham who, by human standards, looks like he is smiling and having a grand time (2002, pp. 25–27). Taking what he experienced before that moment into account, he was, by monkey norms, more than likely terrorized. Fudge explains that the photo of this "animal astronaut" was taken by NASA in 1961 after a test flight that was grueling before taking off, during the flight itself, and even after landing. To signify a successful mission, the photo was circulated across the world, accompanied by a feel-good narrative about the photogenic monkey: "after being recovered from the craft, 'Ham was in good spirits and posed for pictures with the sailors on the recovery ship'" (Fudge, 2002, p. 25). In reality, he was nothing more than a captive lab animal used to experiment on. Since Ham did come back alive, and his reactions could be packaged as glee, he was also used as a jingoistic photo prop to signal space-race superiority. Alongside that, to divert our eyes away from the suffering inflicted upon him and countless other animals used for space exploration, we are made to believe that Ham is a hero, a patriotic monkey that is happy to serve his nation. Fudge asks us:

> So what does the picture of Ham tell us about ourselves? Our desire to interpret the smile of the chimp as evidence of his enjoyment of the experience of space travel reflects, perhaps, our willingness to dismiss in certain contexts some of the differences between humans and animals. Ham looks like he's smiling, happy to receive his reward of an apple, and that's as far as many readers want to go with their interpretation of the image. But if we refuse this anthropomorphism – literally, changing of an animal into a human – then Ham's 'smile' becomes more of a challenge. The bravery of the chimpanzee is translated into something that reflects, perhaps, the terror that he felt.
>
> *(2002, p. 27)*

These sort of clashes between ethnology and ethology mean suffering for animals, and wildlife trophy selfies are just one manifestation of that. Like Ham, although certain animals might seem like they are enjoying canned camera hunting encounters, they are not – and sloth selfies are one example of that.

Although being cute means that humans are more likely to care about an animal, that also puts a target on them, as that means that humans are more likely to want to own or be near them. Sloths are particularly desirable to shoot with a camera because the shape and markings on their face makes it look like they are smiling as we humans do. And since they are extremely slow, they do not show signs of overt resistance and psychosis in ways a human or other animals might – although they can bite, which is why vendors will clip their teeth. Together, many interpret these behaviors as enjoying being held and manipulated by humans that are using them as a photo prop. Exacerbating the situation is that one of the stereotypic behaviors witnessed in sloths being held for selfies is limb stretching (Carder et al., 2018). From an anthropomorphic gaze that could be interpreted as reaching out toward humans, wanting to embrace or hug them – not what stereotypic behaviors really are: signs of stress. Further contributing to the stress of captivity is not allowing sloths to sleep as much as required. Because of these practices it is estimated that those taken from the wild for selfies may die within six months of captivity (World Animal Protection, 2017). While this can be viewed as simple anthropomorphism, we can also see it as something more complex than that: anthropomorphic violence driven by some of the same things that fueled the violence of anthropocentricism described above.

It seems that no matter what a sloth is doing the animal is enjoying life and smiling. For example, even when bloody and clinging to a pole they are seen as happy, and not afraid. As one *New York Post* article described, "Adorable sloth smiles in the face of danger," with the following lead: "This sloth is not afraid of danger. Photos of a sloth in Ecuador clinging onto a traffic barrier for dear life have gone viral, as the tiny creature appeared to be smiling despite its situation."[7] It is not just tourists or lay people on news and social media platforms that animate these narratives. Even zoologists will perpetuate this form of anthropomorphism in an attempt to "rebrand" sloths. In these cases, the anthropomorphic projections switch from one ideological era to another. We are told that sloths should no longer be anthropomorphized as slow, stupid, and useless as they once were by naturalists. Instead, we should anthropomorphize them through a contemporary lens, one that is compatible with neoliberal leisure under capitalist realism: sloths are fun-loving and sweet. For example, the zoologist that some have credited with starting the sloth craze on social media stated the following in a television appearance that featured a sloth from a local zoo: "I fell in love with sloths," Cooke explained. "They're genetically pre-programmed to hug. They smile all the time … I think they're very misunderstood, so I want to make people understand."[8] They are not smiling, that is something humans do. They are not hard-wired to hug; they must be able to hold onto their mother and wrap their arms around trees and limbs to survive. Sloths are still not understood on their own volition here, they have just been repackaged as something culturally legible for consumer capitalism. Albeit implicitly, this perpetuates the idea that the sort of animals that humans can relate to are the ones that are worthy of respect, kind treatment.

Returning to the deviant leisure perspective, we see that sloths are relatable and desirable because they are thought to be the embodiment of leisure, the epitome of a carefree lifestyle. We are made to believe that they too understand the freedom that comes from a life of boundless enjoyment, one uninhibited by work and unencumbered by negative feelings. For example, in an article from the BBC outlet, Discover Wildlife, the popularity of sloths on Instagram is tied to this relatively new perception of them: "Where once sloths were vilified for their apparent laziness, now their languid movements, breezy charm and seemingly carefree attitude have inspired thousands of millennials to pay homage to their 'spirit animals' in the comments sections of the countless viral sloth videos that have flooded TikTok and Instagram."[9] The idea that sloths exist in a state of perpetual pleasure might lead to more harm than vilification. Either way, suffering is justified through whatever the dominant mode of anthropomorphic violence happens to be. Just as humans are interpellated – à la Althusser – as subjects of ideology, we can think of the implications of what "animal interpellation" (Haraway, 2003, p. 17) might mean for them and how they are treated in a particular society. Except, here, they are interpellated as objects of ideology.

Revisiting the Australian nightclub criticized for renting wildlife for promotional purposes, we see that vendors of these encounters also rely on the narrative of enjoyment – it can also be used to justify anthropocentric violence as well. When faced with the reality that he was terrorizing animals for his business, the vendor insisted the animals, like the humans around them, were having fun: "Raymond Hoser who supplied the nightclub with reptiles told Yahoo everyone had a fantastic time. 'The snakes look happy, the people look happy, what the f*** is wrong with that?,' he said."[10]

This narrative is also used for selfies taken with non-captive sloths as well – since they are so happy and playful, they welcome disruptions by tourists and long to be part of their pictures. For example, a popular photo of a selfie with a sloth emphasized how cool, laid back, and affable these animals are, with no mention of fear or harm whatsoever. The photo circulated beyond social media platforms and made its way to popular media outlets as well. One article, "Best selfie ever? Smiling sloth is a total dude," had the following lead:

> This picture of a smiling sloth hanging out in a tree sure has to be a contender for the best selfie ever. A lucky tourist spotted a sloth chilling out up a tree and decided to use his selfie stick to get a picture with it. Not only did the sloth seem happy to oblige, he also apparently managed a smile for the camera. The image was uploaded to Imgur by Nicolas Huskar, and has won the collective heart of the internet. One user wrote: 'A slofie', while another declared it as 'the only reason one should own a selfie stick'. Another joked: 'Smile and wave, just smile and wave. Tourist season is almost over.'[11]

While it may seem that this is completely harmless – particularly since the sloth was seen to be smiling and waving – interrupting sloths attempting to sleep, eat, or defecate in order to get the perfect shot can have consequences for these animals. Although better than canned camera hunting, getting so close to animals in the wild with selfie sticks is still harassment, even if it does not appear that way or done without malicious intentions. This was the case in Australia recently when trophy shot selfies with wombats began to increase in popularity. As a result, campaigns by tourist councils were launched to remind tourists that wombats are wild animals and to respect that by not chasing them with selfie sticks or otherwise harassing them to get a photo.[12] More about these forms of trophy shot selfies are discussed in the next section, as it is important to address how non-captive camera hunting also causes suffering and death.

Open Season for Wildlife Selfies: Life, Death, and Trophy Shot Selfies in the Wild

Being shot with a camera can be just as deadly as being shot with a gun for an animal in the wild, particularly when the thrill of the hunt for the camera hunter is a selfie. Whether sought out directly or captured through chance encounters, trophy shot selfies in natural environments cause harm or death to wildlife in the process. This form of contemporary camera hunting not only legitimizes violence through an anthropocentric and egocentric undercurrent that places selfie culture before animal life, it reproduces harm through the belief that it is perpetually open season for camera hunting in natural places. Although animals in these situations have significantly more agency than those shot by a camera in captivity, camera hunting still threatens them individually and as a community, as open season camera hunting forces wildlife to negotiate the behaviors of camera hunters hoping to obtain a trophy shot selfie with them. From life to death, the harms of these selfie encounters are approached through two broad categories that reveal different dimensions of violence against animals: menacing wildlife and murdering them.

Menacing Wildlife: Ecological Flourishing and Trophy Shot Selfies

While doing no harm to animals encountered in natural spaces is a good ethical baseline, we should ultimately aim for something better – acting in ways that "not only avoids harming others, but positively contributes to society, culture and the environment" (Smith & Raymen, 2018, p. 67). The deviant leisure perspective extends the notion of human flourishing to include matters of ecological justice (Smith & Raymen, 2018, p. 66), which means being concerned with non-human and environmental flourishing simultaneously, what we might call *ecological flourishing*. Focusing specifically on human-animal interactions in natural spaces as one part of that, ecological flourishing directs our attention to both harm and flourishing in relation to not-doings and doings, active absences and presences: not harming wildlife, the active absence of harm. Benefitting wildlife, the active presence of behaviors that allow them and their environment to thrive. Trophy shot selfies threaten ecological flourishing on all fronts. They are not beneficial to wildlife, nor are they simply "harm neutral" (Smith & Raymen, 2018, p. 67). They are harmful. The popularity of "sealfies" – selfies with seals – across the world is one example of this. In places such as Australia, South Africa, England, and the United States camera hunters are harming seals and seal pods, and even hurting themselves, to get a sealfie shot. This is to say nothing of the cumulative damage to flora and fauna wrought by the tourist industry itself.

In the United States, despite efforts of different branches of the NOAA Fisheries organization to discourage the public from taking selfies with seals through press releases, social media, and other outreach materials – notwithstanding signage where these encounters often take place – the exploitative hunt for "sealfies" continues. One issue for the New England/Mid-Atlantic regional division is that sealfies are endangering baby seals, as harbor seal pupping season not only coincides with summer break, it overlaps with a major holiday weekend. This is also the case in the United Kingdom with pup season overlapping with bank holidays in the winter. This combination of events makes encounters with baby seals more likely, and the problem with trophy shot selfies more pronounced.

If alone, approaching a pup could cause stress or cause them to go into the water when they should be resting – this means they are more likely to drown or become more vulnerable to predators if swimming extremely fatigued. If they happen to have an injury, going in the water also makes death more likely. In Wales for instance, an injured seal drowned because they were forced to retreat to the ocean instead of resting, despite signs and flyers posted asking people to stay away.[13] Dealing with harassment is also a waste of energy, and it can hinder proper eating. While these factors are bad on their own, they can also compound. If a mother and pup are harassed, it may cause mothers to abandon their pups out of self-preservation, which significantly increases the likelihood that the pup will die or need to be rescued. In places like California, for example, officials have seen an uptick in the number of selfie-related rescue efforts for seal pups.[14] Individuals and groups beyond the mother-baby dyad are also popular trophy shot selfies – and, like them, they cause stress or diminish their quality of life in other ways.

Where ecological flourishing is concerned, the immediate solution is simple: the most ethical thing camera hunters can do here is nothing at all. Capture a photo from them from far away or simply leave them alone. Or as, NOAA Fisheries West Coast has begged, "Use your zoom and give them room!"[15] Despite these witty nudges, camera hunters seem unable to just do nothing, nor will they allow the animals to do the same. The power of the image, the drive to capture and distribute a trophy shot selfie as a tradable aspect of one's own identity is just too great. Anthropocentric entitlement is certainly one reason for that, but also at play here are forms of anthropomorphism witnessed above. As before, enjoyment is also a factor here.

The idea that seals look like they are having fun or might want to be part of the fun also contributes to harm. For example, in a 2022 media release from the Australian government advising the public to give them space – "Don't be seal-y around seals" – a manager from the Department of Environment and Science offered the following advice:

> "On the Gold Coast, another juvenile fur seal has been seen in the ocean near Coolangatta. The seal was sticking its flippers out of the water and people were concerned it was ill or injured, but that is how seals regulate their temperature. Seeing seals in the wild like this can be exciting, because they always look like they're having fun in the waves or on the beach."[16]

And returning to California, where rescues and deaths of seal pups have increased due to being harassed and picked up, examples of this includes behaviors such as picking up a pup still dependent on their mother and placing them on a surfboard.[17] Playing around like this is likely to lead to a pup's death. Camera hunters that capture these shots might as well shoot the pup with a gun. Yet many, particularly those that consider themselves animal lovers, would never do that. Even so, contemporary camera hunters continue to shoot wildlife in ways that lead to the same outcome. However, there is a double bind here as far as enjoyment is concerned. If seals appear to be having fun, that makes them a target. When they are simply being, not appearing to have fun, that also makes them a target – even in these cases we see how enjoyment-driven anthropomorphism shapes these harmful encounters.

Camera hunters cannot fathom the idea that an animal would like to – or must, in many cases – do nothing. According to cultural theorist Mark Fisher, late capitalism fosters "hedonic depression" (2009), which is akin to, but departs from, the cultural injunction to enjoy here. While framing enjoyment positively is useful – as in, the presence of fun as a cultural imperative – hedonic depression points us to drives that actively avoid the absence of fun. This preoccupation with preventing negative- or non-enjoyment is part of the affective atmosphere of capitalist realism: avoid boredom, seek out or manufacture stimulation. Better yet, remain stimulated by never disconnecting from dominant cultural forms in the first place. Either way, subjects of late capitalism must always be consuming, distributing, and producing culture in order to prevent themselves and others from feeling bored. Or worse, feeling nothing at all. Many not only internalize and perform these cultural ideals, they also project them onto animals. This additional layer of anthropomorphic bias moves us further away from recognizing – nonetheless achieving – ecological flourishing.

For example, the NOAA Fisheries West Coast division has been trying to stop people from harassing marine life that seem to be "doing nothing" according to human standards. In a postcard campaign featuring a seal sleeping on the beach, they remind selfie seekers that "It's normal for me to be onshore. I don't need to go swimming. I'm resting and need my space. I'm not a selfie prop! Please stay back at least 100 feet."[18] Another campaign launched by the NOAA National Marine Sanctuaries division – "There is no selfie stick long enough!" – reminds visitors not to bother animals that seem to be doing "nothing":

> It may be tempting to approach that elephant seal pup lounging on the beach to snap a stellar selfie, but get too close and you put the animal – and yourself – at risk. And while it may seem perfectly harmless to sidle up to a sea turtle that appears to be 'hanging out,' the few seconds it takes to snap a photo can have a compounded and lasting impact on the life of a wild animal.[19]

Doing nothing is something important here, something animals – including humans – need to do to survive and thrive: all animals need time and space to recover and rest. As desire and opportunities to do just that – do nothing – continue to dwindle for us, we also make not-doing untenable for other animals as well.

Murdering Wildlife: Trophy Shot Selfies as Representations of Theriocide

We conclude here with a type of trophy shot selfie that brings us full circle, takes us back to the beginning, as the camera hunter highlighted in this section shoots – and produces a trophy shot – like a traditional gun hunter: both of their shots directly contribute to the death of an animal, both photos depict dead or dying wildlife. In lieu of a bullet, the desire to capture a trophy shot selfie with a wild animal is the very mechanism that kills them, as these camera hunters will do whatever it takes to capture the perfect shot, even if that means killing the animal in the process. To understand how killing or murdering animals – what criminologist Piers Beirne (2018) calls theriocide – might manifest through the pursuit of trophy shot selfies, we turn to another concept situated within the deviant leisure perspective: criminologist Majid Yar's (2012) will-to-representation. As displaying our lives on social media has become the dominant expectation, the drive to represent oneself in popular culture is so strong that many people will do things just to represent them – that is, for the sake of representation itself, including harmful and violent things. In this sense, the very need to represent lends itself to behaviors people might not otherwise engage in to produce and distribute content.

In regard to leisure, Smith and Raymen emphasize how "Yar suggests that the role of the camera, the material's eventual destination of social media and the intensified desire for social recognition play a criminogenic role in these behaviours. We are not talking here about opportunistic or coincidental filming of violence and violent encounters. (…) These are performative acts, enacted for their value on social media and user-generated websites" (2018, pp. 73–74). Similarly, the final encounters described here are not random situations involving animals in trouble, dying animals – they involve camera hunters that concurrently manufacture and document deadly encounters with wildlife. Dolphins, for example, are popular victims of this form of theriocide. Sometimes tourists will take the animal out of the water to get a selfie with them, which can lead to death due to heat exhaustion or their lungs collapsing, as they are not designed to live out of water. For example, in 2016, a rare baby dolphin died in Argentina after being removed from the water and passed around by tourists that wanted to take selfies with the animal. A year later, the same thing happened again. In other cases, tourists have been known to come across a stranded dolphin and handle and take pictures with them in lieu of seeking help for the animal or allowing them to make their way back to the ocean. In 2017, for example, a stranded dolphin died after Spanish beachgoers surrounded the animal to take photos with them. Instances of non-aquatic animals dying at the hand of camera hunters have also circulated in news and social media in recent years. In the United States, for example, a moose drowned in Lake Champlain in Vermont because people crowded the animal to take pictures of them as they were trying to get out of the lake. Eventually the moose retreated back into the lake, then drowned from exhaustion.

Ultimately, despite the innocuous appearance of many, and the positive narratives tied to most, trophy shot selfies are still trophy shots: cultural forms that are produced through – and eventually represent – violence. For traditional trophy shots, that includes narratives of loving and caring about animals and the environment that eventually become images of dead animals or animal parts which further objectify wildlife and deny them of their subjectivity (Kalof & Fitzgerald, 2003). For police trophy shots, that entails narratives around the benevolence of police power that transform into images of control, domination, and accumulation

(Linnemann, 2017). For this newer genre, trophy shot selfies, that involve narratives of enjoyment and harmless fun that manifest into photos of suffering or dying animals. This cultural form is one of many ways of representing animals in contemporary social life, and of abusing them. They remind us to include, but not limit ourselves to, images alone when assessing violence against animals in popular culture. If we are to take harms of animal representation seriously, we must not only represent animals differently, but we must also look at them differently altogether – that is, see them seeing us – see them as subjects: "There are obviously many ways to look at animals, but perhaps the most interesting thing about them is that they look back at us" (Dion, 2012, p. 143).

Notes

1 An earlier version of this paper was presented at the Critical Directions in Green Criminology Conference at the University of Plymouth, UK, April 2, 2019.
2 For example, see: https://www.independent.co.uk/asia/southeast-asia/tigers-abuse-selfies-tourism-thailand-b2368279.html.
3 See: https://www.nationalgeographic.com/animals/article/tigers-in-the-united-states-outnumber-those-in-the-wild-feature.
4 See: https://www.nbcnews.com/news/us-news/miami-school-faces-backlash-bringing-live-tiger-prom-n873846 and https://www.nytimes.com/2018/05/14/us/tiger-prom-florida.html.
5 For example, see: https://www.indy100.com/viral/influencer-event-snakes-crocodiles.
6 See: https://www.curbed.com/2022/12/sloth-encounters-islip-long-island.html.
7 See: https://nypost.com/2016/01/25/adorable-sloth-smiles-in-the-face-of-danger/.
8 See: https://www.today.com/pets/c-c-smiling-sloth-hangs-out-today-1c8709318.
9 See: https://www.discoverwildlife.com/animal-facts/mammals/why-the-sloth-is-instagrams-favourite-animal.
10 See: https://au.news.yahoo.com/investigation-launched-over-controversial-influencer-event-at-melbourne-nightclub-004240951.html.
11 See: https://www.aol.co.uk/2016/07/05/smiling-sloth-selfie-stick-picture-best-ever/.
12 See: https://www.independent.co.uk/travel/news-and-advice/wombats-selfie-australia-tasmania-maria-island-tourists-travel-a8730161.html.
13 See: https://www.bbc.com/news/uk-wales-58629122.
14 See: https://www.foxweather.com/earth-space/marine-mammal-harassment-california.
15 See: https://www.fisheries.noaa.gov/resource/outreach-materials/no-selfies-seals-and-sea-lions.
16 See: https://www.des.qld.gov.au/our-department/news-media/mediareleases/2022/dont-be-seal-y-around-seals.
17 See: https://www.foxweather.com/earth-space/marine-mammal-harassment-california.
18 See: https://www.fisheries.noaa.gov/west-coast/marine-life-viewing-guidelines/share-shore-watch-marine-mammals-responsibly.
19 See: https://sanctuaries.noaa.gov/news/may16/responsible-recreation.html.

References

Bartoli, J. (2023). Missouri Needs stricter koalifications: How Missouri can promote exotic animal welfare. *UMKC Law Review*, *9*(4), 867–908.

Beirne, P. (1999). For a nonspeciesist criminology: Animal abuse as an object of study. *Criminology*, *37*(1), 117–148.

Beirne, P. (2018). *Murdering animals: Writings on theriocide, homicide and nonspeciesist criminology*. Palgrave Macmillan.
Brisman, A. (2008). Crime-environment relationships and environmental justice. *Seattle Journal for Social Justice*, *6*(2), 727–818.
Brisman, A., & South, N. (Eds.). (2013). *Routledge international handbook of green criminology*. Routledge.
Brisman, A., & South, N. (Eds.). (2020). *Routledge international handbook of green criminology* (2nd ed.). Routledge.
Brower, M. (2005). Trophy shots: Early North American photographs of nonhuman animals and the display of masculine prowess. *Society & Animals*, *13*(1), 13–31.
Brower, M. (2010). *Developing animals: Wildlife and early American photography*. University of Minnesota Press.
Carder, G., Plese, T., Carniel Machado, F., Paterson, S., Matthews, N., McAnea, L., & D'Cruze, N. (2018). The impact of 'Selfie' tourism on the behaviour and welfare of brown-throated three-toed sloths. *Animals*, *8*(11), 1–12.
Carr, N., & Broom, D. M. (2018). *Tourism and animal welfare*. CABI.
Cohen, E. (2013). Buddhist Compassion' and 'Animal Abuse' in Thailand's Tiger Temple. *Society & Animals*, *21*, 266–283.
Davis, A. (2011). *Are prisons obsolete?* Seven Stories Press.
Davis, A. (2014). Deepening the debate over mass incarceration. *Socialism and Democracy*, *28*(3), 15–23.
Dion, M. (2012). *Den*. Forlaget Press.
Dion, M. (2017, September 1). Q+A with Mark Dion. Institute of Contemporary Art Boston. https://www.icaboston.org/articles/qa-mark-dion/.
Duffy, R. (2014). Interactive elephants: Nature, tourism, and neoliberalism. *Annals of Tourism Research*, *44*, 88–101.
Dunaway, F. (2000). Hunting with the camera: Nature photography, manliness, and modern memory, 1890-1930. *Journal of American Studies*, *24*(2), 207–230.
Fisher, M. (2009). *Capitalist realism: Is there no alternative?* Zero Books.
Fletcher, R. (2023). Tourism and neoliberalism. *Tourism Geographies*. https://doi.org/10.1080/14616688.2023.2269882
Flynn, M., & Hall, M. (2017). The case for a victimology of nonhuman animal harms. *Contemporary Justice Review*, *20*(3), 299–318.
Ford, W., & Reséndez, A. (2021, March 2). Interview for Theory + Practice Series. *Bomb Magazine*. https://bombmagazine.org/articles/2021/03/02/walton-ford-1/.
Fudge, E. (2002). *Animal*. Reaktion Books.
Glass, T., & Fennell, D. A. (2018). Animal watching in tourism. In A. Butterworth (Ed.), *Animal welfare in a changing world* (pp. 226–234). CABI.
Haraway, D. (1984). Teddy bear patriarchy: Taxidermy in the Garden of Eden, New York City, 1908-1936. *Social Text*, *11*, 20–64.
Haraway, D. (2003). *The companion species manifesto: Dogs, people, and significant otherness*. Prickly Paradigm Press.
Haraway, D. (2008). *When species meet*. University of Minnesota Press.
Kalof, L., & Fitzgerald, A. (2003). Reading the trophy: Exploring the display of dead animals in hunting magazines. *Visual Studies*, *18*(2), 112–122.
Kitson, H., & Nekaris, K. A. I. (2017). Instagram-fuelled illegal slow loris trade uncovered in Marmaris, Turkey. *Oryx*, *51*(3), 394.
Lemieux, A., Bruschi, N., & Slattery, L. (2018, September 1). The overlap between poaching and natural resource extraction [Conference presentation]. In *Eighteenth annual conference of the European Society of Criminology*, Sarajevo, Bosnia and Herzegovina.
Linnemann, T. (2017). Proof of death: Police power and the visual economies of seizure, accumulation and trophy. *Theoretical Criminology*, *21*(1), 57–77.
Lynch, M. J., & Long, M. A. (2022). Green criminology: Capitalism, green crime and justice, and environmental destruction. *Annual Review of Criminology*, *5*, 255–276.

Maher, J., Pierpoint, H., & Beirne, P. (Eds.). (2017). *The Palgrave international handbook of animal abuse studies*. Palgrave Macmillan.
Moorhouse, T., D'Cruze, N. C., & Macdonald, D. W. (2017). Unethical use of wildlife in tourism: What's the problem, who is responsible, and what can be done? *Journal of Sustainable Tourism*, *25*(4), 505–516.
Nagel, M. E., & Nocella, A. J. II (Eds.). (2013). *The end of prisons: Reflections from the decarceration movement*. Brill.
Nocella, A. J. II, Salter, C., & Bentley, J. K. C. (Eds.). (2013). *Animals and war: Confronting the military-animal industrial complex*. Rowman & Littlefield Publishing.
Noske, B. (1989). *Humans and other animals*. Pluto Press.
Nurse, A., & Wyatt, T. (2020). *Wildlife criminology*. Bristol University Press.
Osterberg, P., & Nekaris, K. A. I. (2015). The use of animals as photo props to attract tourists in Thailand: A case study of the slow loris *Nycticebus* spp. *Traffic Bulletin*, *27*(1), 13–18.
Purdy, J. (2015, August 13). Environmentalism's Racist History. *The New Yorker*. https://www.newyorker.com/news/news-desk/environmentalisms-racist-history.
Raymen, T & Oliver Smith. (2019). Introduction: Why Leisure? In T. Raymen & O. Smith (Eds.), *Deviant leisure: Criminological perspectives on leisure and harm* (pp 1–13). Palgrave Macmillan.
Rothe, D. L., & Medley, C. (2019). Beyond state and state-corporate crime typologies: The symbiotic nature, harm, and victimization of crimes of the powerful and their continuation. In M. L. Rorie (Ed.), *The handbook of white-collar crime* (pp. 81–94). Wiley Blackwell.
Rowlands, M. (2002). *Animals like us*. Verso.
Ruggiero, V., & South, N. (2013). Toxic state-corporate crimes, neo-liberalism and green criminology: The hazards and legacies of the oil, chemical and mineral industries. *International Journal for Crime, Justice and Social Democracy*, *2*(2), 12–26.
Shiras, G. III. (1900, June 7). A harmless sport – Hunting with the camera. *The Independent* (52), 1364–1368.
Smith, O. (2019). Luxury, tourism, and harm: A deviant leisure perspective. In T. Raymen, & O. Smith (Eds.), *Deviant leisure: Criminological perspectives on leisure and harm* (pp. 305–323). Palgrave Macmillan.
Smith, O., & Brisman, A. (2021). Plastic waste and the environmental crisis industry. *Critical Criminology*, *29*, 289–309.
Smith, O., & Raymen, T. (2018). Deviant leisure: A criminological perspective. *Theoretical Criminology*, *22*(1), 63–82.
Turner, D. (2018). Managing tourism's animal footprint. In N. Carr & D. M. Broom (Eds.), *Tourism and animal welfare* (pp. 106–111). CABI.
Twine, R. (2013). Addressing the animal-industrial complex. In R. Corbey & A. Lanjouw (Eds.), *The politics of species: Reshaping our relationships with other animals* (pp. 77–91). Cambridge University Press.
van Uhm, D. (2018). The social construction of the value of wildlife: A green cultural criminological perspective. *Theoretical Criminology*, *22*(3), 384–401.
Vasan, S. (2018). Consuming the tiger: Experiencing neoliberal nature. *Conservation and Society*, *16*(4), 481–492.
White, R. (2018). Green victimology and non-human victims. *International Review of Victimology*, *24*(2), 239–255.
White, R. (2019). Loving the planet to death: Tourism and ecocide. In T. Raymen & O. Smith (Eds.), *Deviant leisure: Criminological perspectives on leisure and harm* (pp. 285–297). Palgrave Macmillan.
Winders, D. J. (2017). Captive wildlife at a crossroads – Sanctuaries, accreditation, and humane-washing. *Animal Studies Journal*, *6*(2), 161–178.
Wolch, J. (1998). Zoöpolis. In J. Emel & J. Wolch (Eds.), *Animal geographies: Place, politics, and identity in nature-culture borderlands* (pp. 119–138). Verso.

Wolch, J., & Emel, J. (Eds.). (1998). *Animal geographies: Place, politics, and identity in nature-culture borderlands*. Verso.
World Animal Protection. (2017). *A closeup on cruelty: The harmful impact of wildlife selfies in the Amazon*. World Animal Protection. https://www.worldanimalprotection.us/siteassets/reports-programmatic/a-close-up-on-cruelty-report.pdf.
Yar, M. (2012). Crime, media, and the will-to-representation: Reconsidering relationships in the new media age. *Crime, Media, Culture*, *8*(3), 245–260.

9

FOLLOWING THE CULTURAL TRACES OF NORMALIZED AND LEGITIMIZED VIOLENCE BY ISRAELI KOSHER SLAUGHTERERS TOWARD NONHUMAN ANIMALS

Anat Ben Yonatan

> The kosher-slaughter knife is so sharp and smooth that even if it cuts the *shochet*'s own finger, it will take a few seconds until he feels it. When you slit the animal's throat, it actually does not feel [the pain], and almost does not feel anything. And by the time it feels the first cut … it is no longer with us. In other words, I am not slaughtering it. I am putting it to sleep.
>
> [Elad]

This is how Elad, a 52-year-old Jewish slaughterer (in Hebrew, *shochet*; plural – *shochtim*) describes the nonhuman animal's subjective experience while being slaughtered. Elad, who has been in the business for 30 years, conceptualizes the animal's supposed lack of suffering as a way to illustrate the facilitator's perspective on the violence involved in slaughter.

This chapter explores violence as acts of *physical force* that influence, harm, or kill others – whether human or nonhuman. It deals with violence that is considered legal, which is one of the reasons conventional criminology is often reluctant to investigate it (Sollund, 2021).

This study considers the case of animal slaughter through the facilitator's vantage point and explores the cultural and discursive mechanisms that legitimize and naturalize violence toward animals in Israeli Jewish culture. It argues that while violence toward animals is viewed as natural, destined, or necessary for maintaining social order, when its facilitators become conscious of their actions' violent nature, society provides them with cultural coping mechanisms to deal with its negative effects and sterilize it from its emotional valence.

DOI: 10.4324/9781003441908-12

Violence and Social Order

In certain cases, violence can be not only justifiable or legal, but also ingrained into the very social structure of societies (Fitzgerald & Taylor, 2014), such as in the case of warfare (Howe, 2006). In these cases, the violence is realized by formal entities with the legitimacy to use it. Accordingly, violence can be articulated as a phenomenon to maintain social order, which is realized in all kinds of sovereignty – national, religious, or otherwise. In other cases, such legal violence can be seen as a form of oppression and exploitations of the weak, and social order as consisting of unjust power relations. This has been claimed to be the case with regard to various victims of harm – whether human or nonhuman.

Green criminology, unlike traditional criminology, explores such harm, whether legal or not. Often combined with critical victimology and critical animal studies, green criminology considers various acts of violence committed by institutions, as well as individuals, against humans, the environment and nonhuman animals (Taylor & Fitzgerald, 2018). Critical victimology is incorporated in order to emphasize situations that produce significant victimization through institutions and structural relations, even if they are not defined as such (Chouliaras, 2011). From this perspective, eating practices of humans are of importance to criminological research, since the consumption of nonhuman animals substantially harms them (Peggs, 2023). Critical animal studies additionally interleave asymmetrical power relations between species (Taylor & Fitzgerald, 2018) and challenge oppressive social constructions (Sollund, 2021). Combined, these areas of research deal with harms in a broader sense than the mere focus on its legal aspects.

Nevertheless, much of the work that considers nonhuman animals within green criminology remains anthropocentric (Beirne, 1999; Sollund, 2021). Moreover, its focus is mainly on "wild" animals, and has yet to contribute further to other aspects, such as the industrial animal agriculture. Slaughterhouse work, for example, has been underscored as a crucial component of the livestock chain, exploiting both humans and nonhuman animals. Green criminology, combined with aspects of critical animal studies, has much to offer to the understanding of the power relations and victimization entailed in animal agriculture (Taylor & Fitzgerald, 2018).

Social order intends to prevent chaos, as described by early thinkers like Hobbes (2018), and helps to organize the social space through various categorization efforts (Durkheim & Mauss, 1903; Lévi-Strauss, 1969). Social categorization necessarily presumes a boundary between different social categories, and researchers explore such boundary work in various contexts, including the human–animal boundary (Cartmill, 2001; Russell, 2010). Throughout these negotiations, some recognize that violence is often involved in boundary-crossing processes (Tilly, 2003, 2004), reemphasizing the close connection between social order and violence. Accordingly, when recognized as a key mechanism for arranging social relations, these boundaries may be violently defended if challenged.

This chapter deals with socially sanctioned violence toward animals. While the instrumental aspects of animal slaughter are central to these acts and their legitimacy, animal slaughter is additionally a boundary-maintaining practice that supports the social order that clearly distinguishes between humans and other animals. The very phenomenon of animal slaughter is rooted in the human–animal divide, underscoring it as a way to avoid the potential chaos emanating from mixing up these categories.

Violence in the Human–Animal Context

Like humans, nonhuman animals are exposed to various forms of violence. However, sociological and legal definitions of violence mainly emphasize interhuman variations. According to Wolfe (2013), this is because animals are outside the framework of the law, in that the distinction between humans and animals precedes the law. While many researchers are reluctant to recognize violent acts against animals as violence (e.g., Eisner, 2009; Grossman, 1995), in some cases, existing definitions of violence also apply to nonhuman animals.

The growing discipline of animal studies has incorporated various forms of violence toward animals into the academic discourse. However, the focus has mostly been on violence committed by those conceived of as antisocial individuals toward their "companion animals" (Ascione et al., 2018). These acts of violence are often considered as deviations from social morals and norms in many places around the world.

Animal slaughter is, in contrast, considered to be a morally acceptable social norm. Moreover, it is structural or systematic violence that stems from the very social structure (following Galtung, 1969), and can be seen as closely linked to speciesism (Ryder, 1972), a form of discrimination that promotes structural violence toward nonhuman animals.

Traditionally, structural violence against animals has not received the same level of attention in the literature, perhaps precisely because it is systematic and institutionalized. As such, it is normalized and often deemed necessary, and furthermore, it is more likely to be concealed and defended by stakeholders (Fitzgerald & Taylor, 2014). However, several contemporary scholars have started focusing on these forms of violence (Ben-Yonatan, 2022; Cudworth, 2015; Joy, 2011; McLoughlin, 2019; Nibert, 2013; Pachirat, 2011). Cudworth (2015), for example, calls to incorporate the more-than-human world into the sociology of violence, claiming anthropocentrism leads to the exclusion of systemic and extensive violent social practices against animals.

While violence toward animals takes various forms, animal slaughter is the most prevalent and legitimized. Various researchers have examined the complexities inherent in such widespread institutionalized violence. Cudworth's (2015) research has highlighted the embeddedness of these practices, their connection to structures of authority and the way they are strongly linked to human privilege and domination.

Other studies have illustrated that while slaughter is a form of socially legitimate and certainly legal violence, and is perceived as necessary, this work nevertheless entails unwanted aspects and is often hidden and unwelcome (Elias 1994 [1939]; Pachirat, 2011). Slaughter mostly takes place in remote areas, in order to conceal undesirable facets even from those who consume the products of such violence, that is animal flesh (henceforth: meat) and other animal products (Adams, 1990; Franklin, 1999; Joy, 2011).

What is more, those that perform these routine acts of violence seem to be affected by the contradicting aspects of these practices – their legitimacy combined with their perceived offensiveness. Hamilton and McCabe (2016) explored these emotional issues in their research on meat inspectors in England. Their study found a complex range of conflicting affects such as indifference and apathy on the one hand and compassion and emotional involvement on the other hand. The same has been found among industrial farm workers (Wilkie, 2005).

Further research looked into the ways slaughterhouse workers cope with the killing of animals. These include emotional detachment (Cudworth, 2011; Pachirat, 2011; Wilkie, 2005) and legitimization of the killing by underlining its necessity (Vialles, 1994). Another way to overcome such difficulties is through division of responsibilities, so that no worker is fully responsible for, or even aware of, the complete killing process (Pachirat, 2011). Pachirat further describes the role of the worker that pulls the trigger of the hydraulic gun that actually kills the animal as the most terrifying job in the slaughterhouse. These workers suffer from the most severe emotional consequences (Dillard, 2008; Pachirat, 2011). Baran et al. (2016) similarly show that not only is the job of slaughterers considered "dirty work" and of lower status, it is also associated with lower physical and psychological well-being.

This chapter deals with the very people performing the act of slaughter in the Israeli setting. As will be shown, Jewish slaughter has unique and interesting aspects that differ from other contexts. Unlike previous findings (i.e., Baran et al., 2016; Hamilton & McCabe, 2016; Pachirat, 2011), this setting provides Jewish slaughterers with mechanisms and narratives they employ in order to not only cope with the killing of animals, but also to position themselves as moral and their actions as contributing to the slaughtered animals as well as the Jewish community.

Kosher Slaughter

Consumption of kosher food (i.e., food that conforms to the Jewish dietary law) is an integral part of the Jewish lifestyle. For this reason, religious slaughter is fundamentally different from other forms of animal killing in the meat industry (Grandin, 1988). Specifically, Jewish slaughter (*shechita*) differs from non-religious meat industries, in that Judaism imposes strict laws on the act of slaughter for meat. Thus, animal slaughter is based both on the general instrumental conception of meat as a basic consumer product *and* on the central role of the *Halacha* [Jewish religious law]. Jewish slaughter is considered a complex act, which requires precision, and is one of the oldest and most respected professions among the ultra-Orthodox religious Jewish communities.

The Jewish slaughterer – the *shochet* – is considered to be engaging in sacred work, in a similar manner to other religious practitioners, such as circumcisers. This work is performed only by religious Jewish men, and it requires a special license, that is granted only after comprehensive training, and requires thorough knowledge and continuous study of holy Scriptures.

Traditionally, the act of slaughter can be performed by *shochet* alone, and in the past, before the institutionalization of industrial slaughterhouses, was carried out within the community. However, the modern transition to large meatpacking factories requires the employment of a staff, whose function is to assist in slaughter-related tasks that are not directly related to *kashrut* (kosher) requirements. The *Shochet* usually encounters the animals already restrained by the workers or a special device, and does not need to exercise physical force, except for the movement of the knife as he slits the animal's throat.

In many ways kosher slaughter can be seen as reflecting the complex attitude toward animals in Judaism that incorporates both compassion and subordination. While the *Torah* (old testament) has extensive laws aimed toward minimizing cruelty to animals, it also defines clear boundaries and power relations between humans and other animals, with animals treated as instrumental and as human property. Thus, while Jewish laws generally aim to minimize animal suffering, they nevertheless allow and even encourage their slaughter.

Methods

This chapter derives from research on Jewish slaughterers, conducted between 2014 and 2019. Research tools include in-depth interviews and observations as primary sources of information, as well as informal conversations and textual sources (such as canonical literature, pamphlets, and leaflets) as secondary sources of information.

The findings presented in this chapter are mainly based on 35 semi-structured in-depth interviews with Jewish workers in the animal slaughter industry in Israel. Thirty of these have been trained and actually practice animal slaughter, or at least have in the past. The rest have another direct connection to the industry and the profession (e.g., slaughterhouse *kashrut* supervisors). The ages of the interlocutors range from 24 to 80, and they live mainly in ultra-Orthodox communities in Israel.

Contact with most participants was made by way of snowball sampling, with initial contacts connecting the researcher with other people from the industry. It was not rare to encounter difficulties and a significant level of resistance to participate in this research, though in most cases it was not difficult to form a relationship of trust throughout the interviews.

Interviews were conducted at various locations based on the interviewees' preferences, and lasted two hours on average. The participants provided prior written informed consent, and all study procedures were approved by the Institutional Review Board of Bar-Ilan University. All data were anonymized and analyzed using ATLAS.ti software version 1.5.2.

Narratives of Justification – Slaughter as Normative and Legitimate Violence in Israel

The conversations with the Jewish slaughterers, combined with the participant observations and text analyses, delineate the various mechanisms utilized in order to naturalize and normalize the acts of animal killing and the violence they entail. These conversations are quoted below, with real identities protected. Beyond the variations in the various accounts, several themes became evident.

To begin with, the various components of the violent acts – the slaughter itself, the facilitator (the *shochet*), and the animal victim – are discussed, and the ways through which animal slaughter comes to be perceived of as legitimate, necessary, and even moral, are illustrated. The use of metaphors is subsequently identified, as additional means for justification of violent acts as part of Jewish (kosher) animal slaughter.

Legitimization of the Elements of Slaughter – The Act, the Facilitator, and the Victim

Purposeful and normalized violence can acquire legitimacy from society, depending on the motives and execution of the act itself, the identities of the facilitators of the violence, and those of its victims. The Jewish slaughterers in this study heavily relied on religious Jewish traditions and narratives in order to frame these three elements of slaughter in ways that legitimize the violent aspects of slaughter and even deem them moral and honorable.

As mentioned, within the kosher framework, slaughter is not only perceived as "permitted by law" but also constitutes a sacred duty, rooted in the broader context of *kashrut*. The **act** of slaughter, when incorporated with its social meaning as a holy mission ordered by none other than God himself, manages to obfuscate its violent aspects, as Oz explains:

> I do not consider myself a murderer, God forbid, neither a killer nor an executioner. I see it [slaughtering] as holy work because it emanates from my belief, by which we are obliged. The Lord (blessed is He) himself ordered us to eat kosher food, so I just see it as a mission.
>
> *[Oz, 47, cattle slaughterer]*

This is not to say that the *shochtim* are not aware of these violent aspects, or that in general, Judaism is free from violence. The Bible describes certain violent acts as justified and worthy, and distinguishes between violence committed within the purview of social law and order (i.e., institutionalized and purposeful violence) such as sacrificing animals, and violence for its own sake or personal desires (e.g., pleasure, passion, or revenge). The interviewees also seemed to support this distinction. Seventeen out of 35 interviewees considered animal slaughter to be violent or cruel in a way, but added that the way it is performed within the Jewish legal framework relieves it of its violent aspects:

> It is a violent and a cruel act, but it is permitted by law [...] And if the *Torah* has permitted us [to slaughter], and if we slaughter as per this permission, it [the slaughter] will not create cruelty within us.
>
> *[Zinger, 49, a Rabbi in a poultry slaughterhouse]*

The interviewees did not clearly differentiate between violence and cruelty, but they did differentiate between "cruelty for its own sake" – when the motives for the act may include aspects of pleasure, from "cruelty for a purpose" – which is necessary and beneficial and may entail indifference, or a certain attempt to suppress emotions.

Beyond the act of slaughter itself, the role and identity of the slaughterer is another element that is acknowledged and legitimized in Judaism. The *shochet*, as the **facilitator** of the violent act, can be seen as engaging in morally dubious practices. Indeed, despite this depiction of Jewish slaughter as a sacred duty, there is a constant tension between kosher slaughter's sacred and dirty aspects. Many societies consider slaughter "dirty work," and consequently, those performing it are seen as "dirty" (Ashforth & Kreiner, 2014; Baran et al., 2016; McLoughlin, 2019; Meara, 1974), whether physically, socially, or morally (following Hughes, 1958). Indeed, while the *shochtim* are socially constructed as people performing essential work for the society's benefit, they are nevertheless treated with constant suspicion.

These cultural dictates force the *shochtim* to constantly examine their own positioning within the boundaries of violence, as well as that of their peers, through constantly critiquing their behaviors and practices. Elad underlines such a neutralization technique that transforms the violent and dirty occupation into a clean and righteous one – namely, through being a devout, studious, and virtuous person:

> There are two kinds of *shochtim*—the *shochet*, and the one who works in *shechita*. The one who works in *shechita* knows how to handle a knife and is a master of slaughter, but has no values [...] Therefore, even if he practices kosher slaughter, he may incite violence in his surroundings, not only toward animals but also toward humans. However, the *shochet* is an aspiring student who studies the *Torah* to understand how he should shape his personality to be able to curb his inner violence, yes? [...] Slaughterers who do not study [the *Torah*] are very dangerous people.
>
> *[Elad, 52, slaughterer]*

As facilitators of violence, Jewish culture provides *shochtim* with protective measures against the exposure to this violence and its influences, by appointing them as experts in their profession, and limiting the scope of their violence to specific acts performed for others' benefit. In this way, there is a simultaneous social sanctioning and neutralization of violence. As mentioned in the introduction, from a criminological perspective, this grants these acts legality and normativity, while concurrently cloaking the victimization they entail.

Lastly, the Jewish toolkit provides the *shochtim* with means to contend with the harm inflicted on the **victims** – the animals. While the concern for animals was clearly proclaimed in some of the interviews, at the same time, it was understood that the animals are intended to be used by humans:

> Considering the restrictions the *Halacha* has imposed on me, you cannot say that the animal is suffering. It is not suffering, because they were destined for it, period.
> *[Elad]*

More than that, the supposed interests of animals were often discursively invoked in the interviews as a cultural tool to justify slaughter:

> The Torah has permitted slaughter. On the sixth day of Genesis, when God created the world, he asked the animals, "Are you willing to be slaughtered by humans and receive your *Tikkun* through it?" To which they replied, "We are."
> *[Lior, 29, cattle slaughterer]*

Tikkun refers to forms of action designed to repair or improve something broken or that has gone wrong. In this case, it is believed to be part of a form of reincarnation in which individuals are reborn as animals, so that their souls can be disciplined in order to absolve the sins committed in previous lifetime. This concept - *tikkun* of the animals' soul - was central to the interviewees' stated perceptions regarding the animals' perspective, implying that animal slaughter is a means for these animals to gain redemption.

Moreover, according to this position, animals consent to their slaughter in order to redeem themselves, and this consent further contributes to neutralizing the violent quality of slaughter (following Eisner, 2009). More than that, it enables a complete reorganization and replacement of the meaning of slaughter as violent and cruel with an entirely contrasting one – that of rescue and redemption performed for the animal's benefit.

Discursive Strategies in the Process of Legitimization and Naturalization of Kosher Slaughter

Beyond the consideration of the various components of the slaughter, the interlocutors in this research additionally incorporated the use of discursive mechanisms, namely metaphors, to describe their relationships with the animals, as well as the act of slaughter. According to Lakoff and Johnson (2003), metaphors are embedded in an individual's physical and cultural experiences. Beyond that, culture is the primary factor for metaphors' creation in conjunction with individuals' subjective experiences within a society (Yu, 2008). In our case, metaphors are important tools to understand the process of moral justification the slaughterers engage in, vis-a-vis society, as well as themselves. Unlike the justifications for the three elements of slaughter, which heavily relied on Jewish tradition and were mainly confined to the *shochtim*'s worldview and specific cultural toolkit, the metaphors they employed were based on routine Israeli secular experiences, ones taken from their own personal everyday reality. These metaphors are mainly connected to three specific social realms: surgical medicine, state security, and childrearing.

In various conversations the *shechita* was likened to surgical **medicine**, implying that like surgeons, the *shochtim* cause pain in order to heal the patients' bodies. Invoking this metaphor, Jacob argues that a certain degree of cruelty is indeed demanded in *shechita*, as well as in medicine, although it is necessary and nevertheless beneficial for the recipient:

> When doctors perform surgery and injure the body, do you consider it violent? [...] They cut people [...] A dentist who inflicts pain, sometimes even without anesthesia, what is that? However, it [slaughtering animals] is our purpose in the world; this is the *tikkun* [redemption] of the chicken, which is what we believe in, what our ancestors taught us for generations.
>
> *[Jacob, 66, a Rabbi, senior slaughterer and slaughter certifier]*

The comparison drawn between slaughterers and physicians indicates the *shochtim*'s self-perception as high-status professionals, or at least an attempt to avoid stigma and elevate their status vis-à-vis that of other workers in the meat industry. This metaphor additionally normalizes slaughter by locating it in the physical experience of surgical procedures and shaping it according to cultural understanding (following Yu, 2008). In a similar fashion, the actions of both doctors and *shochtim* are considered to be a legitimate part of their existence as well as unavoidable necessary evils, while portraying suffering as an integral part of life. Interestingly, the interviewees hardly referred to the patients' perspectives while using this metaphor.

Another context that often came up in the use of metaphors to validate the violent aspects of slaughter was that of **war and the military**, where violence is almost immediately identifiable. This metaphor, specifically the analogies between the soldier and the slaughterer, the enemy and the animals, is used to privilege the integrity and security of the state or the nation rather than the individual physical body. The act of slaughter, when detached from its underlying motive, might seem cruel. But, as the slaughterers maintain, the context completely relieves the act from its cruel nature:

> While the objective might be good, the act is cruel. For example, a soldier must act appropriately when faced with a terrorist [...] his actions can be seen as cruel. Similarly, during slaughter, the *shochet* must not listen to the cries of the animal,

> or consider what might be less painful. He must act now—do not think about the pain, think about what is the best location [in the neck] to cut. This is an appropriate attribute for *shochtim* [emphasis]. I need to get information from the terrorist and even if he screams it is for the greater good. What kind of man can do this? A man who does not care about crying, he is doing a *mitzvah* [a religious duty or command].
>
> *[Ze'ev, 52, supervising Rabbi]*

The soldier-enemy metaphor therefore constructs the provision of kosher meat to Jewish communities as a basic need, similar to a state's need for security. Israel's tense military environment allows the *shochtim* to easily invoke a metaphor that is used in the broad context of the need for survival. That is, the greater good demands the justification of cruelty, even if in the case of animal slaughter, the animals do not pose a threat to the community.

The third metaphor is used to create the connection between slaughter and **childrearing**. In the intimate social relationships described through this metaphor, it is inferred that parents must discipline their children in order to teach them acceptable behavior. This, they claim, corresponds with the *shochtim*'s belief that the animal is slaughtered as per its *tikkun* or redemption. Below, Erez describes how he must remain indifferent to the suffering of the slaughtered animals:

> Suppose you are taking your child to the doctor. He may yell or go wild, but you grab him, and make him get the [vaccination] shot because you know that it is for his benefit.
>
> *[Erez, 40, previously a cattle slaughterer abroad, currently does not work as a slaughterer]*

This metaphor presents a hierarchical relationship in which the parent, as the discipliner, knows and decides what is the best for the child. Similarly, *shochtim* assume that they know what is best for the animals and must "discipline" them through slaughter in order to help them achieve their purpose. By using this metaphor, the *shochtim* seem to insist that they have no choice but to perform this violent (albeit legitimate) act, similar to the obligation of parents to restrain their crying children as they are vaccinated.

Discussion

This study illustrates the various mechanisms utilized and cultural forces influencing Jewish slaughterers in Israel in neutralizing, legitimizing, and normalizing violence toward animals necessitated by their practices. The narratives they use strongly draw both from their religious Jewish cultural toolkit, and the more readily available Israeli everyday experiences, in order to present their work not only as legitimate and justifiable, but as something to take pride in.

While the need for emotional detachment is evident in this case, similarly to previous research (Cudworth, 2011; Pachirat, 2011; Wilkie, 2005), as well as the underscoring of the necessity of the slaughter (as in Hamilton & McCabe, 2016), other emotional consequences (Dillard, 2008) or distancing mechanisms are not evident (as in Pachirat, 2011). It seems that the religious justification with its related mechanisms, and the social environment they are submerged in, provide the *shochtim* with sufficient means to deal with their violent practices.

To begin with, the various components of the slaughter were considered – from the act itself, one that might only appear as cruel, but the cultural meaning of being mandated by God releases it from the burden of violence; to the slaughterer as a trained religious professional that does not enjoy the pain he inflicts and can remove himself from the cruel aspects of the slaughter; to the nonhuman victims that are redeemed through the act, fulfilling their purpose and the terms of their contract with God. Therefore, social order is kept through God's commandments as the supreme authority; the constant boundary work between beneficial violence and violence for its own sake, performed by the shochet; and the contract signed between God and the animals, agreeing to their own slaughter in order to rectify their souls. In this form, violence is strongly subdued and connected to the social order, in this case to the religious Jewish cultural one.

The interviews with the *shochtim* further revealed the influence of the Israeli setting, through three categories of metaphorical analogies: doctor–patient, soldier–enemy, and parent–child. Each of these provides a different, yet complementary, justification for the violence performed by those that have power to those that do not. The violent encounter between the *shochet* and the animals about to be slaughtered is described as analogous to other hierarchical or authoritative interactions, in ways that reinforce the social order of the human–animal divide through other forms of social order.

Combining the various mechanisms and narratives, it can be seen that cultural and social contexts influence violence and cruelty by guiding both the actions and the accompanying emotions. However, for *shochtim*, these contexts highlight the very essence and purpose of the practices and their corresponding meaning, and seem to eradicate aspects of cruelty from the violent act. This eradication further alleviates the "dirty" aspects of the trade. Unlike many secular slaughterers around the world, the *shochtim* underscore their professionalism, education, and divine mandate to provide for their people.

It is important to note that these mechanisms are socially sanctioned. Jewish religion and Israeli culture provide the slaughterers with the mechanisms, metaphors, and justifications to endorse their acts and absolve them of any wrongdoing. Consequently, it can be claimed that religion can transform dirty and violent aspects of slaughter into clean and beneficial ones. Religion, in this case, has an important role in regulating and legitimizing violence, and more generally the construction of human–animal hierarchical relations.

This study shows how *shochtim* cope with the violence they inflict on animals. In doing so, it also demonstrates how they subscribe to culturally prevalent ideas about violence as a social phenomenon – as a necessary evil that cannot be omitted, but only directed toward acceptable targets. Justifiable interhuman violence is utilized in order to normalize human–animal violence, so that violence legitimizes violence in what can be seen as an endless cycle.

Violence may be humankind's inherently problematic quality, but it also serves as a way to maintain social order. Nevertheless, from a (green) criminological perspective, it is important not only to explore the ways social order is maintained and reproduced through these mechanisms and narratives, but also to delineate the victimization of the countless animals that are the targets of the violent acts being justified. Such social constructions are oppressive and sustain the asymmetric power relations between humans and other animals. As can be seen in this chapter, social meanings are attributed in ways that are coherent to those that ascribe them – defining what violence is, when it is legitimate and even necessary, when it is cruel, and how the targets of this violence should feel as they are faced with it. But for others, whether outside the specific cultural context, or whether its victims (even being killed under its premises, such as the slaughtered animals in this case), these meanings may feel arbitrary, and may bear significant and ominous consequences.

References

Adams, C. J. (1990). *The sexual politics of meat: A feminist-vegetarian critical theory*. Continuum.

Ascione, F. R., McDonald, S. E., Tedeschi, P., & Williams, J. H. (2018). The relations among animal abuse, psychological disorders, and crime: Implications for forensic assessment. *Behavioral Sciences and the Law*, *36*(6), 717–729. https://doi.org/10.1002/bsl.2370.

Ashforth, B. E., & Kreiner, G. E. (2014). Dirty work and dirtier work: Differences in countering physical, social, and moral stigma. *Management and Organization Review*, *10*(1), 81–108. https://doi.org/10.1111/more.12044.

Baran, B. E., Rogelberg, S. G., & Clausen, T. (2016). Routinized killing of animals: Going beyond dirty work and prestige to understand the well-being of slaughterhouse workers. *Organization*, *23*(3), 351–369. https://doi.org/10.1177/1350508416629456.

Beirne, P. (1999). For a nonspeciesist criminology: Animal abuse as an object of study. *Criminology*, *37*(1), 117–148. https://doi.org/10.1111/j.1745-9125.1999.tb00481.x

Ben-Yonatan, A. (2022). 'Yesterday you slaughtered animals, today you pity them': Ambivalence and resolution among Jewish Israeli slaughterers. *Anthrozoös*, *35*(3), 355–370. https://doi.org/10.1080/08927936.2021.1996023.

Cartmill, M. (2001). Taxonomic revolutions and the animal-human boundary. In R. Corbey & B. Theunissen (Eds.), *Ape, man, apeman: Changing views since 1600* (pp. 97–106). Department of Prehistory, Leiden University.

Chouliaras, A. (2011). The victimological concern as the driving force in the quest for justice for state-sponsored international crimes. *Victimological Approaches to International Crimes: Africa*, 35–63. https://doi.org/10.1017/9781839703089.003.

Cudworth, E. (2011). *Social lives with other animals: Tales of sex, death and love*. Palgrave McMillan.

Cudworth, E. (2015). Killing animals: Sociology, species relations and institutionalized violence. *Sociological Review*, *63*(1), 1–18. https://doi.org/10.1111/1467-954X.12222.

Dillard, J. (2008). Slaughterhouse nightmare: Psychological harm suffered by slaughterhouse employees and the possibility of redress through legal reform. *Georgetown Journal on Poverty Law & Policy*, *15*(2), 391–408.

Durkheim, É, & Mauss, M. (1903). *Primitive classification, 1963*. University of Chicago Press.

Eisner, M. (2009). The uses of violence: An examination of some cross-cutting issues. *International Journal of Conflict and Violence*, *3*(1), 40–59. https://doi.org/10.4119/ijcv-2792.

Elias, N. (1939). *The civilizing process: The history of manners and state formation and civilization, 1994*. Blackwell Publishing.

Fitzgerald, A. J., & Taylor, N. (2014). The cultural hegemony of meat and the animal industrial complex. In R. Twine & N. Taylor (Eds.), *The rise of critical animal studies: From the margins to the centre* (pp. 165–182). Routledge.

Franklin, A. (1999). *Animals and modern cultures: A sociology of human-animal relations in modernity*. Sage Publications (pp. 127–144).

Galtung, J. (1969). Violence, peace, and peace research. *Journal of Peace Research*, *6*(3), 167–191. https://doi.org/10.1177/002234336900600301.

Grandin, T. (1988). Behavior of slaughter plant and auction employees toward the animals. *Anthrozoös*, *1*(4), 205–213. https://doi.org/10.2752/089279388787058335.

Grossman, D. (1995). *On killing*. Back Bay Books.

Hamilton, L., & McCabe, D. (2016). It's just a job: Understanding emotion work, de-animalization and the compartmentalization of organized animal slaughter. *Organization*, *23*(3), 330–350. https://doi.org/10.1177/1350508416629448.

Hobbes, T. (2018). *Leviathan* (J. C. A. Gaskin, Ed.). Oxford University Press.

Howe, B. (2006). Normative war-fighting and the new world order. *Politics and Ethics Review*, *2*(1), 38–61. https://doi.org/10.1177/1743453X0600200104.

Hughes, E. C. (1958). *Men and their work*. Free Press.

Joy, M. (2011). *Why we love dogs, eat pigs, and wear cows: An introduction to Carnism*. Conari Press.

Lakoff, G., & Johnson, M. (2003). *Metaphors we live by*. University of Chicago Press.

Lévi-Strauss, C. (1969). *The elementary structures of kinship*. Beacon Press.

McLoughlin, E. (2019). Knowing cows: Transformative mobilizations of human and non-human bodies in an emotionography of the slaughterhouse. *Gender, Work and Organizations*, *26*(3), 322–342. https://doi.org/10.1111/gwao.12247.

Meara, H. (1974). Honor in dirty work: The case of American meat cutters and Turkish butchers. *Sociology of Work and Occupations*, *1*(3), 259–283. https://doi.org/10.1177/073088847400100301.

Nibert, D. (2013). *Animal oppression and human violence: Domesecration, capitalism and global conflict*. Colombia University Press.

Pachirat, T. (2011). *Every twelve seconds: Industrialized slaughter and the politics of sight*. Yale University Press.

Peggs, K. (2023). Eating animals: A critical criminology of the domestic. In P. Davies & M. Rowe (Eds.), *Criminology of the domestic* (pp. 101–116). Routledge.

Russell, N. (2010). Navigating the human-animal boundary. *Reviews in Anthropology*, *39*(1), 3–24. https://doi.org/10.1080/00938150903548592.

Ryder, R. D. (1972). Experiments with animals. In S. Godlovitch, R. Godlovitch, & J. Harris (Eds.), *Animals, men and morals- an enquiry into the maltreatment of non-humans* (pp. 41–82). Taplinger Publishing.

Sollund, R. (2021). Green criminology: Its foundation in critical criminology and the way forward. *The Howard Journal of Crime and Justice*, *60*(3), 304–322. https://doi.org/10.1111/hojo.12421.

Taylor, N., & Fitzgerald, A. (2018). Understanding animal (ab)use: Green criminological contributions, missed opportunities and a way forward. *Theoretical Criminology*, *22*(3), 402–425. https://doi.org/10.1177/1362480618787173.

Tilly, C. (2003). *The politics of collective violence*. Cambridge University Press.

Tilly, C. (2004). Social boundary mechanisms. *Philosophy of the Social Sciences*, *34*(2), 211–236. https://doi.org/10.1177/0048393103262551.

Vialles, N. (1994). *Animal to edible*. Cambridge University Press.
Wilkie, R. (2005). Sentient commodities and productive paradoxes: The ambiguous nature of human-livestock relations in Northeast Scotland. *Journal of Rural Studies*, *21*(2), 213–230. https://doi.org/10.1016/j.jrurstud.2004.10.002.
Wolfe, C. (2013). *Before The law: Human and other animals in biopolitical frame*. Chicago University Press.
Yu, N. (2008). Metaphor from body and culture. In R. W. Gibbs Jr. (Ed.), *The Cambridge handbook of metaphor and thought* (pp. 247–261). Cambridge University Press.

10

THE ARENA OF CONTROVERSY

Bullfighting and Its Implications in Modern Spanish Society

Patricia Puente-Guerrero

Bullfighting, with its roots at least in the Middle Ages (Badorrey, 2017), is a long-standing tradition in Spain. Beyond the well-known contemporary *corridas* (bullfights carried out in a bullring), which emerged in the eighteenth century as commercialized spectacle (Shubert, 1999), a variety of bullfighting events are deeply integrated into Spanish culture (see Douglass, 1997), particularly in rural communities (Martín-Cabello, 2021).

This industry, however, is marked by institutionalized and highly normalized violence and harm affecting both nonhuman animals and humans, ranging from overt to less apparent forms. Practices like the use of punishment or black banderillas, thicker and longer, in bullfights when "due to his tameness the bovine could not be lanced in the stipulated manner" (Article 75 of the Regulation for Bullfighting Events—Royal Decree [*Real Decreto*] 145/1996—exemplify state-legitimized cruelty, aiming to provoke aggressive reactions from bulls so that the 'show' can continue. Human participants in bullfighting events also face significant risks: within the Valencian Community alone, 9 people died and 984 were injured in 2022 (Giménez, 2023, January 14).

Moreover, bullfighting is intertwined with societal issues such as hegemonic masculinity and the patriarchal system (Hunnicutt, 2020; Vericat, 2019), illustrating how cultural practices can both reflect and reinforce gender hierarchies[1]. Historically, bullfighting has been linked with expressions of virility and traditional masculinity, establishing itself as an arena where the highly celebrated values and attributes associated with manliness predominate (Castro, 2020). Within this context, the active participation of women in *corridas* has been seen as a transgression of the gender roles typically assigned to them, and interpreted as a potential threat to the masculine qualities central to bullfighting, thereby questioning the prevailing gender structure in society (Douglass, 1997). Hence, women have consistently

DOI: 10.4324/9781003441908-13

been attributed a traditional, passive function in bullfighting, often as aesthetic objects in the form of 'beautiful spectators' and/or as suffering mothers or wives of the male bullfighters, positions traditionally associated with feminity (Pink, 1997). At present, the field remains overwhelmingly male-dominated: almost all professional bullfighters are male (98% in 2022—Ministerio de Cultura, 2023), as are, too, most of the supporters, spectators, and active participants in bullfighting events (Fundación BBVA, 2022; Ipsos, 2022; Ministerio de Cultura, 2022). Furthermore, the correlation observed between the San Fermín festival and heightened reports of sexual violence (Zuloaga et al., 2018) invites further examination of how these practices intersect with gender dynamics.

Violence and harm embedded in the bullfighting industry have tended to be denied, ignored, or even praised by invoking culture and tradition as legitimizers (Andrade, 2021), a stance supported by the Spanish Government: in 2011, competences on bullfighting and bullfighting activities were assumed by the Ministry of Culture [*Ministerio de Cultura*] (Royal Decree [*Real Decreto*] 1151/2011), and in 2013 bullfighting was formally declared Spanish cultural heritage (Law [*Ley*] 18/2013). Even the dictionary of the Royal Spanish Academy [*Real Academia Española*] defines bullfighting as "the art of fighting [*lidiar*] bulls". Viestenz (2013) has condemned the employment of such an aesthetic veil to "distract from the bloody and violent nature of the spectacle" (p. 160). As Andrade (2021) asserts, framing bullfighting as an art form shifts the debate from moral considerations to an emphasis on its aesthetic value, thereby sidelining ethical concerns.

This chapter assesses the shifting social support and legal stance toward bullfighting in Spain, revealing a complex landscape that mirrors broader societal attitudes toward nonhuman animals and human-nonhuman animal relationships. Additionally, it delves into the interrelation between harm inflicted on nonhuman animals within the bullfighting industry and certain instances of harm against humans.

Bullfighting in Transition

Bullfighting has long sparked debate in Spanish society, with dissenting voices growing stronger over time in line with social changes (Codina, 2018). This divide is evident in some municipalities and regions openly opposing bullfighting (CAS International, n.d.), while others have declared it an Asset of Cultural Interest or part of the Intangible Cultural Heritage (Martel, August 24, 2021). Key legal battles include Catalonia's 2010 ban of bullfighting events involving the death of the bulls,[2] overturned in 2016 by the Constitutional Court citing interference in state competencies concerning the preservation of bullfighting as Spanish cultural heritage (SCC 177/2016). Similarly, a 2017 Balearic law that prohibited causing injuries or killing the bulls in the bullring was voided in 2018 (SCC 134/2018). The struggle is far from over, with a recent Popular Legislative Initiative aiming to repeal the 2013 law that proclaimed bullfighting as cultural heritage[3] (AnimaNaturalis, January 4, 2024; Mulà, 2024).

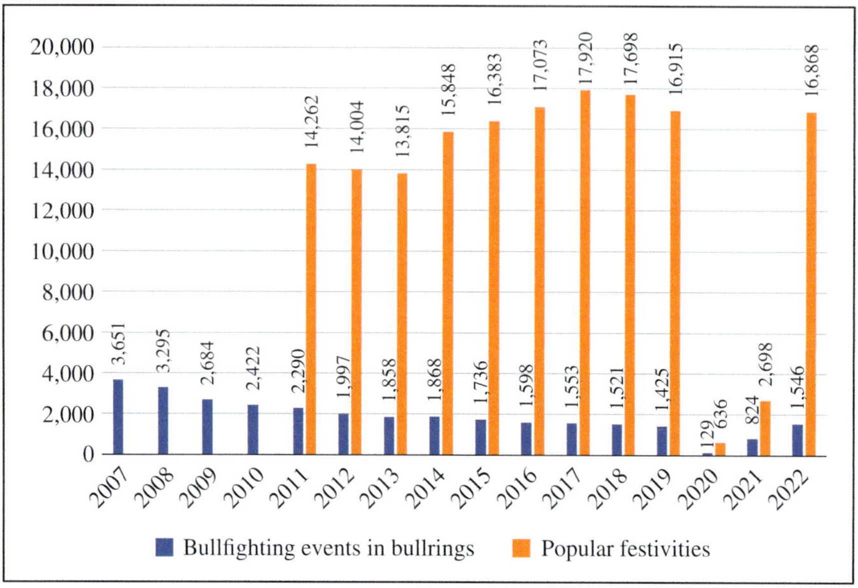

FIGURE 10.1 Evolution in the number of bullfighting events held in Spain 2007–2022.

Source: Own elaboration with data from Statistics of Bullfighting Affairs (Ministry of Culture)

Be that as it may, the popularity of bullfighting has declined overtime, as evidenced by the gradual decrease in bullfighting events held in bullrings[4] (see Figure 10.1) (Ministerio de Cultura, 2023).[5] The number of active registered professional bullfighters has also significantly reduced over time: from 6,703 in 2012 to 4,956 in 2022 (–26.1%) (Ministerio de Cultura, 2023).

Surveys show a minority attendance and interest, decreasing over time (see Tables 10.1 and 10.2). The latest Survey of Cultural Habits and Practices in Spain reported a remarkably low engagement with bullfighting, as only 1.9% of the respondents had attended an event in the previous year. It should be noted that the

TABLE 10.1 Percentage of Respondents of the Survey of Cultural Habits and Practices in Spain Who Attended a Bullfighting Event in the Previous Year

Wave of the survey	*Sample size*	*Global percentage*
2002–2003	12,180	8.6
2006–2007	16,408	9.8
2010–2011	16,408	8.5
2014–2015	16,576	9.5
2018–2019	16,520	8.0
2021–2022	16,464	1.9

Source: Own elaboration with data from the Surveys of Cultural Habits and Practices in Spain (Ministry of Culture)

TABLE 10.2 Percentage of Respondents of the Surveys of Cultural Habits and Practices in Spain by Their Degree of Interest in Bullfighting from Highest (10) to Lowest (0)

Wave of the survey	*10–9*	*8–7*	*6–5*	*4–3*	*2–0*	*Somewhat interested (7–10)*	*Somewhat not interested (0–4)*	*Mean score*
2014–2015	6.8	9.8	13.3	7.9	62.2	16.6	70.1	2.5
2018–2019	5.9	9.3	13.3	6.2	65.3	15.2	71.5	2.3
2021–2022	5.2	8.2	12.3	9.1	65.2	13.4	74.3	2.4

Source: Own elaboration with data from the Surveys of Cultural Habits and Practices in Spain (Ministry of Culture)

period under consideration encompassed the time when the most stringent restrictions were imposed due to the COVID-19 pandemic.

EM-Analytics' 2023 study also reveals growing opposition to bullfighting and its public funding: 45.7% of respondents (N = 1,928) favored a ban, 24% thought bullfighting should be promoted, and 30.3% considered it should be neither banned, nor promoted. Additionally, 69.6% were against granting public subsidies for bullfighting, compared to 25.1% in favor[6] (Electomanía, 2023a). Furthermore, 59.6% were against bullfighting, while only 35.1% supported it[7] (Electomanía, 2023b). This contrasts sharply with a 1995 study by the Sociological Research Center [*Centro de Investigaciones Sociológicas*], where 39.9% of participants (N = 2,487) were "somewhat against" and 45.5% "somewhat in favor" of bullfighting.[8]

Despite a noticeable decline in social support and acceptance over time, many bullfighting events, particularly those associated with local popular festivities, have endured or even increased, as shown in Figure 10.1. However, some have undergone transformations in an attempt to mitigate social criticism. An illustrative case is the controversial Toro de la Vega festival, held every second Tuesday of September in Tordesillas (Valladolid, Castile and Leon). Notably, since 2016, the bull is no longer killed in public, although he still meets his fate in private settings (for an overview, see Cacho, 2022, September 11).

Evolving Attitudes Toward Nonhuman Animals

The observed transformations have taken place within a broader context of growing social sensitivity and empathy toward nonhuman animals. Key findings from a BBVA Foundation study with 2,000 respondents include (Fundación BBVA, 2022):

- A strong acknowledgment of nonhuman animals' capacity to experience physical pain, fear, and emotions akin to humans, with average ratings of 8.7, 7.9, and 7.1, respectively.[9]

- Widespread consensus (average score of 9.1) on the moral imperative to treat nonhuman animals with respect akin to that afforded to humans.
- In terms of speciesism, scores averaged 7.5 for "No species has the right to exploit another species", and 8.9 for "All animals have equal right to life". In contrast, a lower mean score of 4.7 was noted for "Human interests are above animal interests", with 40% somewhat disagreeing (scoring 0 to 4), and 3.1 for "Humans have the right to use animals for our purposes", with 60% scoring 0 to 4.
- An overwhelming 98% of respondents advocated for recognizing rights to nonhuman animals, with divergent views on whether it should parallel human rights.
- A significant 70% of participants found the use of nonhuman animals in bullfights [*corridas de toros*] highly unacceptable (scoring 0 to 2), with an average score of 1.9. Additionally, 76% expressed strong disapproval of using nonhuman animals for entertainment in local festivals, with an average score of 1.4.

A corroborating 2022 IPSOS survey commissioned by the Ministry of Social Rights and the 2030 Agenda (N = 800) disclosed:

- 77% of participants disagreed, to varying degrees,[10] with the idea that "Animals exist to be used by humans".
- A significant 89% strongly agreed[11] that "Animals experience physical pain in a way similar to human beings"; 69% strongly supported equal protection against abuse for both wild and companion animals.
- High levels of concern for nonhuman animal protection (78% responded with scores of 8 to 10), along with 89% advocating for the recognition of nonhuman animal rights.
- A notable 86% of respondents disapproved of using nonhuman animals in bullfights (responses 0 to 5 on the acceptability scale from 0 to 10), with 78% holding the same view regarding popular festivals.
- Merely 11% thought current laws sufficiently penalize nonhuman animal abuse.

Importantly, the Spanish Civil Code was amended in 2021 to acknowledge nonhuman animals as "living beings endowed with sensitivity" (Law [*Ley*] 17/2021). This law mandates owners and guardians of nonhuman animals to exercise their rights over them and duties of care respecting their nature as sentient beings and ensuring their well-being according to the characteristics of the species (Article 333 bis). The law's preamble elaborates on this perspective, asserting that despite nonhuman animals being subjected to ownership and commerce, human-nonhuman animal relationships must be modulated by the sentient nature of nonhuman animals. Accordingly, "rights and faculties over animals must be exercised with attention to the welfare and protection of the animal, avoiding mistreatment, abandonment, and the infliction of a cruel or unnecessary death". Giménez-Candela (2021) cites the shift in social sensitivity toward nonhuman animals, scientific affirmations of nonhuman animal sentience, the global de-objectification movement and EU nonhuman animal welfare legislation as key drivers for this reform.

Navigating Legal Inconsistencies in Responses to Nonhuman Animal Abuse

In Spain, criminal protection for nonhuman animals is relatively recent. The Spanish Criminal Code of 1995[12] (Organic Law [*Ley Orgánica*] 10/1995) initially addressed[13] nonhuman animal abuse under "misdemeanors against the general interests",[14] with fines for those who cruelly mistreated domestic animals or any others in unauthorized shows (Article 632), thus shielding bullfighting.

A significant change occurred in 2003 with Organic Law [*Ley Orgánica*] 15/2003, categorizing serious nonhuman animal abuse as a crime "related to the protection of flora, fauna, and domestic animals"[15]: as stated in Article 337, cases in which domestic animals were mistreated "with undue cruelty [*ensañamiento*] and without justification" resulting in their death or severe physical impairment would be punished with a prison sentence of three months to one year.[16] Cruelly mistreating domestic animals or any others in shows not legally authorized without falling under the cases provided for in Article 337 kept on being defined as a misdemeanor against the general interests, according to Article 632.2, and punishable by a fine or community service. Since the so-called fighting bulls [*toros de lidia* or *toros bravos*] are not legally considered domestic animals, Article 337 did not apply to them either (Blanco, 2010).

Organic law 5/2010 expanded protection to tamed animals[17] and removed the "undue cruelty" requirement.[18] Additionally, mistreatment could be committed "by any means or procedure"[19] and, instead of referring to "severe physical impairment", the new formulation alluded to the more comprehensive concept of "severe health impairment". Article 632.2 remained unchanged.

To heighten the complexity, Organic Law [*Ley Orgánica*] 1/2015 introduced substantial changes. It broadened the scope to include nonhuman animals who are commonly domesticated, those who temporarily or permanently live under human control or any nonhuman animal who does not live in the wild. A prison sentence of three months and one day to one year[20] would be imposed for causing them severe health impairment (Article 337.1). Article 337.2 established some aggravating circumstances which would entail imposing the sentence in its upper half, such as having committed the action with undue cruelty or in the presence of minors.[21] If the abuse had caused the animal's death, Article 337.2 mandated a prison sentence ranging from 6 to 18 months.[22] Additionally, Article 632.2 was repealed, and its stipulations prescribing fines[23] for those who, outside the cases referred to in the previous sections, cruelly mistreated domestic animals or any others in unauthorized shows, were incorporated into Article 337.4. The revised Article 337 theoretically extends to fighting bulls, but bullfighting's designation as Spanish cultural heritage in 2013 effectively exempted it from these provisions.[24]

The latest amendment in 2023 (Organic Law [*Ley Orgánica*] 3/2023) introduced a new Title XVI bis in the Criminal Code, "crimes against animals", thereby acknowledging inherent protection for nonhuman animals (Serrano, 2023). Article 340 bis (1) now establishes a prison sentence of 3 to 18 months or a fine[25] for injuring domestic, tamed, domesticated nonhuman animals or those living under human control needing veterinary care, provided it happens "outside of legally regulated activities". For injuries to vertebrates not covered under these categories, penalties range from 3 to 12 months or a reduced fine, thus offering some protection for wild animals. Any reference to mistreatment or its "unjustified" nature has been removed. Article 340 bis (2) outlines a more extensive list of aggravating circumstances that intensify penalties, such as a profit motive, committing the abuse in a public event or sharing it through information and communication technologies. The presence of particularly vulnerable individuals is also considered, in addition to minors. Under Article 340 bis (3), causing the animal's death leads to steeper sentences: 12 to 24 months of imprisonment[26] in the case of domestic, tamed, domesticated animals or those living under human control; 6 to 12 months or a larger fine for other vertebrates. Aggravating factors concurring result in harsher penalties. Last, Article 340 bis (4) mandates fines or community service[27] when no veterinary treatment is required or no injuries have been caused as a result of serious mistreatment.

Legal reforms addressing nonhuman animal abuse have stemmed from various factors. Prats (2020) identifies key reasons as increased social awareness and sensitivity toward nonhuman animals' plight both in Spain and globally, reinforced by legislative changes granting them rights and protection in other countries and amplified media exposure of egregious nonhuman animal abuse in Spain. This exposure "has further fueled social animosity in favor of a legal improvement in the conditions of animals, and particularly a toughening of penalties for those guilty of such acts" (p. 982).

Reflecting these changes, Law [*Ley*] 7/2023 on the Protection of Animal Rights and Wellbeing has recently been enacted. Notably, its preamble emphasizes the increasing citizen awareness of the need to protect all nonhuman animals, especially those in human environments, recognizing them as sentient beings with rights. It also underscores regulations developed by local and regional governments to enhance nonhuman animal protection and welfare and to deter mistreatment. Nevertheless, Article 1.3(a) of this law excludes animals used in bullfighting[28] from its scope. Paradoxically, Article 25(e) of the law explicitly forbids the use of nonhuman animals "in public spectacles or artistic, tourist, or advertising activities that cause them distress, pain, or suffering (…)", but this applies only to companion animals and wild animals held in captivity. Similarly, Article 75(g) considers "the use of animals in prohibited activities, particularly in cultural and festive activities (…)" a very serious infraction, punishable with a fine from 50,001 to 200,000 euros (Article 76).

Children's Exposure to Bullfighting

Spanish legislation reveals a complex mosaic of regulations regarding minors' exposure to nonhuman animal abuse, illustrating clear inconsistencies that hinge on whether such exposure is legally sanctioned or not. Regional laws on the participation or attendance of minors at bullfighting events or bullfighting schools vary significantly: the Balearic Islands prohibit under-18s from attending bullfights (Law [*Ley*] 9/2017), whereas the Basque Country allows those under 16 to attend general bullfighting events with an adult. This region also forbids under-16s' active participation in popular festivities (Decree [*Decreto*] 183/2008). They can attend, however, as spectators. Children over 12 can enroll in bullfighting schools, but they are not allowed to participate in bullfighting events or train with live bovines until the age of 16. In Catalonia, under-14s cannot participate in traditional festivities with bulls but can be spectators (Law [*Ley*] 34/2010), while Galicia disallows under-12s from attending bullfights (Law [*Ley*] 10/2017). Navarre bans under-16s from participating in traditional events but allows them as spectators, with bullfighting school students participating in practical sessions from 14 (Foral Decree [*Decreto Foral*] 249/1992). Aragon forbids the participation of under-16s in popular festivities (Decree [*Decreto*] 226/2001) but allows bullfighting school enrollment from 12 and practical classes with bovines from 14 (Decree [*Decreto*] 16/2003). Extremadura requires under-18s in bullfighting schools to have the authorization of their legal representative (Decree [*Decreto*] 61/2006).

This regional variation contrasts with the Criminal code, which considers nonhuman animal abuse in the presence of minors as an aggravating circumstance, and deviates from the United Nations Committee on the Rights of the Child's 2018 recommendation against under-18s participating or spectating in bullfighting events to prevent harmful effects on them. According to Mulà (2021), it follows from this recommendation that bullfighting represents one of the violent practices from which children should be shielded, regardless of the cultural heritage bullfighting events may entail. Moreover, it also conveys that such a prohibition does not violate the right to access culture, "since the child's paramount interest in not being exposed to violence takes precedence over any other legitimate interest that might arise or come into conflict, such as the right to freely participate in cultural life" (Mulà, 2021, p. 4). Indeed, the Committee has recently stated that Article 19 of the Convention on the Rights of the Child includes the children's right to "be protected from all forms of physical and psychological violence and from exposure to violence, such as (…) violence against animals" (United Nations Committee on the Rights of the Child, 2023, August 22).

In this regard, Graña et al. (2004) studied the psychological effects of viewing bullfights in 240 Madrid-based children aged 8 to 12. They first administered a questionnaire to examine children's attitudes toward bullfighting and then used videotapes of bullfights with different commentaries to assess the emotional impact of these events, as well as that of narrative explanations on aggression and anxiety levels. Their findings revealed mixed attitudes toward bullfighting, with 57%

disliking it and 65.1% considering it violent. Despite that, most children (65.6%) were fine with their age group watching bullfights, although 24.5% thought it is wrong for children to watch bullfights. Common emotional responses to watching a bullfight were grief (52.8%) and indifference (36.8%). Only 10.4% asserted feeling happiness. When asked about their thoughts during a bullfight, 58.3% reported thinking it is wrong, 28.7% were neutral, and 13% thought it acceptable. More than half of the sample indicated that no harm should be done to the bull, and a quarter considered bullfighting a clear example of nonhuman animal abuse. Regarding personal harm caused by viewing a bullfight, 49.6% felt it had a negative impact, whereas 50.4% perceived no negative effect.

With respect to the influence of the type of justification given to the bullfight (festive, aggressive, or neutral) on anxiety levels, aggression, and emotional impact, Graña and colleagues (2004) found that children who had been exposed to the festive justification—which focused on the descriptive elements of the bullfight and ignored its negative consequences—scored significantly higher on the aggression and anxiety tests than those who viewed the neutral video. The aggressively justified bullfight—which emphasized negative or violent descriptions—caused greater negative emotional impact on children as compared to the neutral scene. The effects were greater in young boys.

Conclusion

Despite its entrenched historical roots and cultural significance, bullfighting in Spain is facing increasing scrutiny due to growing ethical concerns over nonhuman animal abuse and the societal ramifications of such practices. This chapter outlines a significant shift in public perception of bullfighting in Spain, marking a transition from a deep-seated cultural tradition to an increasing awareness of nonhuman animal welfare and rights. The declining popularity of bullfighting, evidenced by reduced attendance and escalating opposition, mirrors this societal move toward a more empathetic and ethical gaze on nonhuman animals. As Spain navigates this transformation, the tension between preserving cultural heritage and advancing nonhuman animal rights remains a critical issue.

This struggle is reflected in the contradictions of Spain's current legislation. On the one hand, it acknowledges nonhuman animals' sentience and advocates for their wellbeing and protection against abuse. On the other hand, it shields bullfighting as cultural heritage, exempting it from these protective provisions. Such an exemption creates a pronounced legal inconsistency, evincing a hierarchy of victims based on an anthropocentric viewpoint (Wyatt, 2021). Unlike some other nonhuman animals, the so-called 'fighting bulls' are thus regarded as *justified* victims (Sollund, 2015). The differing legal consideration of qualitatively similar acts and the associated justifications raised for criminalizing and/or punishing some acts but not others are relevant subjects of study for green criminology (Sollund, 2015).

This normative framework also resonates with practices like *greenwashing* (de Freitas Netto et al., 2020; Miller, 2017), *welfare washing* (Bjørkdahl & Syse, 2021), or *humane-washing* (Vila, 2021; Winders, 2017). Inasmuch as policies and legislation aimed at protecting nonhuman animals from abuse become merely symbolic and/or are selectively applied, they may be considered instances of what could be termed *nonhuman animal protection washing*. Thus, the current legal landscape, with its incongruities and concessions for bullfighting, casts doubt on the efficacy, coherence, and intentions behind nonhuman animal protection laws, pointing to the need for a more harmonized and ethically consistent approach.

Given this scenario, it is imperative for Spain to reassess its cultural traditions in light of ethical considerations and emerging societal values. The growing public opposition to bullfighting, along with increasing sensitivity toward the wellbeing and protection of nonhuman animals, and even the recognition of their rights, demand a reevaluation of these activities which have been proven to cause suffering and inflict harm on both nonhuman and human animals. Future legal reforms should seek greater consistency and ensure that cultural practices are not exempted from ethical scrutiny. Additionally, it is essential to promote education and raise awareness about the ethical implications, the harm caused, and the broader societal impacts of bullfighting, particularly among young people. These initiatives are crucial in cultivating a culture that respects and protects all sentient beings.

In this regard, presumed interspecies differences, physical, social, and mental distance, and denial have been deemed key elements in the understanding of speciesism and nonhuman animal abuse (Sollund, 2008). Elaborating on Sollund's (2012) application of Sykes and Matza's techniques of neutralization to the acceptance of nonhuman animal abuse and exploitation, in the case of bullfighting, denial of injury and denial of the victim, as well as the condemnation of the condemners—the critics of these practices—and the appeal to higher loyalties—the preservation of culture and tradition as legitimizers—are commonplace techniques of denial. Social distance also allows individuals to overlook the fact that behaviors like the ones exhibited by bulls in the arena are not natural, but rather forced responses (Sollund, 2022). In the context of the dominant speciesist ideology, empathy for nonhuman animals tends to be ignored, denied, or suppressed. Therefore, Sollund (2022) advocates for enhancing empathy as a fundamental component of progress. Importantly, and in line with what was stated in the introduction, this author underscores how the various forms of 'isms'—sexism, racism, speciesism, etc.—are influenced by similar patterns of discrimination and prejudice. As a result, examining one type of abuse illuminates the understanding of another (Sollund, 2015). Furthermore, Sollund asserts that those who are oppressed by these 'isms' have an equal right to concern and recognition as victims.

This chapter highlights the necessity of an ongoing dialogue that carefully balances the preservation of cultural heritage with advancement in (non)human animal protection and rights in Spanish society. Emphasizing this point, Barry (2002, p. 258) asserts that culture is no excuse: "if there are sound reasons against doing something, these cannot be trumped by saying—even if it is true—that doing it is part of your culture". This argument is critical when considering the deeply ingrained tradition of bullfighting in Spain. Andrade (2021), drawing on Cohen's (2011. Cited by Andrade, 2021) concept, argues that, while there is merit in preserving cultural elements of value, this does not justify the continuation of practices like bullfighting that are exploitative and unjust due to severe nonhuman animal abuse. Andrade suggests that, although the harmful aspects of bullfighting should not be maintained, Spain could preserve its valuable aspects in non-exploitative forms, such as in museums or through complementary art forms like flamenco. Reflecting on these insights, it becomes apparent that the path forward requires a nuanced understanding and reimagining of cultural traditions, aiming to honor historical heritage without compromising ethical standards. This endeavor necessitates not only legislative changes but also a significant cultural shift toward acknowledging the intrinsic value of all beings.

Acknowledgments

I would like to express my profound gratitude to the editors of this compilation for their thorough review and invaluable insights on this chapter. Their feedback has not only broadened the scope of the analysis but also significantly enhanced the overall quality of the work presented. I am deeply appreciative of their dedicated efforts throughout the entire editing process, which have indubitably enriched the final outcome.

Notes

1 An extract from the lyrics of the song "The miniskirt", launched in 1971 by the popular Spanish Singer Manolo Escobar (1931–2013) exemplifies this: "I don't like it when you wear a miniskirt to the bullfight. I don't like you to go to bullfight with miniskirt. People look up because they want to see your face and they want to see your knees. Those annoying brats won't stop staring at you. I rebel and rebel, and I have to confront them, and don't even watch the bullfight. So now you know: don't wear a miniskirt, because I want to see the bullfight this afternoon without quarreling with anyone".
2 Thereby contentiously excluding "correbous".
3 The text (in Spanish) can be consulted in https://pdf.animanaturalis.org/files/Texto-Iniciativa-Legislativa-Popular-NoEsMiCultura.pdf.
4 As shown in Figure 10.1, the evolution of popular festivities (those in which bovine individuals are played with or run according to the traditional customs of the locality) has been more irregular. Data on these events started to be recorded in 2011.
5 The 2022 resurgence in bullfighting events has been attributed to the post-pandemic context, where "the bullfighting sector has been bolstered by public subsidies to ensure its survival, as well as a significantly reduced price in bulls, due to the unsold excess 'stock' in farms during the pandemic years" (AnimaNaturalis & CAS International, 2023, p. 7).
6 5.3% chose "don't know/no answer".
7 5.3% responded "don't know/no answer".
8 11.5% chose "don't know" and 3.2% opted for not responding.

9 Measured in a scale from 0 (completely disagree) to 10 (completely agree).
10 Responses 0–5 in a scale from 0 (completely disagree) to 10 (completely agree).
11 Responses 8–10.
12 The Criminal Code of 1995 is the one currently in force, although it has undergone several and significant legal reforms over time. The most relevant regarding our subject of study will be discussed in this section.
13 Non-human animal abuse had been first introduced in the Spanish criminal legislation in 1928, in the Article 810.4° of the Criminal Code (Real decreto-ley aprobando el proyecto de Código Penal, que se inserta, y disponiendo empiece a regir como Ley del Reino el día 1° de enero de 1929). It was defined as a misdemeanor against the general interests and the population regime. Those who publicly mistreated domestic animals or forced them to an "excessive fatigue" would be punished with a fine. However, this code was in force only until 1932, when the Second Republic was proclaimed. The Criminal Code of 1928 was then immediately replaced by the Criminal Code of 1870. No provisions on nonhuman animal abuse were included in the subsequent Criminal Codes until 1995.
14 As Serrano (2019) elucidates, the legal protection provided in this article safeguards human sensibilities that are affronted by the mistreatment of nonhuman animals, instead of nonhuman animals themselves. Thus, their protection under this statute is indirect, operating from an anthropocentric standpoint. It primarily regards the impact on human feelings and moral judgments, rather than addressing the intrinsic rights or wellbeing of the nonhuman animals themselves.
15 These crimes were included in the broader category of "crimes related to territorial planning and the protection of historical heritage and the environment".
16 Additionally, a sentence of special disqualification for a period of one to three years for the practice of any profession, trade, or business involving nonhuman animals would be imposed.
17 In any case, the so-called 'fighting bulls' kept on being excluded (Blanco, 2010).
18 This requirement was quite problematic and made it difficult to apply the article: if the deliberate increase of the nonhuman animal's suffering was not observed, the behavior could even result in impunity (for a discussion, see Arregui, 2022).
19 This phrase was absent in the article's previous formulation.
20 As well as the sentence of special disqualification for the practice of any profession, trade, or business involving nonhuman animals, and, as a novelty, for their ownership.
21 Individuals under the age of 18.
22 The sentence of special disqualification to be imposed was also potentially longer: for a period of two to four years.
23 The judge may additionally impose a special disqualification sentence of three months to one year for the practice of any profession, trade, or business involving nonhuman animals, and for their ownership.
24 As Boiso (2021) observes, the existence of legislation that grants special protection to these activities (see also Law 10/2015) allows bullfighting to be considered as justified under Article 337, thereby circumventing punishment.
25 Together with the already mentioned sentence of special disqualification for a period of one to three years.
26 In addition to a sentence of special disqualification in the same terms as in prior Article 337.3: of two to four years.
27 In addition to the special disqualification sentence of three months to one year.
28 Along with others, such as nonhuman animals who live in the wild, so-called 'production animals' and, having generated substantial controversy since they are commonly the victims of neglect, abuse, and abandonment, "hunting dogs, packs of hunting dogs [*rehalas*], and auxiliary hunting animals" (Article 1.3).

References

Andrade, G. E. (2021). A response to cultural arguments in the renewed disputes over the ethics of bullfighting. *Sport, Ethics and Philosophy*, *16*(1), 50–65. https://doi.org/10.1080/17511321.2021.1887332.

AnimaNaturalis. (January 4, 2024). *Registran ILP para derogar la protección que tiene la tauromaquia declarada como patrimonio cultural*. Retrieved January 10, 2024, from https://www.animanaturalis.org/n/46611/Registran-ILP-para-derogar-la-proteccion-que-tiene-la-tauromaquia-declarada-como-patrimonio-cultural.

AnimaNaturalis & CAS International. (2023). *Investigación fiestas crueles. Financiación pública de las fiestas populares con toros en España.* AnimaNaturalis. https://www.animanaturalis.org/n/46546#mainForm.

Arregui, R. (2022). Análisis jurídico sobre la concurrencia de ensañamiento en el delito de maltrato animal. *dA. Derecho Animal (Forum of Animal Law Studies)*, *13*(1), 6–18. https://doi.org/10.5565/rev/da.595.

Badorrey, B. (2017). *Otra historia de la tauromaquia: toros, derecho y sociedad (1235-1854)*. Boletín Oficial del Estado. https://www.boe.es/biblioteca_juridica/abrir_pdf.php?id=PUB-DH-2017-36.

Barry, B. (2002). *Culture and equality: An egalitarian critique of multiculturalism*. Harvard University Press.

Bjørkdahl, K., & Syse, K. V. L. (2021). Welfare washing: Disseminating disinformation in meat marketing. *Society & Animals*, *32*(1), 37–55. https://doi.org/10.1163/15685306-BJA10032.

Blanco, I. (2010). Artículos 334 a 337. In M., Gómez Tomillo (Dir.). *Comentarios al Código Penal* (pp. 1299–1316). Lex Nova.

Boiso, M. (2021). Análisis del delito de maltrato animal (art. 337 CP). *dA. Derecho Animal (Forum of Animal Law Studies)*, *12*(1), 82–111. https://doi.org/10.5565/rev/da.519.

Cacho, R. (2022, September 11). Historia del Toro de la Vega: entre polémicas, defensores y detractores y tradición. El Español. Retrieved January 15, 2024, from https://www.elespanol.com/castilla-y-leon/region/valladolid/20220911/historia-toro-vega-polemicas-defensores-detractores-tradicion/701430109_0.html.

CAS International. (n.d.). *Municipios antitaurinos*. Retrieved January 2, 2024, from https://www.stieren.net/es/municipios-antitaurinos/.

Castro, D. (2020). La exclusión social y jurídica de las toreras: un símbolo de la consolidación del orden de género establecido (1895-1910). *Arenal. Revista De Historia De Las Mujeres*, *27*(1), 199–218. https://doi.org/10.30827/arenal.v27i1.6235.

Centro de Investigaciones Sociológicas. (1995). *Estudio: 2203 – Perfiles Actitudinales en la Sociedad Española.* https://www.cis.es/detalle-ficha-estudio?origen=estudio&idEstudio=1193.

Codina, J. I. (2018). *Pan y toros: breve historia del pensamiento antitaurino español.* Plaza y Valdés. http://dx.doi.org/10.5565/rev/da.391.

de Freitas Netto, S. V., Sobral, M. F. F., Ribeiro, A. R. B., & da Luz Soares, G. R. (2020). Concepts and forms of greenwashing: A systematic review. *Environmental Sciences Europe*, *32*(19), 1–12. https://doi.org/10.1186/s12302-020-0300-3.

Decreto 226/2001, de 18 de septiembre, del Gobierno de Aragón, por el que se aprueba el Reglamento de los Festejos Taurinos Populares. *Boletín Oficial de Aragón*, *122*, 2001, October 17. https://www.boa.aragon.es/cgi-bin/EBOA/BRSCGI?CMD=VERDOC&BASE=BOLE&SEC=BUSQUEDA_AVANZADA&SEPARADOR=&&DOCN=000065813.

Decreto 16/2003, de 28 de enero, del Gobierno de Aragón, por el que se aprueba el Reglamento de las Escuelas Taurinas de Aragón. *Boletín Oficial de Aragón*, *17*, 2003, February 12. https://www.boa.aragon.es/cgi-bin/EBOA/BRSCGI?CMD=VEROBJ&MLKOB=385307434040&type=pdf.

Decreto 61/2006, de 4 de abril, por el que se establecen normas para la ordenación de las Escuelas Taurinas en la Comunidad Autónoma de Extremadura. *Diario Oficial de Extremadura*, *43*, 2006, April 11. https://doe.juntaex.es/eli/es-ex/d/2006/04/04/61/dof/spa/pdf.

Decreto 183/2008, de 11 de noviembre, por el que se aprueba el Reglamento de Espectáculos Taurinos. *Boletín Oficial del País Vasco*, *243*, 2008, December 19. https://www.euskadi.eus/bopv2/datos/2008/12/0807047a.pdf.

Decreto Foral 249/1992, de 29 de junio, por el que se aprueba el Reglamento de Espectáculos Taurinos. *Boletín Oficial de Navarra*, *80*, 1992, July 4. https://www.lexnavarra.navarra.es/detalle.asp?r=28674.

Douglass, C. B. (1997). *Bulls, bullfighting, and Spanish identities*. The University of Arizona Press. https://doi.org/10.2307/j.ctv2jhjw0x.

Electomanía. (2023a). *ElectoPanel (1S): polarización sobre prohibir y subvencionar la tauromaquia.* Retrieved January 10, 2024, from https://electomania.es/electopanel-1s-polarizacion-sobre-prohibir-y-subvencionar-la-tauromaquia/.

Electomanía. (2023b). *ElectoPanel (31Ag): la mayoría de los españoles contra la tauromaquia.* Retrieved January 10, 2024, from https://electomania.es/ep-toros-1electopanel-31ag-la-mayoria-de-espanoles-contra-la-tauromaquia/.

Fundación BBVA. (2022). *Visión y Actitudes hacia los Animales en la Sociedad Española.* Departamento de Estudios Sociales y Opinión Pública. Retrieved January 10, 2024, from http://cea.unizar.es/TRANSPARENCIA/BBVA%20Estudio%20Vision%20y%20actitudes%20hacia%20los%20animales%20en%20la%20Sociedad%20Espa%F1ola%202022.pdf.

Giménez, M. (2023, January 14). *Nueve muertos y 984 heridos en los 'bous al carrer' en el año en el que menos municipios valencianos han realizado actos taurinos.* elDiario.es. Retrieved January 2, 2024, from https://www.eldiario.es/comunitat-valenciana/nueve-muertos-984-heridos-bous-carrer-ano-municipios-valencianos-han-realizado-actos-taurinos_1_9863289.html.

Giménez-Candela, M. (2021). The legal status of animals in the Civil Code: The long-awaited de-objectification of animals. *dA. Derecho Animal (Forum of Animal Law Studies)*, *12*(2), 7–22. https://doi.org/10.5565/rev/da.582.

Graña, J. L., Cruzado, J. A., Andreu, J. M., Muñoz-Rivas, M. J., Peña, M. E., & Brain, P. F. (2004). Effects of viewing videos of bullfights on Spanish children. *Aggressive Behavior*, *30*(1), 16–28. https://doi.org/10.1002/ab.20005.

Hunnicutt, G. (2020). Gender violence in ecofeminist perspective. *Intersections of animal oppression, patriarchy and domination of the earth.* Routledge. http://dx.doi.org/10.4324/9781351026222.

Ipsos. (2022). *Sensibilización de la sociedad española respecto a los derechos de los animales. Informe de Resultados.* Retrieved August 8, 2024, from https://www.mdsocialesa2030.gob.es/derechos-animales/estudios/docs/estudio-ley-encuesta.pdf

Ley 34/2010, de 1 de octubre, de regulación de las fiestas tradicionales con toros. *Boletín Oficial del Estado*, *257*, 2010, October 23. https://www.boe.es/eli/es-ct/l/2010/10/01/34/con.

Ley 18/2013, de 12 de noviembre, para la regulación de la Tauromaquia como patrimonio cultural. *Boletín Oficial del Estado*, *272*, 2013, November 13. https://www.boe.es/eli/es/l/2013/11/12/18.

Ley 10/2015, de 26 de mayo, para la salvaguardia del Patrimonio Cultural Inmaterial. *Boletín Oficial del Estado*, *126*, 2015, May 27. https://www.boe.es/eli/es/l/2015/05/26/10/con.

Ley 10/2017, de 27 de diciembre, de espectáculos públicos y actividades recreativas de Galicia. *Boletín Oficial del Estado*, *60*, 2023, March 11. https://www.boe.es/eli/es-ga/l/2017/12/27/10/con.

Ley 9/2017, de 3 de agosto, de regulación de las corridas de toros y de protección de los animales en las Illes Balears. *Boletín Oficial del Estado*, *223*, 2017, September 15. https://www.boe.es/eli/es-ib/l/2017/08/03/9.
Ley 17/2021, de 15 de diciembre, de modificación del Código Civil, la Ley Hipotecaria y la Ley de Enjuiciamiento Civil, sobre el régimen jurídico de los animales. *Boletín Oficial del estado*, *300*, 2021, December 16. https://www.boe.es/eli/es/l/2021/12/15/17.
Ley 7/2023, de 28 de marzo, de protección de los derechos y el bienestar de los animales. *Boletín Oficial del Estado*, *75*, 2023, March 29. https://www.boe.es/eli/es/l/2023/03/28/7.
Ley Orgánica 10/1995, de 23 de noviembre, del Código Penal. *Boletín Oficial del Estado*, *281*, 1995, November 11. https://www.boe.es/eli/es/lo/1995/11/23/10/con.
Ley Orgánica 15/2003, de 25 de noviembre, por la que se modifica la Ley Orgánica 10/1995, de 23 de noviembre, del Código Penal. *Boletín Oficial del Estado*, *283*, 2003, November 23. https://www.boe.es/eli/es/lo/2003/11/25/15.
Ley Orgánica 5/2010, de 22 de junio, por la que se modifica la Ley Orgánica 10/1995, de 23 de noviembre, del Código Penal. *Boletín Oficial del Estado*, *152*, 2010, June 23. https://www.boe.es/eli/es/lo/2010/06/22/5.
Ley Orgánica 1/2015, de 30 de marzo, por la que se modifica la Ley Orgánica 10/1995, de 23 de noviembre, del Código Penal. *Boletín Oficial del Estado*, *77*, 2015, March 31. https://www.boe.es/eli/es/lo/2015/03/30/1.
Ley Orgánica 3/2023, de 28 de marzo, de modificación de la Ley Orgánica 10/1995, de 23 de noviembre, del Código Penal, en materia de maltrato animal. *Boletín Oficial del Estado*, *75*, 2023, March 29. https://www.boe.es/eli/es/lo/2023/03/28/3/con.
Martel, I. (August 24, 2021). *La situación de la tauromaquia en España: ¿dónde están prohibidas las corridas de toros?* ABC.es. Retrieved January 5, 2024, from https://www.abc.es/cultura/toros/abci-situacion-tauromaquia-espana-donde-prohibidas-corridas-toros-nsv-202108241122_noticia.html.
Martín-Cabello, A. (2021). Desruralización y prácticas culturales: el caso de los espectáculos taurinos en España. *Aposta. Revista de Ciencias Sociales*, *88*, 8–29. http://dx.doi.org/10.31921/doxacom.n33a1482.
Miller, T. (2017). *Greenwashing culture*. Routledge. https://doi.org/10.4324/9781315659251.
Ministerio de Cultura (2022). *Encuesta de Hábitos y Prácticas Culturales en España.* https://www.cultura.gob.es/servicios-al-ciudadano/estadisticas/cultura/mc/ehpc.html.
Ministerio de Cultura (2023). *Estadística de Asuntos Taurinos. Resultados.* https://www.cultura.gob.es/servicios-al-ciudadano/estadisticas/cultura/mc/culturabase/asuntos-taurinos/resultados-asuntos-taurinos.html.
Mulà, A. (2021). Los menores de edad y su exposición a la violencia contra los animales: análisis sobre las observaciones finales del Comité de los Derechos del Niño. *Boletín INTERcids de derecho Animal (BIDA)*, AOL/21/G3, mayo-junio, 1–7. Retrieved April, 7, 2024, from https://intercids.org/menores_edad_exposicion_violencia_animales/.
Mulà, A. (2024). La Ley 18/2013, de 12 de noviembre, que declaró la tauromaquia como patrimonio cultural: Antecedentes, irregularidades e implicaciones de su aprobación. La Iniciativa Legislativa Popular para su derogación. *Boletín INTERcids de derecho Animal (BIDA)*, AOL/24/G1, enero-febrero, 1–18. Retrieved April 7, 2024, from https://intercids.org/ley_tauromaquia-patrimonio_cultural_antecedentes_irregularidades_implicaciones_iniciativa_legislativa_popular/.
Pink, S. (1997). *Women and Bullfighting. Gender, Sex and the Consumption of Tradition.* Routledge. https://doi.org/10.4324/9781003087403.
Prats, E. (2020). El delito de maltrato animal en España. *Revista Jurídica de Catalunya*, *4*, 965–1002.
Real Decreto 145/1996, de 2 de febrero, por el que se modifica y da nueva redacción al Reglamento de Espectáculos Taurinos. *Boletín Oficial del Estado*, *54*, 1996, March 2. https://www.boe.es/eli/es/rd/1996/02/02/145/con.

Real Decreto 1151/2011, de 29 de julio, por el que se modifican el Real decreto 1132/2008, de 4 de julio, por el que se desarrolla la estructura orgánica básica del Ministerio de Cultura y el Real decreto 1181/2008, de 11 de julio, por el que se modifica y desarrolla la estructura orgánica básica del Ministerio del Interior. *Boletín Oficial del Estado*, *209*, 2011, August 31. https://www.boe.es/eli/es/rd/2011/07/29/1151.

Real decreto-ley aprobando el proyecto de Código Penal, que se inserta, y disponiendo empiece a regir como Ley del Reino el día 1° de enero de 1929. *Gaceta de Madrid, 257,* 1928, September 13. https://www.boe.es/buscar/doc.php?id=BOE-A-1928-8856.

Sentence of the Constitutional Court (SCC) 177/2016, 2016, October 20. *Boletín Oficial del Estado*, *285*, 2016, November 25. https://www.boe.es/diario_boe/txt.php?id=BOE-A-2016-11124.

Sentence of the Constitutional Court (SCC) 134/2018, 2018, December 13. *Boletín Oficial del Estado*, *13*, 2019, January 13. https://www.boe.es/diario_boe/txt.php?id=BOE-A-2019-459.

Serrano, M. D. (2019). El maltrato de animales. *Revista De Derecho Penal Y Criminología*, (2 Extra), 501–526. Retrieved August 7, 2024, from https://revistas.uned.es/index.php/RDPC/article/view/24885.

Serrano, M. D. (2023). Lección 23. Delitos contra los recursos naturales y el medio ambiente. Delitos contra la flora y la fauna. Delitos contra los animales. In M. D. Serrano (Coord.), *Derecho Penal: parte especial*. Tirant lo Blanch. http://dx.doi.org/10.2307/j.ctv1k03rks.7.

Shubert, A. (1999). *Death and money in the afternoon. A history of the Spanish bullfight.* Oxford University Press. http://dx.doi.org/10.1093/oso/9780195095241.001.0001.

Sollund, R. A. (2008). Causes for speciesism: Difference, distance and denial. In R. A. Sollund (Ed.), *Global harms: Ecological crime and speciesism* (pp. 109–130). Nova Science Publishers. http://dx.doi.org/10.4324/9781315084589-5.

Sollund, R. A. (2012). Speciesism as Doxic Practice Versus Valuing Difference and Plurality. In R. Ellefsen, R. Sollund, & G. Larsen (Eds.), *Eco-global Crimes. Contemporary Problems and Future Challenges* (pp. 91–114). Routledge. https://doi.org/10.4324/9781315578651-6.

Sollund, R. A. (2015). Introduction: Critical Green Criminology – An Agenda for Change. In R.A. Sollund (Ed.), *Green Harms and Crimes. Critical Criminology in a Changing World* (pp. 1–26). Palgrave Macmillan. https://doi.org/10.1057/9781137456267_1.

Sollund, R.A. (2022). *¿Solo un animal?* Universidad Antonio Nariño.

United Nations Committee on the Rights of the Child. (2018, March 5). *Concluding observations on the combined fifth and sixth periodic reports of Spain.* CRC/C/ESP/CO/5-6. Retrieved January 10, 2024, from https://www.ohchr.org/en/documents/concluding-observations/crccespco5-6-concluding-observations-combined-fifth-and-sixth.

United Nations Committee on the Rights of the Child. (2023, August 22). *General comment No. 26 (2023) on children's rights and the environment, with a special focus on climate change*. CRC/C/GC/26. Retrieved April 7, 2024, from https://docstore.ohchr.org/SelfServices/FilesHandler.ashx?enc=6QkG1d%2FPPRiCAqhKb7yhsqIkirKQZLK2M58RF%2F5F0vHrWghmhzPL092j0u3MJAYhyUPAX9o0tJ4tFwwX4frsfflPka9cgF%2FBur8eYD%2BEeDmuoVnVOpjkzwB9eiDayjZA.

Vericat, M. (2019). El patriarcado en la corrida de toros. *dA. Derecho Animal (Forum of Animal Law Studies)*, *10*(4), 112–122. https://doi.org/10.5565/rev/da.447.

Viestenz, W. (2013). Sins of the flesh: Bullfighting as a model of power. In J. R. Resina (Ed.), *Iberian modalities. A relational approach to the study of culture in the Iberian Peninsula* (pp. 143–161). Liverpool University Press. http://dx.doi.org/10.5949/liverpool/9781846318337.003.0010.

Vila, M. (2021). Humane-washing: la publicidad engañosa respecto al bienestar animal en las explotaciones ganaderas. *Boletín INTERcids de derecho Animal (BIDA)*, AOL/21/G1, enero-febrero, 1–18. Retrieved April 7, 2024, from https://intercids.org/

humane-washing-la-publicidad-enganosa-respecto-al-bienestar-animal-en-las-explotaciones-ganaderas-bida-aol21-g1/.

Winders, D. J. (2017). Captive wildlife at a crossroads – Sanctuaries, accreditation, and humane-washing. *Animal Studies Journal*, *6*(2), 161–178. https://ro.uow.edu.au/asj/vol6/iss2/9.

Wyatt, T. (2021). Construcciones verdes de las categorías de víctima y daño. In H. Mol, D. Rodríguez Goyes, N. South, & A. Brisman (Eds.), *Introducción a la Criminología Verde. Conceptos para nuevos horizontes y diálogos socioambientales* (pp. 77–96). Editorial Temis, Universidad Antonio Nariño. http://dx.doi.org/10.1007/s10612-018-9397-8.

Zuloaga, L., Francés, P., Alemán, E., García, L., Tirapu, X., & Jabat, E. (2018). *Agresiones y abusos sexuales en Sanfermines. Estudio diagnóstico de las dimensiones y de los posicionamientos mediáticos e institucionales ante el problema. Informe de resultados*. Universidad Pública de Navarra – Nafarroako Unibertsitate Publikoa. http://dx.doi.org/10.15581/001.3.33897.

PART IV

Legal Collusion and the Violence of "Animal Capital"

11

HORSERACING AS REGULATED CRUELTY

A Nonhuman Animal Victimology Perspective

Melanie Flynn and Angus Nurse

Nonhuman animal victimology (Flynn, 2024; Flynn & Hall, 2017) posits that nonhuman animals should be viewed, and responded to, as victims of crime. This may be from a legal perspective, in relation to policy and procedures, or in terms of research and scholarly activity. This approach has criticized victimology for being speciesist, after its almost complete neglect of the experiences of, and responses to, nonhuman animals who are harmed by humans, echoing similar criticisms that have been made of criminology more broadly (see, for example, Beirne, 2009; Sollund, 2015). In keeping with victimology, this calls for a shift in focus from crimes (and offenders) to victims, in this case those nonhuman animals who are harmed by the activities of humans. Flynn (2024) proposes that attention should be paid to patterns in extent and characteristics of nonhuman victims and the environments and situations in which their victimization occurs, the consequences of victimization for nonhuman animals, the relationships between nonhuman animal victims and offenders, and interactions with institutions – notably the criminal justice system. Having its origins in critical conceptions, the victimization of nonhuman animals is viewed not just in relation to criminal offences, such as animal abuse or illegal trafficking, but from a zemiological or social harm perspective (Hillyard, 2005). This broader scope highlights a greater range of behaviors that require the attention of victimological scholarship beyond those currently prohibited by criminal law. Indeed, as Beirne (2009) points out, most harms resulting from human behavior that are experienced by the vast majority of nonhuman animals are not subject to criminal legislation. These harmful practices may even be regulated *into* the animal-industrial complex.

The horseracing industry can be viewed as part of the animal-industrial complex. It involves interactions between breeders, trainers, jockeys, racecourses, regulatory bodies, the betting industry, and (some of) the general public, all benefitting

DOI: 10.4324/9781003441908-15

from the exploitation of large numbers of sentient beings born and raised for the sole purpose of providing entertainment and profit. As well as the physical harms that result from being used for racing, we also therefore acknowledge the harms inherent in exploiting nonhuman animals for entertainment.

In relation to harms occurring throughout the animal-industrial complex, criminology is mostly silent, with research on nonhuman animal harm focusing mostly on wildlife and companion animals. Yet the more critical and radical elements of the discipline are predicated on ideas of challenging state definitions of crime and victimization and orthodox assumptions, which are most helpful when considering nonhuman animal victimization (Sollund, 2015). Critical victimology and green criminology recognize "inconvenient victims" (Kearon & Godfrey, 2007, p. 31) who are excluded by political and economic considerations influenced by power inequalities (Ruggerio & South, 2010). Relatedly, green criminological discourse often draws on treadmill of production theory and critiques of capitalism in relation to environmental harm (see, for example, Lynch et al., 2020). Therefore, the foundations for viewing and challenging the animal-industrial complex as a creator of animal victimization are there if criminologists can remove their anthropocentric blinkers.

In an effort to address this, in this chapter, we apply a nonhuman animal victimology perspective to horseracing using analysis from the United Kingdom, while also drawing on data from the United States and Australia. We argue that horses used for racing should be considered victims, discussing both horseracing generally and use of the whip (a regulated behavior) more specifically. To reach this conclusion, we draw on elements of the three key areas of victimology: (1) the extent and patterns of horse victimization in racing; (2) the consequences of victimization for horses made to race; and (3) responses to harm in horseracing and the extent to which horses' interests are considered in these processes and outcomes.

The difficulties of adopting nonspeciest language have been rehearsed by several others (Beirne, 2009 summarizes the issues well). While we recognize the critiques of the term, as nonhuman animal victimology explicitly emphasizes the focus on those who are neglected for not being human, this term is considered the most appropriate. When referring to horses in the racing industry, following Beirne (2009) these are horses used for racing (not racehorses). When considering violence experienced by horses, we mean both physical and psychological or emotional harm, though physical harm is far more documented thus dominates our discourse.

Horseracing as Regulated Activity

Use of nonhuman animals for human entertainment has long attracted criticism for its inherent cruelty, combined with the notion that such exploitation is illustrative of problematic, anthropocentric attitudes (Young, 2014). Adherents of traditional fieldsports distinguish between unnecessary suffering of nonhuman animals for no reason and the bravery and combative nature of "sport" that has value to all participants (Hawley, 1993; Huggs, 1993; Matz, 1984). But opponents of "sports" involving nonhuman animals highlight the imbalance in power relationships between humans and nonhuman animals and the inability of nonhuman animals to consent to their involvement. They also note the vulnerability of nonhuman animals within sport and a requirement for humans to safeguard the welfare of nonhuman participants (Dashper, 2017). We might also add that many nonhuman animals used in sports are bred for that very purpose, knowing no other life than to be a commodity from the day they are (usually forcefully) conceived. In this regard, notions of "ownership" and breeding practices reinforce anthropocentric conceptions on nonhuman animal use and the allegedly "benign" nature of exploitation for sport.

Animal welfare issues in using horses for racing are based, at least in part, on the harm caused by training methods and inherent issues within a sport based largely on the endurance of its nonhuman participants and their capacity to push themselves to near exhaustion. However, as "horseracing" is legal, these issues are largely dismissed or ignored within criminological discourse. Horseracing as legal activity broadly considers any harms to be incidental, accidental, or an occupational hazard (Davies et al., 2022).

Horseracing is administrated as a licensed sport primarily with regulatory boards controlling its operation, licensing its participants, and setting the sport's rules. In the United Kingdom, the Jockey Club, founded in 1752, shares regulation with the British Horseracing Authority (BHA; Vamplew, 2003), which "is responsible for the governance, administration and regulation of horseracing and the wider horseracing industry in Britain" (BHA 2024a, Our Mission section). The Jockey Club runs 15 racecourses, retaining some regulatory and disciplinary functions for jockeys. The BHA operates a Judicial Panel chaired by a legally qualified Chair and comprising: a Disciplinary Panel; a Licensing Committee; and an Appeal Board.[1] The Disciplinary Panel holds enquiries and has the power to impose penalties for breaches of horseracing Rules. This panel also hears appeals from Stewards' decisions. The Licensing Committee is responsible for regulation of licensed, permitted, and registered individuals, including all jockeys, trainers, other racing staff and "racehorse owners". The Appeals Board considers all relevant appeals. While the BHA appoints and remunerates the Judicial Panel it is arguably an independent regulatory body.

In the United States, the Horseracing Integrity and Safety Authority (HISA) regulates thoroughbred horseracing, holding jurisdiction over safety and anti-doping measures. The HISA's functions were included in the Horseracing Integrity and Safety Act 2020 representing a change in US thoroughbred racing regulation following safety concerns (Liebman, 2021). Historically, individual states regulated horseracing but arguably this failed to protect horses or horseracing, accordingly federal control became a necessity (McAfee, 2023, p. 784).

Horseracing regulatory boards in various jurisdictions (including the United States, United Kingdom, and Australia) generally allow whipping of horses subject to regulatory control over "excessive use" and type of whip. In 2011, the BHA produced "A Review of the Use of the Whip in Horseracing", concluding that "the use of the whip in Racing – providing strict controls are effectively enforced – remains appropriate and necessary for the safety of both jockeys and horses. Use of the whip to focus and concentrate a horse, and to encourage it to perform at its best, also remains appropriate" subject to some constraints (BHA, 2011, p. 5). More recently, the independently chaired Horse Welfare Board (HWB) produced a strategic plan for the welfare of horses bred for racing.[2] This acts as a "single, overarching strategy for equine welfare in the racing industry" (Horse Welfare Board, 2020, p. 7). The HWB purports to be objective, draw on a range of views, and adopt an evidence-based approach. In the context of the welfare of horses used for racing, this proposed strategy may be seen as broadly progressive, notably in relation to recommendations on collective lifetime responsibility (addressing some of the appalling treatment of horses after they are "retired"). With regard to use of the whip, they recommended the BHA should consult on this topic, and regardless of the outcomes of this consultation, review and increase penalties for existing breaches of whip regulations. Though couched in language to reassure they are not making an explicit recommendation, the report also reflects some concern regarding the use of the whip for encouragement (as opposed to safety). They suggest this should be considered as part of the consultation.

Following consultation, new restrictions on whip use were subsequently introduced in the United Kingdom in 2023, although a recommendation that use of the whip for encouragement should be limited to the "backhand position only" was adjusted following objections from jockeys. However, the new rules did increase penalties for offences, including doubled suspension in major races and the introduction of a disqualification sanction for offences where the whip has been used four times or more above the permitted level.[3] Though excessive use of the whip is clearly discouraged through enhanced sanctions, whether whip use per se is a form of nonhuman animal victimization remains unexamined.

A key question in relation to both training methods and use of the whip is the extent to which these measures amount to nonhuman animal harm, and whether they potentially contravene legislation (Nurse, 2013). In addition to industry regulation, animal law applies to horseracing creating legal obligations to consider horse welfare. UK law primarily provides for animal welfare via the Animal Welfare Act 2006 (AWA) and associated devolved legislation.[4] Section 4 of the AWA redefines the offence of causing "unnecessary suffering" to include both acts and omissions. In R (on the application of Gray and Another) v Aylesbury Crown Court [2013] EWHC 500 (Admin) the courts clarified that the correct interpretation of section 4(1)(a) was whether a person knew or ought to have known that their actions were "likely to cause unnecessary suffering whether by negligent act or omission" (Nurse, 2021, p. 454). UK law's definitions of animal abuse and unnecessary suffering now extend beyond direct intentional infliction of pain as is the case in the United States where anti cruelty laws are primarily designed to "protect animals from the intentional and gratuitous infliction of pain and suffering at the hands of humans" (Schaffner, 2011, p. 22). The UK conception of unnecessary suffering includes consideration of whether the suffering could reasonably be avoided as well as the mental component of the suffering caused by acts or omissions. Section 9 of the AWA imposes a positive duty to consider the needs of companion animals and provide for animal welfare. Section 9(2)(e) includes obligation to protect from "pain, suffering, injury and disease". Thus, training methods and excessive use of the whip during racing arguably give rise to animal welfare offences requiring consideration of the extent to which horseracing as private regulated activity provides for adequate prevention of harm and imposes appropriate sanctions where such abuses occur.

The Harms of Horseracing

Welfare harms associated with horseracing are well documented through undercover NGO investigations (e.g., PETA, 2023) as well as in academic discourse on animal-based entertainment (McManus, 2020). Previous research identified that three-quarters of Australians responding to a poll did not support the whipping of horses in racing (McGreevy & Fawcett, 2018). But use of the whip continues as a safety tool and encouragement to horses, albeit one that by definition constitutes animal harm, especially where excessive whip use breaches regulations (Hood et al., 2017; Nurse, 2013). We therefore contend that (regulated) violence is an integral and accepted part of horseracing.

Further, the evidence suggests that breaches of the industry's regulations continue, resulting in horse victimization. Lee (2015, p. 27) highlighted how drug use was historically common within horseracing. Due to the fragmented nature of US horse racing regulation, a trainer caught running horses on steroids in one jurisdiction, could simply move to another (e.g., a neighboring state) where steroid use might be tolerated. Evidence suggests that doping of horses is relatively commonplace, with evidence of injured horses being drugged and forced to run (Brooks et al., 2013). Limited reliable data exist on the extent of injuries and deaths, but it has been estimated that "for every 1,000 horses that start a race in the United States, over five either will not finish the race or will finish with severe, oftentimes fatal, injuries" (Thole, 2012, p. 560). Studies examining the sudden death of racehorses have concluded that age, distance, race type, season, and number of starts in the period leading up to a race are factors associated with death. The implication is that horses are being raced to exhaustion with the consequence that disproportionately high cases of sudden death are suspected in horses used for racing (Lyle et al., 2012).

This is also evident in the incidence of exercise-induced pulmonary hemorrhage (EIPH). This involves bleeding in the airways and is a result of over-exertion. A phenomenon not often seen in other species, it is reportedly common in horses used for racing and the damage is cumulative (Oke, 2022). Lyle et al.'s (2012) study, which examined the risk of sudden death in 201 case race starts and 705,712 control starts, concluded that EIPH is more commonly a cause of death in shorter races, whereas longer races may cause more severe metabolic derangements thought to contribute to fatal ventricular arrhythmias. The "encouragement" provided by whip use seeks to push horses even harder when they may already be suffering due to the strenuous high-intensity exercise they are being forced to undertake. In essence, horse racing risks forcing horses beyond their natural limits and the high level of injuries and deaths suggests a sport with poor levels of welfare adherence (Physick-Sheard, 2022). This further highlights the inability of regulation to prevent horse victimization.

Whether whipping causes harm, and the extent of this, is a contested issue. While there is a general lack of scientific evidence that whipping causes significant pain, McGreevy et al. (2013) observed that 83% of whip strikes viewed in slow motion caused visible indentation. Whip use is also associated with falls, with one study concluding that horses being whipped and progressing through a race were at greater than seven times the risk of falling compared to horses that were not being whipped (Pinchbeck et al., 2010). However, irrespective of the science on physical harm, the use of the whip as "encouragement" serves to push horses to race harder, which, when already near exhaustion, clearly constitutes animal harm.

The Regulatory Response

In essence, horseracing self-regulates, through the BHA (UK) and even in respect of statute-driven regulation by the HISA (USA). A key issue in relation to the critique of self-regulation is the extent to which the regulatory approach is driven by, and simply reinforces, *human* and industry perspectives and norms of racing. The public interest justification for self-regulation is primarily based on fulfilment of three conditions; that the activity (here, horseracing) is afflicted by market failure, that private law instruments are inadequate or too costly to correct the failure, and that self-regulation is a better (and cheaper) method of solving the problem than conventional regulation (Ogus, 1995). It is argued that industries and specialist associations are better able to regulate practices than public regulators because of their relevant expertise and technical knowledge (Christopher Hodges et al., 2012, p. 145).

From a welfare perspective it may be possible to mount a defense of self-regulation in the case of those reputable breeders, owners, and jockeys who operate responsibly. However, it is less effective in the case of rogue operators and others who see horseracing purely in economic terms, and who may consider operating lower standards of animal welfare as a means through which to secure greater profits. The extent to which players fall into these two categories should also be subjected to debate. Self-regulation also fails to adequately address the kind of rigorous inspection, monitoring, and critical reflection necessary to identify poor and outdated practices, incompetence in engaging with animal welfare standards and, crucially from our perspective, anthropocentric views that construct horses as exploitable resources for human use and entertainment. Improvements to the regulatory framework with standardized rules have arguably failed to achieve the envisaged improvements (Russolello, 2009).

One aspect of the challenges facing the regulatory landscape is the extent to which the industry resists efforts to maintain and enforce standards of ethical animal welfare. For example, narratives that dispute whether horses feel pain or not (from whipping) are prevalent in horseracing discourse (Graham & McManus, 2016). Justifications and neutralizations are regularly used to divert attention away from perceived cruelty in training and racing and to challenge the need for further regulation. Before the HISA came into force in the United States several law suits were launched in an attempt to prevent the new regulatory regime, including National Horsemen's Benevolent & Protective Association v. Black (Black I) and Oklahoma v. United States (Oklahoma I), No. 5:21-CV-104-JMH, 2022 US Dist, cases that sought to argue that HISA violates Private Non-delegation and that the delegation of regulatory authority was unconstitutional (McAfee, 2023, pp. 792–793). Such legal challenges represent an appeal to higher loyalties, seeking to retain control of racing in the hands of sympathetic state boards who will understand and preserve the historical nature of the sport. They also represent a condemnation of the condemners, those who do not understand horseracing and seek to meddle in its affairs through unnecessary and unwelcome regulation.[5] Similar discourses are apparent throughout the literature and policy reports, where the public (outsider) is often presented as having little understanding of the horseracing industry, thus justifying reliance on insider decision-making regarding what is purportedly in the best interests of horses. Such neutralization techniques, combined with those arguing that regulatory attention (and resources) would be better spent elsewhere, represent a mechanism through which wrongdoing or harm can be marginalized (Sykes & Matza, 1957) and, thus, discounted.

The evidence suggests that, while on paper, the regulatory approach deployed in horseracing has potential to deal with infractions, the extent to which it does so to reduce horse victimization in any meaningful way is questionable. Hood et al. (2017) conducted a peer-reviewed characterization of whip rule breaches and their regulatory outcomes. Their research analyzed breaches recorded by stewards from 2013 and 2016 in New South Wales and the Australian Capital Territory. Their initial analysis for 2013 "identified a breach or breaches being reported in 332 of the 5,604 races (5.92%) that took place at 785 meetings" (Hood et al., 2017, p. 4).

Current data from the United Kingdom, from the introduction of the new whip rules to 2nd March 2024, recorded a total number of 87,735 rides with 779 referrals and 676 breaches equating to 0.77% of rides (BHA, 2024b). However, when recalculated based on the number of recorded "runners" (horses) that were raced in 2023 the whip victimization rate equates to approximately 3.63%.[6] For comparison, the most recent police recorded violent crime victimization rate (for England and Wales, year ending September 2023) was 2.31% (excluding stalking and harassment) and national victimization survey data (Crime Survey England and

Wales) results in an estimated 1.28% of the population (aged 16 and over) experiencing violent crime during the same period (ONS, 2024a, 2024b).

The main whip offence types recorded in the United Kingdom were whipping Above Permitted Level (434); Without time to respond (112); Incorrect place (37); and Above Shoulder Height (29) (BHA, 2024b). BHA data also show that:

- Fully professional riders accounted for 392 offences (57.99% of total offences at an offence rate of 0.58%).
- Apprentice riders accounted for 111 offences (16.42% of total offences at an offence rate of 1.02%).
- Amateur riders accounted for 60 offences (8.88% of total offences at an offence rate of 2.53%).

(BHA, 2024b, Offences by License Type section)

Evidence suggests that even where such breaches occur, disciplinary action is minimal with the typical sanction for a whip infraction being a short-term suspension. UK rules provide for suspensions, referrals to the Judicial Panel and ultimately race disqualification where riders breach whip rules (BHA, 2024b. Rules and Limits on use section).

In the United Kingdom, suspensions typically last for two or three races while Hood et al.'s Australian analysis concluded:

> Overall, breach outcomes or penalties were low in our view. Reprimands were the most common outcome for first and second breaches, with less than a third of first breaches resulting in fines, and even then the total was a modest $25,600 (median value $200). Further, all Code 6 first breaches (Excessive, unnecessary, or improper whip use) resulted in cautions or reprimands only.
>
> *(Hood et al., 2017, p. 18)*

Evidence of jockeys being permanently banned and of sanctions having an effect seem minimal. BHA data show evidence of repeat offending by some riders yet in the United Kingdom there have been only four disqualifications since the new whip rules came into effect equating to "one disqualified winning ride from 10,024 races run under the new rules" (BHA, 2024b, Offences by license type section). While the BHA seem to present this as evidence there is little problem with repeated whip use breaches, an alternative interpretation is that the most "severe" sanctions are rarely used. This is plausible given that US analysis of the Animal Welfare Act and the Horse Protection Act indicates that they have low enforcement rates, with most enforcement activity resulting in the lowest levels of formal social control (Lynch & Genco, 2021). Such a pattern has also been identified in criminal disposals for other nonhuman animal victimizations (e.g., Flynn, 2016; Nurse & Harding, 2023). We conclude, therefore, that the evidence suggests a *failure* of regulatory bodies to uphold animal welfare principles, with limited meaningful sanctions for breaches and horse victimization and poor safety standards arguably normalized within the sport (Chapman & Thompson, 2016).

Discussion

Looking through a species justice lens, use of the whip and the harms that have been documented herein are clearly forms of horse victimization. These can be summarized as harms related to breeding and sexual violence, enslavement and lack of autonomy, being forced to race (often to the point of exhaustion and physical harm) for profit and human entertainment, and physical and psychological harm. Taken together these harms clearly result from the commodification of horses. This can be no more clearly seen than in this quote from the HWB (2020) strategy: "lifetime duty of care starts with the decision to breed a Thoroughbred in the first place, where there is a need to ensure sustainable and responsible *production*" (p. 10, emphasis added). Perhaps attempting to normalize this treatment of horses, the industry also talks about them retiring or having second careers, language suggestive of horses being employees. However, we maintain that the horseracing industry is built on enslavement, not employment, as with many other parts of the animal-industrial complex, from the initial decision to create sentient life for the purpose of producing horses for human gain. Similarly, we contend that the practice of horse breeding is a form of sexual violence and coercion, as explored in relation to other nonhuman animals in greater depth elsewhere in this collection (Banwell and Walliss, 2024).

Claims are sometimes made that horses enjoy racing. However, we have established there is evidence indicating that, physiologically, running at this intensity is likely not natural for horses and that their lives are characterized by harmful, and often painful, practices. They are whipped and pushed beyond their limits. Further, those used for racing have known no other life. They have been bred, "broken", trained, and raced regardless of their own desires and needs (see also Puente in this collection). Were we to treat humans this way, not only would there be a public outcry, but several criminal offences would have been committed. Indeed, we have demonstrated that the rate of violent horse victimization in the United Kingdom, through use of the whip alone, is higher than self-reported or police-recorded interpersonal violence.

We recognize that many people involved in this industry genuinely believe they have the welfare of horses in mind, but welfarism has been demonstrated throughout the animal rights literature to be a poor response to animal victimization, not least because conceptualizations of "welfare" come from an anthropocentric perspective, where it is permissible to commodify animals and use them for our ends, as long as this is done "humanely" (e.g., Regan, 2004). This approach is apparent in the language of "a life well lived" and claims throughout the industry that those involved genuinely care for horses and want what is best for them. We contend that those who make such claims simply do not recognize the harms caused by enslavement, commodification, forced breeding, intensive training, over-medication, and injury risk because of institutionalized and socially accepted speciesism. For the same reasons, welfare legislation and industry self-regulation have also been shown to be ineffective in substantially reducing horse victimization.

Turning to the use of the whip more specifically, the evidence regarding the harm this causes may be challenged, but we contend that from a species justice perspective, the whip should be viewed as problematic even if it does not cause physical harm. In horseracing, it can be argued that the whip is the most visible symbol of enslavement and oppression. It is an overt demonstration of the power imbalance between the "participants" (jockey and horse). Therefore, even if it could be demonstrated (or adapted) to not cause physical harm, its presence is totemic (HWB, 2020). This may also explain why the public do not seem to support the whip and what it signifies. It might also be an unacknowledged reason why jockeys are so keen to defend its continued use.

To conclude, our analysis identifies insufficient regulation of animal welfare concerns and that enforcement of whip use, and regulation breaches, warrant further review and a strengthening of sanctions. We therefore also suggest a need for better integration of species justice concerns and independent scrutiny. Ultimately, however, we challenge those who claim to have the best interests of horses at heart to consider whether they can ever have a life well lived while they are treated as commodities who exist only for human entertainment. Additionally, we also challenge criminologists to recognize the extent of harm and suffering experienced by nonhuman animals, such as horses, as part of the animal-industrial complex and to consider meaningfully engaging in criminological research within this area.

Notes

1 At time of writing, the chair is a barrister, Sarah Crowther KC.

2 The Horse Welfare Board (HWB) was set up in 2019 to provide a greater focus on welfare for horses used for racing. The HWB (2020) explains that this was necessary because "many racecourses and participants routinely exceed [the regulatory body] standards" (p. 7) the BHA's remit mostly relates to the active period of racing (not life before and after). This proposed strategy cuts across different members and roles within the industry and during the life of horses. Though purportedly independent and with membership contributing a wide range of views, this is dominated by those within the industry: BHA members, horsemen, racecourse representation, with two independent members: a veterinarian and a former Minister for Sport.

3 There are different whip thresholds for flat races and jump races, but the disqualification sanction applies in all races.

4 Animal welfare is a devolved issue with separate legislation in Northern Ireland and Scotland. The Welfare of Animals Act (Northern Ireland) 2011 and the Animal Health and Welfare Act (Scotland) 2006, respectively provide for animal welfare in the regions with responsibility devolved to administrations and legislatures in Scotland, Wales, and Northern Ireland.

5 The constitutional arguments were partially successful in that the US Court of Appeals for the Fifth Circuit, held that HISA unconstitutionally granted unchecked governmental power to a private entity, namely the Authority. However, the Act was amended in 2022 to provide for greater Federal Trade Commission (FTC) supervisory power over the Authority, addressing the constitutional claims.

6 Based on a published figure of 18,630 "runners" for 2023 (BHA, n.d.). It is acknowledged this may be a slight over-estimation as the number of distinct "runners" for the full period to March 2024 is not known.

References

Banwell, S. and Walliss, J. (2024). "They have literally given up on life;" A review of the experiences of nonhuman animals subject to reproductive violence and coercion on factory and puppy farms. In G. Hunnicutt, R. Twine, & K. Mentor (Eds.), *Violence and harm in the Animal Industrial Complex: Human-animal entanglements*. Routledge.

BHA. (n.d.). *Racing and Industry Statistics 2019-2023*. Retrieved March 20, 2024, from https://media.britishhorseracing.com/bha/Racing_Statistics/Racing_Data_Packs_Full_Year/Full_year_2023.pdf

BHA. (2011). *Responsible regulation: A review of the use of the whip in horseracing*. British Horseracing Authority. https://www.britishhorseracing.com/wp-content/uploads/2014/03/WhipReview.pdf.

BHA. (2024a). *What we do*. British Horseracing Authority. Retrieved February 19, 2024, from https://www.britishhorseracing.com/about/the-bha/what-we-do/.

BHA. (2024b). *The whip*. British Horseracing Authority. Retrieved March 16, 2024, from https://www.britishhorseracing.com/regulation/the-whip-2/.

Beirne, P. (2009). *Confronting animal abuse: Law, criminology, and human-animal relationships*. Rowman & Littlefield.

Brooks, G., Aleem, A., & Button, M. (2013). Fraud and corruption in horse racing. In G. Brooks, A. Aleem, & M. Button (Eds.), *Fraud, corruption and sport* (pp. 107–122). Palgrave Macmillan. https://doi.org/10.1057/9781137290717_8.

Chapman, M., & Thompson, K. (2016). Preventing and investigating horse-related human injury and fatality in work and non-work equestrian environments: A consideration of the workplace health and safety framework. *Animals*, *6*(5), 33. https://doi.org/10.3390/ani6050033.

Dashper, K. (2017). Listening to horses: Developing attentive interspecies relationships through sport and leisure. *Society & Animals*, *25*(3), 207–224. https://doi.org/10.1163/15685306-12341426.

Davies, E., McConn-Palfreyman, W., Parker, J. K., Cameron, L. J., & Williams, J. M. (2022). Is injury an occupational hazard for horseracing staff? *International Journal of Environmental Research and Public Health*, *19*(4), 2054. https://doi.org/10.3390/ijerph19042054.

Flynn, M. (2016). *Sentencing wildlife trade offences in England and Wales: Consistency, appropriateness and the role of sentencing guidelines*. WWF. https://www.wwf.org.uk/sites/default/files/2017-01/WWF-UK%20Report%20-Sentencing%20wildlife%20trade%20offences%20in%20England%20and%20Wales.pdf.

Flynn, M. (2024). Proposing a nonhuman animal victimology. In L. Levitt, D. B. Rosengard, & J. Rubin (Eds.), *Animals as crime victims* (pp. 113–130). Edward Elgar Publishing.

Flynn, M., & Hall, M. (2017). The case for a victimology of nonhuman animal harms. *Contemporary Justice Review*, *20*(3), 299–318. https://doi.org/10.1080/10282580.2017.1348898.

Graham, R., & McManus, P. (2016). Changing human-animal relationships in sport: An analysis of the UK and Australian horse racing whips debates. *Animals*, *6*(5), 32. https://doi.org/10.3390/ani6050032.

Hawley, F. (1993). The moral and conceptual universe of cockfighters: Symbolism and rationalization. *Society and Animals*, *1*(2), 159–168.

Hillyard, P. (2005). Crime obsessions: Crime isn't the only harm. *Criminal Justice Matters*, *62*(1), 26–46. https://doi.org/10.1080/09627250508553097.

Hodges OBE, C., Peysner, J., & Nurse, A. (2012). *Litigation funding: Status and issues*. Oxford Legal Studies Research Paper No. 55/2012. http://dx.doi.org/10.2139/ssrn.2126506.

Hood, J., McDonald, C., Wilson, B., McManus, P., & McGreevy, P. (2017). Whip rule breaches in a major Australian racing jurisdiction: Welfare and regulatory implications. *Animals*, *7*(1), 4. https://doi.org/10.3390/ani7010004.

Horse Welfare Board. (2020). *A life well lived: A new strategic plan for the welfare of horses bred for racing, 2020-2024*. Retrieved March 11, 2024, from https://media.britishhorseracing.com/bha/Welfare/HWB/WELFARE_STRATEGY.pdf.

Huggs, K. C. (1993). Blood sport: A look at the underground world of illegal dogfighting. *Times of Acadiana*, *14*, 20–27.

Kearon, T., & Godfrey, B. S. (2007). Setting the scene: A question of history. In S. Walklate (Ed.), *Handbook of victims and victimology* (pp. 17–36). Willan Publishing. https://doi.org/10.4324/9780203118207.

Lee, K. (2015). Transgressing trainers and enhanced equines: Drug use in racehorses, difficulty assigning responsibility and the need for a National Racing Commission. *Journal of Animal and Natural Resources Law*, *11*, 23–50. https://www.animallaw.info/sites/default/files/jouranimallawvol11.pdf.

Liebman, B. (March 1, 2021). Introducing the Horseracing Integrity and Safety Act and a new era of racing regulation. *Entertainment, Arts and Sports Law Journal*, *64*(1). https://ssrn.com/abstract=3780750.

Lyle, C. H., Blisitt, K. J., Kennedy, R. N., McGorum, B. C., Newton, J. R., Parkin, T. D. H., Stirk, A., & Boden, L. A. (2012). Risk factors for race-associated sudden death in thoroughbred racehorses in the UK (2000–2007). *Equine Veterinary Journal*, *44*(4), 459–465. https://doi.org/10.1111/j.2042-3306.2011.00496.x.

Lynch, M. J., & Genco, L. (2021). Cruelty against animals' welfare (CAAW) violations: A study of Animal Welfare Act and Horse Welfare Act enforcement actions in the US, 2010–2014. *Sociological Spectrum*, *41*(3), 255–272. https://doi.org/10.1080/02732173.2021.1893239.

Lynch, M. J., Stretesky, P. B., Long, M. A., & Barrett, K. L. (2020). Expanding treadmill of production analysis within green criminology by integrating metabolic rift and ecological unequal exchange theories. In N. South & A. Brisman (Eds.), *Routledge international handbook of green criminology* (2nd ed., pp. 79–94). Routledge. https://doi.org/10.4324/9781315207094.

Matz, K. S. (1984). *The pit bull fact and fable*. De Mortmain Publishing.

McAfee, L. (2023). The rise and fall of the Horseracing Integrity and Safety Act: How Congress could save the "Sport of Kings." *Vanderbilt Journal of Entertainment and Technology Law*, *25*(4), 783–812. https://scholarship.law.vanderbilt.edu/jetlaw/vol25/iss4/4.

McGreevy, P., & Fawcett, A. (2018, February 21). *Poll says most people support a ban on whips in Australian horse racing*. The Conversation. https://theconversation.com/poll-says-most-people-support-a-ban-on-whips-in-australian-horse-racing-91917.

McGreevy, P. D., Hawson, L. A., Salvin, H., & McLean, A. N. (2013). A note on the force of whip impacts delivered by jockeys using forehand and backhand strikes. *Journal of Veterinary Behavior*, *8*(5), 395–399. https://doi.org/10.1016/j.jveb.2013.03.003.

McManus, P. (2020). Animal-based entertainment industries, animal death and Social Licence to Operate (SLO): An analysis of 'The Final Race' and the 2019 Melbourne Cup. *Social & Cultural Geography*, *24*(7), 1242–1261. https://doi.org/10.1080/14649365.2022.2053194.

Nurse, A. (2013). *Animal harm: Perspectives on why people harm and kill animals*. Ashgate. https://doi.org/10.4324/9781315567174.

Nurse, A. (2021). Green criminological perspectives on dog-fighting as organised masculinities -based animal harm. *Trends in Organized Crime*, *24*, 447–466. https://doi.org/10.1007/s12117-021-09432-z.
Nurse, A., & Harding, N. (2023). *Policing wildlife: The nature of wildlife crime and its public policy response*. International Fund for Animal Welfare (IFAW). https://d1jyxxz9imt9yb.cloudfront.net/resource/1465/attachment/original/ntu_policing_wildlife_research_report_revised_Oct-Nov_2022.pdf.
Ogus, A. (1995). Rethinking self-regulation. *Oxford Journal of Legal Studies*, *15*(1), 97–108. https://doi.org/10.1093/ojls/15.1.97.
Oke, S. (2022, October 28). The bleeding edge of EIPH in horses, *The Horse*. https://thehorse.com/1105376/the-bleeding-edge-of-eiph-in-horses/.
ONS (Office for National Statistics). (2024a). *Crime in England and Wales: Year ending September 2023*. Retrieved March 20, 2024, from https://www.ons.gov.uk/peoplepopulationandcommunity/crimeandjustice/bulletins/crimeinenglandandwales/yearendingseptember2023.
ONS (Office for National Statistics). (2024b). *Crime in England and Wales, Appendix Tables – Year ending September 2023*. Retrieved March 20, 2024, from https://www.ons.gov.uk/file?uri=/peoplepopulationandcommunity/crimeandjustice/datasets/crimeinenglandandwalesappendixtables/yearendingseptember2023/appendixtablessep23final.xlsx.
PETA. (2023). *Horse racing*. Retrieved July 21, 2023, from https://www.peta.org/issues/animals-in-entertainment/horse-racing-2/.
Physick-Sheard, P. (2022). EIPH – could there be links to sudden death and pulmonary haemorrhage? *Trainer.* Retrieved March 23, 2024, from https://trainermagazine.com/european-trainer-articles/tag/Bleeding+from+the+lungs.
Pinchbeck, G. L., Clegg, P. D., Proudman, C. J., Morgan, K. L., & French, N. P. (2010). Whip use and race progress are associated with horse falls in hurdle and steeplechase racing in the UK. *Equine Veterinary Journal*, *36*(5), 384–389. https://doi.org/10.2746/0425164044868387.
Regan, T. (2004). *Empty cages: Facing the challenge of animal rights*. Rowman & Littlefield Publishers.
Ruggerio, V., & South, N. (2010). Critical criminology and crimes against the environment. *Critical Criminology*, *18*, 245–250. https://doi.org/10.1007/s10612-010-9121-9.
Russolello, A. (2009). *Avoiding a triple frown: The need for a National Horse Racing Commission*, Animal Legal & Historical Centre. https://www.animallaw.info/article/avoiding-triple-frown-need-national-horse-racing-commission.
Schaffner, J. (2011). *An introduction to animals and the law*. Palgrave Macmillan.
Sollund, R. A. (2015). Introduction. In R. A. Sollund (Ed.), *Green harms and crimes: Critical criminology in a changing world*. Palgrave Macmillan. https://doi.org/10.1057/9781137456267_1.
Sykes, G., & Matza, D. (1957). Techniques of neutralization: A theory of delinquency. *American Sociological Review*, *22*, 664. https://doi.org/10.2307/2089195.
Thole, R. J. (2012). Stop beating a dead horse: How attempts to prop up the horseracing industry have led to increased horse fatalities, fewer bettors, and less tax revenue for local governments. *Gaming Law Review and Economics*, *16*(1), 560–586. https://doi.org/10.1089/glre.2012.16105.
Vamplew, W. (2003). Reduced horse power: The Jockey Club and the regulation of British horseracing. *Entertainment Law*, *2*(3), 94–111. https://doi.org/10.16997/eslj.130.
Young, K. (2014). Toward a less speciesist sociology of sport. *Sociology of Sport Journal*, *31*(4), 387–401. https://doi.org/10.1123/ssj.2014-0140.

12

NON-HUMAN ANIMALS AS PROPERTY

What This Means When Companion Animals Are Stolen

Daniel Allen and Tanya Wyatt

Non-human animals are often legally defined as 'things'—property of their human owners, with little material difference or increased legal protection than an inanimate object. Yet, non-human animals are not the same as a piece of furniture or a laptop and we suggest their status as property contributes to their invisibility as victims of the range of crimes that humans perpetrate against them. One crime that is overlooked and has yet to be explored from a criminological perspective is that of 'pet theft'.

More animals are farmed than in any time in history. Whilst the animal-industrial complex is most frequently talked about in terms of farming animals for food, there is a significant portion of this industry dedicated to the farming of animals for the companion animal industry. Global 'pet' ownership is at an all-time high and this is true in the UK as well. An estimated 62% of UK households have a companion animal—this includes 13 million dogs and 12 million cats (PFMA, 2022). During the pandemic many people acquired dogs to combat social isolation and loneliness. This demand saw prices rise, and the number of reported thefts increase (Home Office, 2021). DogLost, a UK organisation that helps search for lost and stolen dogs, reported a 170% increase in the abduction of dogs between 2019 (172 reported) and 2020 (465 reported); DirectLine estimated that 2,438 dogs were reported stolen to the police in 2020, the equivalent of seven dogs a day (Blue Cross a, No date). Although the proportion of thefts per dog population is very small, that is not to say the victims of this crime are not important. 'Like many crimes, dog theft is driven by supply and demand' (Crimestoppers, No date). Money is made by reselling stolen dogs; breeding and selling puppies; obtaining a reward for returning a dog as part of a scam; and extortion. The societal consideration and response to the abduction of non-human companion animals raise several points as to how non-humans are treated in our multi-species societies and how they can be better represented and protected in our anthropocentric systems.

DOI: 10.4324/9781003441908-16

This chapter lays some foundation to address these points. First, we provide some background and examples of legal systems that define non-human animals as property. This includes an exploration of the implications of the theft for the non-human animal who is stolen/abducted. We then go on to detail the situation in England and Wales. Our theoretical framework section outlines two aspects that are relevant to the situation. First, we introduce species justice as the approach to justice that society should be moving towards as it recognises that society is made up of multiple species who are entangled. Next, we outline situational crime prevention (SCP), which is the current response to pet theft in England and Wales. Our aim is to assess whether SCP is up to the task of reducing pet theft and whether it can contribute to achieving species justice. We then outline our methodology for gathering data on the nature and extent of pet theft in England and Wales as well as the criminal justice system response to the crime. This is followed by a discussion of the findings and an analysis from a species justice perspective. We conclude by detailing the ongoing efforts to improve the protection of non-human companion animals from theft in England and Wales.

Background

Non-Human Animals as Property

For the most part, legal systems are centred around protecting the property rights of people (Nurse & Wyatt, 2020). This is relevant to non-human animals since their status as property means the priority of the legal system is to provide redress for the harm or loss to the human owner rather than to address the suffering of the non-human animal. The non-human animal is not the victim of the crime (in this case being abducted from their home); the human owner is the victim of a property crime (an object they own has been stolen from them). That is not to say that non-human animal welfare is completely ignored. Countries around the world have legislation that protects non-human animals from cruelty and injury. Yet, as Nurse and Ryland (2013) note, animal welfare legislation can also be anthropocentric in that it benefits farmers and 'livestock' holders among others who maximise their profit by owning healthier non-human animals. The non-human animal welfare regime has also been criticised for only providing marginal improvements to the lives of non-humans whilst perpetuating exploitation and impeding non-human animals from being afforded rights (Francione, 2007).

Being property rather than a subject of the law with rights and interests means that non-human animals cannot legally be recognised as crime victims (Nurse & Wyatt, 2020). The criminal justice system then is speciesist. Speciesism is where members of another species are not given the same consideration as a member of one's own species (Ryder, 1971). In the case of the legal and criminal justice systems, non-human animals' suffering is not directly engaged with in investigations, prosecutions, and/or court processes (Beirne, 2007). Moore (2005, p. 91) argues that '[non-human] animals who have been abused do not have their interests represented in court. Instead, the state alone is able to prosecute crimes against [non-human] animals'. Being property and not a victim means that non-human animals do not necessarily have their interests represented (Moore, 2005). This is part of the speciesism of the system because in the context of human crime victims, they would be able to advocate on their own behalf. Through victim impact statements or direct engagement with investigators or prosecutors, human crime victims can share their suffering and have the harm that they have endured heard. In the context of non-human animals who are (unrecognised) crime victims, their suffering may play no part in the evidence of the case, or they are reliant on humans to speak on their behalf. This brings us to the suffering and harm of an abducted non-human animal.

The Harms of Non-Human Animal Abduction

Pet theft is known to be a gateway to animal cruelty and extortion, whilst the emotional impact of forced absence can be traumatic for both humans and animals. Pets are generally abducted by non-family members for economic profit, although family disputes over ownership are also reported to the police and recorded as a crime. Many stolen pets are sold on to unsuspecting families in the UK and Ireland; with some, notably working dogs, also having a market in mainland Europe. Some dogs and cats are abducted as a relatively quick way for criminals to claim the reward money, others are used to extort money from families as ransom (Allen et al., 2022; Home Office, 2021). Whilst economic motivations to commit can be driven by poverty, this may not always be the case.

Recent research by Allen et al. (2022) explored the experiential significance of dog theft on human victims. They found that 'the theft of a pet, from the realisation of loss through to ongoing searches to reunite, is filled with emotions that reflect a traumatic experience' (Allen et al., 2022, p. 529). The impact of this absence on human lives can also be influenced by the everyday role of absent pets beyond mere companionship. For example, stolen dogs have included emotional support dogs for children and adults; assistance dogs for the visually impaired; also working dogs central to farming livelihoods.

Scholars have overlooked the physical and experiential impacts on stolen pets, but media stories provide useful insights. Assuming there is a good pre-existing relationship and environment for the animal, all abducted dogs and cats experience the harms of forced separation from their human guardians and home. This includes the disappearance of human family members, daily routines of walks (for dogs), meals, and play. This would lead to varying degrees of anxiety and fear related to species, age, duration of separation, and new environment(s) of the stolen pet. When abducted, different pets will have different experiences. Some dogs, for example, have their coats shaved or dyed to hide their identity; on rare occasions, microchips have also been gouged out.

Unweaned puppies have been taken, risking their lives, and distressing their mother; some abducted pets also require medication to survive. There have also been cases where abducted puppies and adult dogs have had their ears cropped—this painful illegal surgery has become increasingly fashionable in certain breeds.

Some abducted female dogs are being placed in puppy farms, where they are kept in poor conditions and forced to reproduce year after year. Abducted dogs are also being used in illegal dog fighting. Beyond profit, there are also various reports of pets (notably cats) being taken by individuals and cruelly tormented and killed. Yet despite these traumatic experiences, companion animals remain property.

The English and Welsh Context

In England and Wales, stolen pets are regarded as stolen property under the Theft Act 1968. With no specific offence for pet theft, the crime is covered as: theft from the person; theft in a dwelling; theft in breach of trust; theft from a motor vehicle; robbery; and burglary. Sentencing is dependent on monetary or retail value of stolen property, with the courts also assessing the offender's culpability, harm to the human victim, and any aggravating or mitigating factors. The Sentencing Council (2022) provides sentencing guidelines which courts must follow to determine the offence category (see Figure 12.1). Owing to the economic value of

Category 1	Very high value goods stolen (above £100,000) **or** high value with significant additional harm to the victim or others
Category 2	High value goods stolen (£10,000 to £100,000) **and** no significant additional harm **or** Medium value with significant additional harm to the victim or others
Category 3	Medium value goods stolen (£500 to £10,000) **and** no significant additional harm **or** Low value with significant additional harm to the victim or others
Category 4	Low value goods stolen (up to £500) **and** little or no significant additional harm to the victim or others

	A	B	C
Category 1 Adjustment should be made for any significant additional harm factors where very high value goods are stolen.	**Starting point** 3 years 6 months' custody	**Starting point** 2 years' custody	**Starting point** 1 year's custody
	Category range 2 years 6 months'–6 years' custody	**Category range** 1–3 years 6 months' custody	**Category range** 26 weeks'–2 years' custody
Category 2	**Starting point** 2 years' custody	**Starting point** 1 year's custody	**Starting point** High level community order
	Category range 1–3 years 6 months' custody	**Category range** 26 weeks'–2 years' custody	**Category range** Low level community order–36 weeks' custody
Category 3	**Starting point** 1 year's custody	**Starting point** High level community order	**Starting point** Band C fine
	Category range 26 weeks'–2 years' custody	**Category range** Low level community order–36 weeks' custody	**Category range** Band B fine–Low level community order
Category 4	**Starting point** High level community order	**Starting point** Low level community order	**Starting point** Band B fine
	Category range Medium level community order–36 weeks' custody	**Category range** Band C fine–Medium level community order	**Category range** Discharge–Band C fine

FIGURE 12.1 England and Wales theft sentencing guidelines

stolen pets, their theft is generally treated as a category 3 (goods up to £10,000) or category 4 offence (goods up to £500). Offenders can also be charged with moving stolen property when selling stolen pets (Nurse & Ryland, 2014). Harris (2018, p. 6) argues, 'the law trivializes the emotional harm that can be caused by the theft of a companion animal'; this trivialisation extends to the individual animals whose experiences receive no consideration with this offence.

Regardless of the pet and/or the offender, the chances of being arrested and charged for stealing pets are very low. Using Freedom of Information (FOI) data (see Methodology for details) for 43 police forces in England and Wales, Allen et al. (2019) found the percentage of charges for offences including dog theft to be around 4% in 2015, 3% in 2016, and 2% in 2017. Selby-Fell and Allen (2021) suggested that less than 1% of recorded dog thefts for 33 forces resulted in a charge in 2020.

Recognising the importance of the human-animal bond, the devastating impact that pet theft has on people and pets, and the need for pet abduction as a specific offence, Dr Daniel Allen and the Stolen and Missing Pets Alliance launched the Pet Theft Reform campaign in January 2018. This evidence-based campaign harnessed public and political support through government e-petitions, traditional and social media, and lobbying as a coalition. Although initial calls to revise the sentencing guidelines in relation to theft were rejected by the Sentencing Council (Lord Justice Holroyde, 2021), the campaign led to the creation of the government's Pet Theft Taskforce in 2021 (Home Office, 2021), and the addition of the Pet Abduction offence to the Animal Welfare (Kept Animals) Bill 2021. This Bill was withdrawn on 8 June 2023. Anna Firth MP introduced the Pet Abduction Bill to create offences of dog abduction and cat abduction on 6 December 2023. The Pet Abduction Act 2024 gained Royal Assent on 24 May 2024, with the law coming into force on 24 August 2024. As animal welfare is a devolved matter, the Act applies to England and has been extended to Northern Ireland; the Welsh government chose not to go forward with similar measures at this time. In Scotland, the Proposed Dog Abduction (Scotland) Bill (2023) 'would take account of considerations such as the emotional impact and dog welfare'.

This change to the legislation in England but not Wales brings us to consider what is species justice and whether the current approach to tackling companion animal theft contributes to species justice or could the approach be adapted.

Theoretical Framework

We propose that non-human animals could and should be more central in the systems and processes when crimes are directed at them. To achieve justice for all species, the legal and criminal justice systems need to challenge speciesism and adopt a species justice approach to justice. This would include consideration of whether the non-human animal's interests are central to proposed SCP strategies, which as we discuss here, are the main strategies for preventing dog theft. Some thefts, for example, could be liberations if the person has grounds to believe the current owner is abusive.

Species Justice

Species justice has two levels (White, 2008), each of which needs to be addressed to improve the lives of non-human animals. The first level of harm is that endured and inflicted upon individual non-human animals (White, 2008). In the context of pet theft, this means the individual cat or dog who has been forcibly taken from their home and family. That individual non-human has experienced harm and trauma, and this singular form of suffering deserves redress by society as part of achieving justice. The second level of harm is the institutional or systematic exploitation of whole species or all non-humans (White, 2008). Factory farming where pigs and other non-human animals are routinely exploited is one example of many of this level of harm. Regarding pet theft, institutional exploitation is evident in at least two ways. The first way relates to the issue of non-human animals only having property status. The commodification of sentient beings through their marketisation as products that can be bought and sold is proof of institutional, systematic exploitation. The second way that non-human animals are systematically exploited is their status as 'pets' or more recently 'companions'. Not all 'owners' of non-human animals treat them well or provide good lives for them. Whilst many people think of their pets as members of the family, this may not always be the case (see Greenebaum, 2004, among many others). There are also more radical ideas that non-human animals perhaps should never be kept as pets and that the relationship is inherently exploitative, with distinctions being made between breeding and rescuing (Francione, 2007; Meijer, 2019). Thus, the very practice of having companion animals is one that needs to be analysed with the aim of ensuring that the systematic keeping of non-human animals in our homes is not one that inadvertently or directly harms them. This is an example of placing the non-human at the centre of consideration and such an approach would also change the development of crime-prevention strategies.

Situational Crime Prevention

Clarke (1980) proposed that in societal efforts to reduce crime, the more realistic—yet undervalued—way of preventing crime would be to consider and disrupt the choices and decisions made by people who commit crime. The situation or environment around people could be changed in ways to prevent crime. These changes would reduce the physical opportunities for committing crime as well as increase the chances of the person being caught (Clarke, 1980). This 'situational crime prevention' is linked closely with Cohen and Felson's (1979) theorising about a routine activity theory of offending behaviour. They suggested that there are three spatial and temporal elements to criminal events—a potential offender, who is capable of committing crime, a suitable target or victim, and the absence of a capable form of protection or a guardian. Felson and Clarke (1998) theorise that the target is suitable to an offender if the target has value, there is inertia to obtain the target, the target is visible, and is accessible (VIVA—value, inertia, visible, accessible). When these three elements overlap (and the VIVA criteria are met), crime occurs. Thus, disrupting any of these elements will contribute to crime prevention. Following on from this theorising, scholars proposed specific ways for this disruption to occur. Perhaps the most influential are the 25 techniques drafted by Cornish and Clarke (2003). Their categories are grouped into: increase the effort, increase the risks, reduce the rewards, reduce provocations, and remove the excuses (see Figure 12.2). Under each of these five groups are five techniques.

In the case of pet theft, we suggest that SCP is at the heart of official responses. In the forthcoming analysis section, we provide evidence of the SCP strategies recommended to the public. Whilst these efforts may indeed prevent dogs and other companion animals from being abducted, they do not address the underlying causes or motivations for the abduction outlined earlier. As Wortley (2010) notes, SCP approaches do not address social reform or offender rehabilitation and are only concerned with the immediate circumstances of the crime.

Methodology

To further understanding of the nature and scope of pet theft in England and Wales and the formal responses to the crime, we conducted a mixed-methods study. The first method was a literature review of the relevant academic and grey literature. Next, a series of FOI requests were made to the 43 regional police forces in England and Wales to gather data on the annual number of dog theft crimes in 2015–2021 (Allen, 2020; Allen et al, 2022). FOI data for cat theft crimes from 2015 to 2021 were collected by voluntary group Pet Theft Awareness and published as the Cat Theft Report 2022. This chapter brings together dog and cat theft data for the 43 regional police forces in England and Wales for the first time. Following Selby-Fell and Allen (2021), we approach these data with caution. There is likely under-reporting to the police; police systems capture pet theft under a range of crime types; and the parameters/search techniques that each force has used to collate their figures are not clear; also, various police forces could not provide comparable data for recorded dog and cat thefts for the requested timeframe.

Finally, we gathered information from government, police, and charity sources advising the public on how to reduce the risk that their non-human companion animal will be stolen. The content of these materials was analysed under the framework of species justice and SCP. The next section sets out the findings of the data collection.

Increase the Effort	Increase the Risks	Reduce the Rewards	Reduce Provocations	Remove the Excuses
1. Target hardening • Steering column locks • Anti-robbery screens • Tamper-proof packaging	***6. Extend guardianship*** • Take routine precautions • 'Cocoon' neighbourhood watch	***11. Conceal targets*** • Off-street parking • Gender-neutral phone directories • Unmarked bullion trucks	***16. Reduced frustrations and stress*** • Efficient queues and polite service • Expanded seating • Soothing music/muted lights	***21. Set rules*** • Rental agreements • Harassment codes • Hotel registrations
2. Control access to facilities • Entry phones • Electronic card access • Baggage screening	***7. Assist natural surveillance*** • Improved street lighting • Defensible space design • Support whistleblowers	***12. Remove targets*** • Removable car radio • Women's refuges • Pre-paid phone cards for pay phones	***17. Avoid disputes*** • Separate enclosures for rival soccer fans • Reduce crowding in pubs • Fixed cab fares	***22. Post instructions*** • 'No parking' • 'Private property' • 'Extinguish camp fires'
3. Screen exits • Ticket needed for exit • Export documents • Electronic merchandise tags	***8. Reduce anonymity*** • Taxi driver IDs • 'How's my driving?' decals • School uniforms	***13. Identify property*** • Property marking • Vehicle licensing and parts marking • Cattle branding	***18. Reduce emotional arousal*** • Controls on violent pornography • Enforce good behaviour on soccer field • Prohibit racial slurs	***23. Alert conscience*** • Roadside speed display boards • Signatures for customs declarations • 'Shoplifting is stealing'
4. Deflect offenders • Street closures • Separate bathrooms for women • Disperse pubs	***9. Utilise place managers*** • CCTV for double-decker buses • Two clerks for convenience stores • Reward vigilance	***14. Disrupt markets*** • Monitor pawn shops • Controls on classified ads • License street vendors	***19. Neutralise peer pressure*** • 'Idiots drink and drive' • 'It's OK to say no' • Disperse troublemakers at school	***24. Assist compliance*** • Easy library check-out • Public lavatories • Litter bins
5. Control tools/weapons • 'Smart' guns • Disabling stolen mobile phones • Restrict spray paint to juveniles	***10. Strengthen formal surveillance*** • Red light cameras • Burglar alarms • Security guards	***15. Deny benefits*** • Ink merchandise tags • Graffiti cleaning • Speed humps	***20. Discourage imitation*** • Rapid repair of vandalism • V-chips in TVs • Censor details of modus operandi	***25. Control drugs and alcohol*** • Breathalysers in pubs • Server intervention • Alcohol-free events

FIGURE 12.2 Situational crime prevention measures (Cornish & Clarke, 2003).

TABLE 12.1 Results of FOI Requests Regarding Pet Thefts 2015–2021

Of the 43 regional police forces in England and Wales, 28 provided comparable data for total dog/cat theft offences** for 2015–2021.*

	2015	*2016*	*2017*	*2018*	*2019*	*2020*	*2021*
Dog theft offences	1119	1157 (+3%)	1315 (+14%)	1345 (+2%)	1264 (−6%)	1358 (+7%)	1700 (+25%)
Cat theft offences	119	186 (+56%)	230 (+24%)	300 (+30%)	352 (+17%)	383 (+9%)	528 (+38%)

*15 forces could not provide complete datasets.

**This data does not show the total number of stolen animals.

FOI police force data requested by Dr Allen (for dogs) and the Pet Theft Awareness team (for cats).

Findings

The figures in Table 12.1, as mentioned, are only part of the picture. Not all forces provided data, not all people who have a pet stolen report it, and pet theft may get lost during the commission of other crimes taking place at the same time. For cats in particular, people may under-report as it is often more difficult to know if a cat has been stolen or if they are temporarily or permanently living somewhere else since most cats in the UK are let outdoors (Nurse & Ryland, 2014). The agency of cats who tend to be freely roaming makes it even more difficult to know if they have been stolen as compared to dogs who tend to be under continual control. As we indicated earlier, whilst the overall percentage compared to the population of pets in the UK may seem small, it is a form of harm to both companion animals and people. Police forces seem to recognise this as they produce dedicated material to contribute to preventing pet theft.

We collected 12 guides from 2022 or undated about pet theft from police constabularies and charity organisations. These are not an exhaustive list of guides, but rather the most prominent ones identified from Internet searches and our work on the Pet Theft Reform campaign. The guides analysed have a significant overlap in their content as seen in Table 12.2. Of importance is that all guides are focused on dog theft and not on other species kept as companion animals.

Other advice contained within some of the guides included not walking your dog alone, not walking your dog in quiet areas, and not letting children walk your dog (Blue Cross b, No date; Dog Lost, No date; Suzy Lamplugh Trust, No date). Blue Cross b (No date) also recommended getting your dog spayed or neutered as this may be why some dogs are targeted for abduction (breeding). They also suggest carrying a personal alarm, having your dog's collar fitted with a GPS tracker, and having CCTV on your property. The Suzy Lamplugh Trust (No date) advises not to have visible signs through your house windows that you have a dog and that when going for a walk, people should share their live location with a friend.

TABLE 12.2 Common Preventative Measures in Dog Theft Guides

Category	*Actions*
Registration	Microchip your dog
	Have a collar with an identification tag on your dog
	Have up to date photographs of your dog
At Home	Never leave your dog in your garden unattended
	Have a secure garden or kennel where your dog cannot get out
	Have a security system with alarms linked to gates on your property and kennel
Out and About	Never leave your dog alone in the car
	Never leave your dog outside shops etc.
	Make sure your dog has good recall (they come back when you call them)
	Mix up your walks
	Be suspicious of strangers asking a lot of questions about your dog
	Be careful what you post about your dog on social media
	Check that dog walkers and kennels are reliable
If Stolen	Call the police
	Call the Local Authority/Dog warden
	Call the microchip company
	Place information about your stolen dog at veterinarians and local notice boards
	Visit places with lots of dogs to look for your dog and tell other dog 'owners'
	Post on 'missing animal' social media groups about your stolen dog

Sources: (Battersea, No date; Blue Cross b, No date; Crimestoppers, No date; Dog Lost, No date; Dogs Trust, No date; Metropolitan Police, 2022; Neighbourhood Watch, 2022; Norfolk Police, No date; PAWS, 2022; RSPCA, 2022; Sussex Police, 2022; Suzy Lamplugh Trust, No date)

Analysis

Gaining a more accurate picture of pet theft would be beneficial to better understand the crime as well as in developing prevention measures. One way to improve the accuracy of the data is to have a specific offence as described earlier. A specific crime for pet theft could then be recorded by each of the constabularies. Another way of achieving this is to adopt the International Classification of Crime for Statistical Purposes (ICCS) created by the United Nations Office on Drugs and Crime, which, in addition to dozens of crime classifications, has a specific classification for pet theft (50229—UNODC 2015). This classification does, however, also list pets as property, but would at least contribute to more statistics. In turn, better data can inform crime-prevention measures.

Currently, SCP (see Figure 12.2) is *the* approach to preventing pet theft. A majority of the crime-prevention measures offering pet theft—really dog theft—guidance fall into the category of 'Increase the Effort' particularly '1. Target hardening' and '2. Control access to facilities' (see Cornish & Clarke, 2003). Securing gardens and not leaving your dog unattended at home or in the community fit in these categories. Other recommended preventative actions 'Increase the Risks'. By having alarms, CCTV, or walking with other people, this '6. Extends guardianship', '7. Assist natural surveillance', and '10. Strengthen formal surveillance'. Hiding evidence that you have a dog and again, not having them unattended, fall into the category of 'Reduce the Reward' by '11. Conceal target' and '12. Remove target'. All the actions in Table 12.2 under registration—microchipping, having a collar with a tag, and having photographs—also fall under 'Reduce the Reward'. These three actions serve to '13. Identify property'. This reinforces the status of dogs as property as does the overall guidance to prevent their theft since it is the same guidance that would be given to protect inanimate objects that are property.

In relation to the three spatial and temporal elements that contribute to crime commission (Cohen & Felson, 1979), the actions suggested under these SCP measures predominantly address the target and to a lesser degree the guardian or protection. The actions also only address particular aspects of the target as proposed by Felson and Clarke's (1998) VIVA. Increasing the effort actions challenge the inertia behind some crimes. For instance, by making it more difficult to snatch a dog from a garden by having secure gates and alarms, an opportunistic kidnapping of a dog might be thwarted. Similarly, not leaving evidence of having a dog in the windows of your house reduces their visibility and thus the likelihood that your house would be targeted by a dog-napper. Also, regarding targets, accessibility to stealing dogs is reduced when they are not left alone in gardens, cars, and outside shops. However, the SCP measures do not appear to relate to the last element of targets—value. Monetary value for dogs is a complex mix of their material and emotional worth to humans. Arguably though, the value within VIVA does not completely capture value beyond financial value. Having a family member abducted has much different value implications than having 'property' stolen.

For the most part, whilst the actions proposed might disrupt the offender's behaviour temporarily, overall—as criticised by Wortley (2010) as mentioned earlier—the suggestions do not address why dogs and other species are targeted specifically for theft. Thus, for theft of a sentient family member, the recommendations to protect companion animals do not unpack their commodification and exploitation. Further understanding of both elements would seem to be crucial to not only prevent pet abduction in the moment but also reduce it from taking place at all. The suggestions for preventing pet theft then do not contribute to achieving species justice. The status of pets is not challenged, only dogs are the focus of guidance, and that guidance limits the lives of dogs. Throughout the SCP measures, dogs continue to be treated as property. The actions to protect them, as mentioned, are not different to actions recommended to protect other forms of property, except for perhaps recall and neutering. As repeated, the guidance says nothing of other companion animals. This perhaps speaks to some amount of speciesism that dogs are the sole focus rather than advising people on how to protect all non-human animals from abduction. The guidance also limits the lives of dogs and/or subjects them to particular procedures. Never being able to leave your dog alone in a garden or to wait outside of a shop alters the way people and non-human animals go about their routines. Whilst admittedly these changes are not harmful, it is worth examining the need to change behaviours to prevent potentially being a victim. Some actions are more invasive, like microchipping. Again, this may not be all that physically harmful, but may contribute to reinforcing non-human animals' status as property.

An interesting insight from the dog theft guides is the advice after a dog has been stolen. The usual advice is given—call the police—but also people are expected to help solve the crime themselves. This is evident in the advised outreach to local authorities, vets, notice boards, dog parks, and social media. There is no obvious equivalent or parallel to other types of crime where the victim would be advised to become so involved in recovering stolen 'property'. We can only speculate that this is because the police and society recognise that companion animals mean more to us than other 'property', but that police resources do not exist or will not be allocated to dedicate more time to finding abducted pets—which itself conveys anthropocentrism.

Conclusion

A new approach to pet theft is needed both in terms of legislative status and protection and in terms of crime prevention. In England and Wales, companion animals' legal status and the approach to their protection is speciesist and has not taken into consideration the suffering of the individual animal or their systematic exploitation. The changes to legislation in England are a step in the right direction of recognising that pets are not property but family members - this will not be the case in Wales. It is not clear that there are additional SCP approaches that could be incorporated which would place the animal at the centre of consideration. The SCP framework overall seems to be anthropocentric (disrupting human behaviour in the moment) and this does not seem to be conducive to more species inclusive actions. Whilst SCP can be useful in disrupting opportunistic crimes and these efforts may indeed prevent some pet theft, SCP needs to be supported by other crime-prevention approaches that address the speciesism and anthropocentrism of society. Education of young people about animal suffering, for instance, is something that may contribute to better treatment of non-humans in general.

With or without a new approach to pet theft, there still needs to be improvement of data collection about pet theft. This can be better recording of existing statutes or a more standardised approach such as that proposed by UNODC (2015). In the context of England and Wales, an offence code for pet theft/abduction could be created through Home Office Counting Rules to provide 'a national standard for the recording and counting of "notifiable" offences recorded by police forces in England and Wales' (Home Office, 2022). Either way more accurate data will give a better understanding of the nature and scale of pet theft. What we do know about pet theft currently paints a bleak picture of non-human animal and human suffering for money and blood sports. The non-human animals who we bring into our homes deserve protection.

References

Allen, D. (2020). Written evidence from Dr Daniel Allen (PEE0001) 12 October 2020 (HC 546). Available at: https://committees.parliament.uk/writtenevidence/12901/default/. Accessed 14 September 2022.

Allen, D., Arathoon, J., & Selby-Fell, H. (2022). Experiences of dog theft and spatial practices of search/ing. *The Geographical Journal*, *188*(4), 518–533. Available at: https://doi.org/10.1111/geoj.12474. Accessed 12 September 2022.

Allen, D., Peacock, A., & Arathoon, J. (2019). Spatialities of dog theft: A critical perspective. *Animals*, *9*(5), 209. Available at: https://doi.org/10.3390/ani9050209. Accessed 12 September 2022.

Animal Welfare (Kept Animals) Bill. (2021–2022). Available at: https://publications.parliament.uk/pa/bills/cbill/58-03/0002/220002.pdf. Accessed 12 September 2022.

Battersea (No date). *Dog theft advice*. Available at: https://www.battersea.org.uk/pet-advice/dog-advice/dog-theft-advice?gclid=EAIaIQobChMIjrjl09e–AIVioBQBh2NggL-EAAYAyAAEgIsLvD_BwE. Accessed 2 August 2022.

Beirne, P. (2007). Animal rights, animal abuse and green criminology. In P. Beirne, & N. South (Eds.), *Issues in green criminology* (pp. 55–83). Willan.

Blue Cross a. (No date). *Take the lead on dog theft*. Available at: https://www.bluecross.org.uk/campaign/take-the-lead-on-dog-theft. Accessed 2 August 2022.

Blue Cross b. (No date). *Ways to avoid dog theft*. Available at: https://www.bluecross.org.uk/advice/dog/how-to-protect-against-dog-theft?gclid=EAIaIQobChMIjrjl09e–AIVioBQBh2NggL-EAAYAiAAEgK_c_D_BwE. Accessed 2 August 2022.

Clarke, R. (1980). 'Situational' crime prevention: Theory and practice. *British Journal of Criminology*, *20*(2), 136–149.

Cohen, L., & Felson, M. (1979). Social change and crime rate trends: A routine activity approach. *American Sociological Review*, *44*, 588–608.

Cornish, D., & Clarke, R. (2003). Opportunities, precipitators and criminal decisions: A reply to Wortley's critique of situational crime prevention. In M. Smith, & R. Clarke (Eds.), *Theory for practice in situational crime prevention (crime prevention studies)* (pp. 41–96). Lynne Rienner Publishers.

Crimestoppers. (No date). *Dog theft*. Available at: https://crimestoppers-uk.org/campaigns-media/campaigns/dog-theft. Accessed 2 August 2022.

DogLost. (No date). *Dog theft prevention campaign pamphlet*.

Dogs Trust. (No date). *Pet theft: How to prevent it and what to do if your dog is stolen*. Available at: https://www.dogstrust.org.uk/help-advice/advice-for-owners/dog-theft-advice. Accessed 2 August 2022.

Felson, M., & Clarke, R. (1998). *Opportunity makes the thief. Practical theory for crime prevention* (Police Research Series, Paper 98). London: Home Office, Policing and Reducing Crime Unit.

Francione, G.L. (2007). *Animals, property and the law*. Temple University Press.

Greenebaum, J. (2004). It's a Dog's life: Elevating status from pet to 'Fur Baby' at yappy hour. *Society & Animals*, *12*, 117–35.

Harris, L. (2018). Dog theft: A case for tougher sentencing legislation. *Animals*, *8*(5), 78–86. Available at: https://doi.org/10.3390/ani8050078. Accessed 12 September 2022.

Home Office. (2021). *Pet theft taskforce: Policy paper*. Available at: https://www.gov.uk/government/publications/pet-theft-taskforce-report. Accessed 12 September 2022.

Home Office. (2022). *Counting rules for recoded crime*. Available at: https://www.gov.uk/government/publications/counting-rules-for-recorded-crime. Accessed 24 November 2022.

Lord Justice Holroyde. (2021). Written response from Chair of Sentencing Council to Tom Hunt MP and Siobhan Baillie MP – 19 February 2021.

Meijer, Eva. 2019. *When animals speak: Toward an interspecies democracy*. New York University Press.

Metropolitan Police. (2022). *Protecting your dog from theft*. Available at: https://www.met.police.uk/police-forces/metropolitan-police/areas/campaigns/2021/protecting-dog-theft/. Accessed 2 August 2022.

Moore, A. (2005). Defining animals as crime victims. *Journal of Animal Law*, *1*(1), 91–108.

Neighbourhood Watch. (2022). *Dog theft*. Available at: https://www.ourwatch.org.uk/protectyourpooch. Accessed 2 August 2022.

Norfolk Police. (No date). *Dog theft*. Available at: https://www.norfolk.police.uk/sites/norfolk/files/dog_theft.docx.pdf. Accessed 2 August 2022.

Nurse, A., & Ryland, D. (2013). *Cats and the law: A research report for the feline advisory bureau*. Middlesex University/University of Lincoln.

Nurse, A., & Ryland, D.. (2014). *Cats and the law: A plain English guide*. Available at: http://www.thecatgroup.org.uk/pdfs/Cats-law-web.pdf. Accessed 2 August 2022.

Nurse, A., & Wyatt, T. (2020). *Wildlife criminology*. Bristol University Press.

Pet Abduction Bill. (2023). Available at: https://bills.parliament.uk/bills/3549/stages. Accessed 14 January 2024.

Pet Food Manufacturers' Association (PFMA). (2022). *Pet population 2022*. Available at: https://www.pfma.org.uk/news/new-pfma-pet-population-data-highlights-pet-peak-but-the-number-of-owners-giving-up-their-pet-is-huge-concern-. Accessed 8 August 2022.

Proposed Dog Abduction (Scotland) Bill. (2023). Available at: https://www.parliament.scot/bills-and-laws/proposals-for-bills/proposed-dog-abduction-scotland-bill. Accessed 3 April 2024.

Royal Society for the Prevention of Cruelty to Animals (RSPCA). (2022). *Preventing pet theft*. Available at: https://www.rspca.org.uk/adviceandwelfare/pets/theft. 2 August 2022.

Ryder, R. (1971). Experiments on animals. In S. Godlovitch, R. Godlovitch, & J. Harris (Eds.), *Animals, men and morals* (pp. 41–82). Victor Gollanz.

Selby-Fell, H., & Allen, D.. (2021). *Dog theft: What can we infer from the evidence so far?* Available at: http://www.pettheftreform.com/research. Accessed 12 September 2022.

Sentencing Council. (2022). *Theft – general*. Available at: https://www.sentencingcouncil.org.uk/offences/magistrates-court/item/theft-general/. Accessed 12 September 2022.

Sussex Police. (2022). *Operation collar*. Available at: https://www.sussex.police.uk/police-forces/sussex-police/areas/campaigns/campaigns/op-collar/. Accessed 2 August 2022.

Suzy Lamplugh Trust. (No date). *Dog theft and personal safety*. Available at: https://www.suzylamplugh.org/dog-theft-personal-safety. Accessed 2 August 2022.

Theft Act. 1968. Available at: https://www.legislation.gov.uk/ukpga/1968/60/contents. Accessed 12 September 2022.

United Nations Office on Drugs and Crime (UNODC). (2015). International Classification of Crime for Statistical Purposes (ICCS) Version 1.0). United Nations.

White, R. (2008). *Crimes against nature: Environmental criminology and ecological justice*. Willan.

Wortley, R. (2010). Critiques of situational crime prevention. In B. Fisher and S. Lab (Eds.), *Encyclopedia of victimology and crime prevention*. Sage.

13

"THEY HAVE LITERALLY GIVEN UP ON LIFE"

A Review of the Experiences of Nonhuman Animals Subject to Reproductive Violence and Coercion on Factory and Puppy Farms[1]

Stacy Banwell and John Walliss

In this chapter, we argue that reproductive violence and reproductive coercion take place on factory and puppy farms. Reproductive violence refers to violence that violates a person's reproductive autonomy or to violence that is directed against an individual due to their reproductive capabilities (Grey, 2017). Allied to this is reproductive coercion. This behavior includes, among other things, coercing someone to become pregnant, interfering with birth control and/or coercing an individual to continue or terminate a pregnancy. It is typically regarded as a type of intimate partner violence (Miller & Silverman, 2010). However, in this chapter we demonstrate that this type of violence is committed by humans against nonhuman animals (hereafter NHA) on a daily basis. In terms of factory farms, while both acts of violence and coercion are present across the meat, egg, and dairy industries, due to space constraints, we will limit our discussion here to the dairy industry. And given our focus on NHA we will draw on Kelty-Huber (2015, p. 4) who uses the term reproductive violence to refer to "the systematic exploitation; physical and mental violence, and trauma experienced by female farmed animals" whose reproductive systems are controlled within the animal-industrial complex. Noske (1997) coined the term animal-industrial complex to refer to the industrial domination and exploitation of NHA as commodities in capitalist societies. For Noske (1997, p. ix) the animal-industrial complex is the result of "human domestication of animals," with current industrial production demonstrating "poignantly the extent to which humans have come to dominate and exploit animals as mere resources." The case studies reviewed in this chapter take place within the broader animal-industrial complex.

To capture the gravity of the violence inflicted on NHA across factory and puppy farming, following Banwell (2023), we frame these acts of reproductive violence and coercion as crimes against humanity. Large-scale, serious crimes that violate international law – that is, core international crimes and atrocity crimes –

DOI: 10.4324/9781003441908-17

include crimes against humanity. Atrocity crimes "…are considered to be the most serious crimes against humankind"; acts believed to "…affect the core dignity of human beings" (Framework of Analysis for Atrocity Crimes, 2014, p. 1). To be clear, we are not comparing the experiences of humans with NHA; rather, we are arguing that routine animal husbandry practices and breeding practices on factory and puppy farms, respectively, are often shrouded by euphemistic language when they ought to be framed and taken as seriously as crimes against humanity. Furthermore, and perhaps most crucially, we will demonstrate how NHA meet the criteria outlined within the following crimes against humanity: rape, forced pregnancy, and sexual slavery. We will return to this in due course. First, a note on the position taken in this chapter. From a species justice perspective, both authors advocate for the elimination of all violence and exploitation of NHA. We believe that all vertebrate and invertebrate NHA with brains and nervous systems should be regarded as sentient beings. We contend that NHA should be granted passive legal personhood. Moving beyond the welfarist approach – which seeks to improve the welfare and treatment of NHA (see for example Singer, 1975) – we support the rights-based approach, which contests the property status of NHA (see for example Regan, 1983; Wise, 2000). For our purposes in this chapter, we will narrow our focus to Dunayer's (2013) notion of new speciesism and her blueprint for a nonspeciesist law. Among other things, this law would "…accord all sentient beings a legal right to liberty – physical freedom and bodily integrity" (Dunayer, 2013, p. 37). As a result, NHA would be released from captivity. To put it bluntly: It would be illegal to hold NHA captive. This standpoint is pertinent to the argument that we will present in due course.

The chapter proceeds as follows. We start by outlining the three key elements of crimes against humanity before explaining why we believe NHA, within factory and puppy farming, are victims of rape, forced pregnancy, and sexual slavery. After outlining each of these crimes in detail we then move on to our case study examples to illustrate how bovine and canine bodies are subject to these acts of reproductive violence and reproductive coercion. In both cases, the "widespread" and "systematic" nature of these crimes, as well as the physical and emotional harm experienced by NHA, is laid bare. Drawing on empirical research on puppy farms in the UK, we close by reinforcing our alignment with the animal rights-based approach.

Reproductive Violence and Coercion as Crimes against Humanity

To reiterate, reproductive violence and coercion are acts of violence that target and interfere with bodies capable of becoming pregnant and bodies capable of reproduction through semen production. Both take place on factory and puppy farms. The nature, scale, and gravity of these acts lead us to argue that these can be understood as crimes against humanity. The crimes against humanity that we will address are rape, forced pregnancy, and sexual slavery. Crimes against humanity do not have to be carried out during an armed conflict, they can also occur in peacetime societies. It is worth pausing here to (1) outline our reasoning for framing these acts as crimes against humanity and (2) unpack the specific crimes we are interested in before examining them in the context of factory and puppy farming.

Crimes against humanity contain three key elements: the *physical*, *contextual*, and *mental element*. Listed as a *physical element*, and relevant to our discussion here, is "grave forms of sexual violence and other humane acts." The *contextual element* refers to acts that are "committed as part of a widespread or systematic attack directed against any civilian population." This includes, but is not limited to, "rape, sexual slavery…forced pregnancy… *or any other form of sexual violence of comparable gravity*" (see Element g of Article 7 of the Rome Statute of the International Criminal Court, 1998, p. 4). Finally, the *mental element* stipulates that: "[T]he perpetrator must also act with knowledge of the attack against the civilian population and that his/her action is part of that attack" (UN Office on Genocide Prevention and the Responsibility to Protect – Crimes against Humanity, n.d., para 9). Following Banwell (2023), we regard NHA as civilians. In her recent monograph, *The War Against Nonhuman Animals: A Non-speciesist Understanding of Gendered Reproductive Violence*, Banwell (2023), as the title suggests, argues that humans are currently engaged in a species war against NHA. The book offers practical and operational guidelines on how we might protect NHA from the violence(s) of war. Specifically, Banwell provides an original analysis of non-international armed conflict as outlined in International Humanitarian Law (IHL). Currently, IHL protects NHA during armed conflict when they are classified as property. Article 53 of the Geneva Conventions – *The Protection of Civilian Persons in Time of War* – prohibits destruction to the personal property of "protected persons." This includes NHA (see Roscini, 2017, p. 8). To put it another way, the destruction of NHA, when they are classified as the private property of protected persons, is considered a War Crime and a breach of the Geneva Conventions (Roscini, 2017). Rather than view NHA as property, Banwell (2023) makes the case that NHA should be treated as civilians and that existing protections within IHL should be applied to them as protected persons not the *property* of protected persons.

Why Frame These Acts as Crimes against Humanity?

The violence(s) inflicted on NHA, within factory and puppy farming, meet the *physical*, *contextual*, and *mental* elements of crimes against humanity described earlier. Both the *physical* and *contextual* elements will become clear when we review the *extent* and *nature* of reproduction within factory and puppy farming. The *mental* element is best explored in relation to animal husbandry practices. For Cusack (2013, p. 25), animal husbandry is a euphemism for "…rape and sexual slavery/trafficking." It involves repeated nonconsensual penetration, either with hands or objects of the NHA, and nonconsensual insemination for the purposes of reproduction (Bourke, 2020; Cusack, 2013). Cusack (2013) believes that the term rape describes these processes more accurately and should replace the euphemistic phrase "animal husbandry." For us, this euphemistic phrasing suggests prior knowledge of the attack. Furthermore, the overuse of antibiotics to curtail disease and illness among NHA facilitates industrialized reproduction within the animal-industrial complex. Put simply, the measures put in place to increase meat, egg, and dairy production, as well as minimize the risk of disease and illness, reveal knowledge of the attack (Banwell, 2023). In the following, we outline the three crimes against humanity that form the basis of this chapter: rape, forced pregnancy and sexual. Following this, we demonstrate how they are present within factory and puppy farming.

Rape

While the language of rape laws varies, in the US, Australia, and across EU member states, all refer to penetration by a penis/sex organ or an object. In the UK, however, rape refers to nonconsensual penetration with a penis. In this instance we might use the term *assault by penetration.* This is when a person penetrates another person's vagina or anus with any part of the body other than a penis, or by using an object, without the person's consent (Sexual Offences Act, 2003). Whichever definition or term we use – rape, assault by penetration – the key is the nonconsensual nature of the act.

Forced Pregnancy

There are three cumulative elements contained within the crime of forced pregnancy. First, "…the victim must be unlawfully confined by the perpetrator…" Second, "…the victim must have been forcibly made pregnant." And third, "the perpetrator acted with one of two specific intents: To affect the ethnic composition of a population, *or* to carry out other grave violations of International Law" [emphasis added] (Amnesty International, 2020, pp. 8–9; see also Boon, 2001). We will unpack each of these in more detail in the following.

Unlawful Confinement

Forced pregnancy requires that a woman be both forcibly made *and* kept pregnant, often through confinement. The International Criminal Court (ICC) definition does not specify the exact duration of the unlawful confinement of the woman. However, it is reasonable to assume that the timeframe we are concerned with here is around 9 months (Boon, 2001).

The Forcible Impregnation of a Woman

There have been recent calls to replace the term "woman" with the term "pregnant person" to include all individuals capable of becoming pregnant. As pointed out by Amnesty International, "…if the essence of the crime is the denial of autonomy over a pregnancy by means of unlawful confinement… then the crime must be applied to all pregnant persons subjected to such treatment" (2020, pp. 14–15).

The Intention of the Perpetrator

Forced pregnancy, like rape, *can* constitute the crime of genocide. However, as per the definition, the intention can also be based on "carrying out other grave violations of international law." Indeed, under article 7 of the Rome Statute of the ICC, crimes against humanity also refer to: "Other inhumane acts of a similar character intentionally causing great suffering, or serious injury to body or to mental or physical health" (Rome Statute of the International Criminal Court, 1998, p. 5).

Sexual Slavery

The key elements of sexual slavery, as defined by the ICC *Elements of Crime* (2013, p. 6), are as follows:

> The perpetrator exercised any or all of the powers attaching to the right of ownership over one or more persons, such as by purchasing, selling, lending or bartering such a person or persons, or by imposing on them a similar deprivation of liberty. The perpetrator caused such person or persons to engage in one or more acts of a sexual nature.

Now that we have outlined the key characteristics of the crimes against humanity under review, we will apply these to the experiences of bovine and canine bodies within factory and puppy farming respectively. Our review of the dairy industry draws on empirical research conducted on dairy farms, auctions yards, and sanctuaries in the US (see Gillespie, 2014; 2018); as well articles and research by PETA (n.d.) and Compassion in World Dairy Farming (n.d.). While our analysis of puppy farming draws on Walliss' (2024) recent research with animal caregivers in the UK who have adopted ex-breeding dogs.

The Rape and Forced Impregnation of NHA

As noted earlier, under nonspeciesist law (Dunayer, 2013), the confinement of NHA is illegal. NHA on factory and puppy farms are held in confinement during the 9-month timeframe outlined earlier. Therefore, we make the case that NHA meet the first criterion of the ICC definition of forced pregnancy. In terms of the second criteria – the forcible impregnation of a "pregnant person" – if we accept that NHA are legal persons; biologically capable of becoming pregnant, then they too, as "pregnant persons," can be victims of this crime (Banwell, 2023). Taking this on board we contend that NHA meet the second criteria of the definition of forced pregnancy. Finally, as noted earlier, forced pregnancy is not always used as an act of genocide. It is used to carry out other serious crimes. Based on this we adopt the following revised definition of forced pregnancy put forward by Banwell (2023, p. 97):

> The unlawful confinement of a person forcibly made pregnant, with the intent of carrying out inhumane acts that cause great suffering or serious injury to body or to mental or physical health. Rape, sexual slavery, and other forms of sexual violence are used in the commission of this act. All of which are grave violations of International Law.

Drawing on this interpretation of "other grave violations of International Law" (see definition of forced pregnancy mentioned earlier), we argue that NHA meet this third criteria of forced pregnancy. This is illustrated here in our examination of factory and puppy farming.

The Dairy Industry

The production of milk for agribusiness is the result of the repeated rape and forcible impregnation of dairy cows (Brown, 2016; Cusack, 2013; Gillespie, 2014; 2018). Nonconsensual invasive methods of penetration (both vaginal and rectal) are used to impregnate dairy cows (Cusack, 2013). This includes the use of the "rape rack." This is the device used to forcibly restrain the cow with chains while she is forcibly impregnated artificially or by a bull (Bourke, 2020; Cusack, 2013; Shuchat, 2016). The process of artificially inseminating cows is referred to as recto-vaginal rape (Vandermark, Salisbury & Boley 1951 as cited by Cusack, 2013, p. 27). During this process the worker inserts one arm into the rectum of the cow in order to locate the cervix. The farmer then uses the other arm to insert the artificial insemination gun into the cow's vagina. The cow is restrained by the rape rack during this process. The rod-like gun contains bull semen. It is forced into the cow until it reaches her cervix where the semen is injected into her uterus (Shuchat, 2016). Eighty percent of dairy cows in North America and Europe are subject to forced impregnation through artificial insemination. "Therefore," as Gillespie (2018, p. 174) highlights, "to speak about dairy production is to speak primarily about artificial insemination."

Puppy Farming

As with factory farming, repeated rape and sexual violence are also key characteristics of illegal dog-breeding establishments. These are often referred to as "puppy farms," "puppy mills," or "puppy factories." However, whereas insemination in the former occurs through artificial means, in the latter it ostensibly occurs naturally, with a sire/stud dog penetrating a bitch/dam with his penis when she is in season, thereby impregnating her. That is not to say, though, that the process is completely "natural." The time, location, and mating pair are all selected by the puppy farm operatives themselves. Neither the breeding bitch or the sire are able to give any real consent to the process. Not only are both confined within the puppy farm, but the former are also reportedly trapped in partly closed doors – makeshift "rape racks" – to ensure that they are unable to escape from the insemination process. In this way, insemination within puppy farms, while ostensibly occurring as part of a "natural" process, is clearly an act of coerced sexual violence for both the bitch and sire. This violence is experienced as an act of rape by the bitch and as an act of forced sex by the sire.

The Physical and Emotional Impact on NHA

The Dairy Industry

In addition to the physical violence inflicted on the cow, her emotional and spiritual integrity is diminished when her offspring are taken from her almost immediately after she has given birth. Within the first 12 hours of being born, 97% of calves are taken from their mothers (Brown, 2016). Mothers will cry and bellow for days, sometimes weeks, after their calves have been removed from them (Cusack, 2013; Gillespie, 2014; Shuchat, 2016). This separation of mother and child is seen as necessary within the dairy industry to meet the demands of consumers and ensure maximum profitability. Once the calf is removed, the milk can then be used for human consumption. This forced separation is a form of reproductive coercion, specifically, control of pregnancy outcomes. We regard the involuntary separation of mother and calf as an assault on bovine motherhood. Here it is instructive to draw on Halbmayr (2010, p. 30) who posits that: "…violent acts can be understood as sexualized if they are directed at the most intimate part of a person and, as such, against that person's physical, emotional, and spiritual integrity." Based on this, we frame this act of reproductive coercion as "any other form of sexual violence of comparable gravity," as outlined in element g of 7, article 7 of crimes against humanity. As a reminder, crimes against humanity include, but are not limited to "rape, forced pregnancy… *or any other form of sexual violence of comparable gravity*" [emphasis added] (Rome Statute of the International Criminal Court, 1998, p. 4).

Puppy Farms

Similarly, puppy farming produces physical and behavioral harms for both the sire and the dam. However, it is also important to recognize the harms incurred by the puppies themselves. Two recent studies from the US and the UK, respectively, found that dogs who were bred on puppy farms were significantly more likely to exhibit aggression or fear responses toward strangers, other dogs, and nonsocial stimuli when compared with dogs bought from noncommercial breeders. They also exhibited more touch sensitivity and separation-related problems (McMillan et al., 2013). The UK study also found that dogs bred on puppy farms were also more likely to suffer from both genetic disorders and infectious diseases, such as parvovirus (Wauthier & Williams, 2018).

Several factors have been highlighted within the literature that explain how puppy farms produce these negative health and behavioral outcomes in dogs (McMillan, 2017). For example, where little attention is paid to either the health or temperament of the breeding pair, any negative traits will be passed onto their offspring. Likewise, physical, and psychological stressors experienced by the pregnant dog within a puppy farm – as a consequence of her confinement and poor treatment – will also be experienced by the puppy developing inside her. This, in turn, will negatively affect the latter's development. Once born, severe stressors during the puppy's early life, coupled with inadequate socialization, are also likely to produce negative outcomes for dogs in later life. Puppies on puppy farms also have their development negatively impacted by being separated too early from their mother during weaning. Not only does this lead to extra stressors for the puppy, but the separation from its caregiver and littermates also negatively affects their socialization. The latter is attributed to the loss of the ability to learn through observing and interacting with others (McMillan, 2017). So far, we have considered rape and forced pregnancy, as well as the physical and emotional impact these crimes have on NHA. In the following section we will address sexual slavery – the third crime against humanity considered in this chapter.

Sexual Slavery on Factory and Puppy Farms

It is important to clarify the meaning of the phrase "acts of a sexual nature" as outlined within the ICC definition of sexual slavery. Sexual in the context of what we are discussing here does not relate to sexual desire or intent on the part of the human actor, as in the case of bestiality. Rather, what we are referring to here is the sexual reaction of the NHA. Before we continue, it is worth noting the difference between animal husbandry and bestiality: The goal of animal husbandry is industrialized reproduction while bestiality is rooted in sexual desire (Bourke, 2020; Cusack, 2013). Furthermore, animal husbandry is a legal practice, justified on clinical and scientific grounds. Conversely, bestiality is prohibited in most countries, including the UK and the US. This is because animals cannot consent to sex with humans, therefore sexual acts rooted in human desire are considered cruel and violent for the NHA. Let us return to the sexual reaction and experience of the NHA. Here we draw on Halbmayr's (2010) notion of sexualized violence outlined earlier. We believe that the experiences of NHA within the animal-industrial complex can be placed within this framework of sexualized violence, specifically, as demonstrated earlier, the impact on the physical, emotional, and spiritual integrity of these NHA.

Bovine bodies are subject to sexual slavery as the result of the global intensification of dairy farming. We have already outlined the cycle of rape and forcible impregnation inflicted upon dairy cows. These acts of sexual violence are carried out by those who claim ownership over the bodies of these NHA. Based on her empirical research, Gillespie (2014, pp. 1326–7) highlights how the bodies of dairy cows are treated as disposable commodities within agribusiness. Here the emphasis is on productivity, global markets, international trade, and maximizing profit. Describing the experiences of dairy cows, she emphasizes the continuous nature of the violence, stating:

> This cycle of artificial insemination, birth, and milking [is] repeated for several years until lameness, mastitis, infertility, and/or declining milk production set in. At this point, the farmer make[s] a careful calculation of her profitability as a milk producer weighed against the cost of maintaining her.

In terms of bulls, they were previously kept on their home dairy farms. However, to ensure maximum economic efficiency they are now "housed on breeding farms that specialize in semen production [which] store large tanks of each bull's semen in cold storage for later sale and use" (Gillespie, 2014, p. 1328). Indeed, "[b] ull semen," Gillespie continues, "has become an internationally traded commodity and semen is shipped from suppliers for use in impregnating cows on dairy and beef operations around the world…" (Gillespie, 2014, p. 1328). This commodity is advertised in semen catalogs where bulls are commodified for their reproductive prowess, appearance, genetic heritage, and the quality and virility of their semen.

We can look to the intensification of dairy farming for evidence of the "widespread" and "systematic" nature of the crimes outlined earlier. This adds to our claim that these acts need to be framed as crimes against humanity. Intensification refers to the increase in dairy cows per hectare of land alongside the decrease in dairy farms. For example, in 1970 there were an estimated 640,000 dairy farms in the US. This figure was 40,000 in 2017 (see Clay et al., 2020). A similar trend can be found in the UK, where dairy farms decreased from 35,000 in 1995 to 13,000 in 2017, while the number of dairy cows tripled. And in terms of milk production, "[i]n the US, the average dairy cow produces more than 7.5 gallons of milk per day. If she was producing just enough to feed her calf, a dairy cow would only produce about one gallon of milk per day" (Compassion in World Dairy Farming, n.d., para 3). To put it another way: Despite the decrease in the number of cows on dairy farms in the US in 2017, compared with 1950 (an estimated 12 million fewer), milk production has increased from "116 billion pounds of milk per year in 1950 to 215 billion pounds in 2017" (PETA, n.d., para 5). Indeed, to meet this increase in demand, cows are subject to genetic manipulation, as well as antibiotic and hormonal treatment. Their natural diets of grass are substituted with diets that are unnaturally high in protein to fulfill the demand for dairy (PETA, n.d., para 5).

We frame the "widespread" and "systematic" reproduction and sale of canine bodies within puppy farming as sexual slavery. As the operative words "farm," "mill," and "factory" suggest, the primary business model of such establishments is to produce puppies on a mass, almost industrial scale. The aim is to maximize the number of litters/puppies that are produced and then sold for profit. Within puppy farms, breeding bitches are treated merely as "breeding stock;" sexual slaves whose entire existence is defined by their ability to reproduce. While the exact size of the puppy trade in the UK is unknown, it is clear that it is an extremely lucrative business: puppies from some breeds routinely exchanging hands for several thousand pounds each. In their report for the Scottish government, for example, Wyatt et al. (2017) estimated a conservative value of the puppy trade in that country alone at £13 million. Extrapolating from this to the rest of mainland UK would place a value of £130 million on the trade (Maher & Wyatt, 2019). Similarly, a 2015 report by the BBC estimated the value of the trade in Northern Ireland at £160 million. By comparison, the value of the *illegal* trade is harder to discern. However, information gleaned from cases of dealers found guilty of the illegal importation of puppies from oversees puppy farms, estimates that dealers can earn up to £40,000 per week from this illegal trade (Yeates & Bowles, 2017). Indeed, the importation of puppies bred overseas appears to be a more lucrative trade than puppy farming *per se*. This is because it does not include any of the overheads incurred when breeding dogs (see Dogs Trust, 2018). Consequently, in recent years, criminal gangs have been increasingly attracted to smuggling puppies bred overseas, knowing that (1) there is an extremely buoyant and lucrative market for puppies and (2) that the risk of detection, and possible penalties if they are caught, are significantly smaller compared with importing drugs or weapons into the country (Maher & Wyatt, 2021).

Discussion

Earlier we noted our alliance with the animal rights-based approach. Here, drawing on Walliss' (2024) empirical research on puppy farms in the UK, we concretize our position by outlining the limitations of existing welfare approaches within the animal industrial complex. While commercial dog breeding within the UK is governed by several pieces of legislation, puppy farms routinely operate outside of the law to maximize their profits (see Maher & Wyatt, 2019). For example, the Animal Welfare Act (2006), and its equivalents in Scotland and Northern Ireland, place a duty of care on humans to ensure that they meet the welfare needs of their NHA and prevent any "unnecessary suffering." The Animal Welfare Act defines these needs as including the following: A suitable environment, a suitable diet; the ability to exhibit normal behavioral patterns; and appropriate housing (with, or apart from, other animals). Humans should also ensure that the NHA in their care are protected from pain, suffering, injury, and disease. Puppy farms contravene this legislation by failing to meet the basic needs of breeding dogs and their offspring, and in some cases, by causing their unnecessary suffering. Obvious signs of neglect were reported by individuals who had rescued or adopted ex-breeding dogs. Interviewees described the poor physical condition of the dogs in their care and attributed their (undiagnosed and untreated) chronic illnesses to the constant breeding they had been subject to on the puppy farms. Additionally, interviewees relayed the emotional impact on ex-breeding dogs. One interviewee described how her ex-breeding dog "just purely lived on fear" and how "her eyes were dull and dead…as you came near her, you could always see she was looking for any way to run, to see where she could escape away." While, in many cases, they would appear to overcome this fear, this is not always the case and some dogs, as one interviewee stated: "Never really come back from [their] time in the puppy farm."

Conclusion

We end by returning to Noske and the human domestication of NHA. Nibert (2013) prefers the term *domesecration*. He uses it to describe "…the systemic practice of violence in which social animals are enslaved and biologically manipulated, resulting in their objectification, subordination, and oppression." This process, he explains, "violates the sanctity of life of the sentient being involved" (2013, p. 12). This, as we have demonstrated in this chapter, is what takes place on factory and puppy farms. The acts of reproductive violence and reproductive coercion reviewed across both industries – rape, forced pregnancy, and sexual slavery – were framed as crimes against humanity. As both case studies illustrate, the nature and scale of the violence(s) meet the *physical*, *contextual*, and *mental element* of crimes against humanity.

We hope that this work responds to and enriches existing work within criminology that addresses the abuse of NHA. For example, in 1999 Beirne in his article, *For a Nonspeciesist Criminology: Animal Abuse as the Object of Study,* urged criminologists to develop and theorize the category of animal abuse. He concluded that "a criminology that ignores animal abuse will be a speciesist discourse utterly irrelevant to the understanding of much harm and suffering inflicted by humans on nonhuman, but nevertheless valuable, forms of life" (p. 140). In this chapter, we have highlighted animal abuse across two industries within the animal industrial complex: factory and puppy farming. We theorized that as sentient beings, NHA, precisely because of their reproductive capabilities, are subject to rape, forced pregnancy and sexual slavery. We argued that these crimes affect their core dignity and bodily integrity. In a more recent article, Flynn and Hall (2017) have called for a victimology of NHA. Enriching their argument we have extended the focus of their thesis to include core international crimes and atrocity crimes. Specifically, we have proposed a nonspeciesist interpretation of crimes against humanity as outlined by the ICC. This requires acknowledging that any being who can suffer and feel pain can become the target and victim of certain crimes against humanity. In sum, we make the case for reframing crimes against humanity as crimes against sentient beings. We apply our notion of crimes against sentient beings – rape, forced pregnancy, and sexual slavery – to bovine and canine bodies capable of becoming pregnant and capable of reproduction through semen production.

Note

1 The quote from the title is taken from an interview with "Mark" who runs a rescue center in South Africa. He is describing the state of fear that some of the ex-breeding dogs he has rescued have at the sight of humans. This fear stems from their previous experience of living on a puppy farm. He states: "When they see a human they fall down and they start shaking, no matter where they are. So, there's no flight anymore. They have literally given up in life…They accept their fate." This interview forms part of a larger empirical research project conducted by Walliss (2024). It includes 20 interviews with individuals who have either adopted or fostered a female ex-breeding dog. For more details about the methodology, see the relevant section of the forthcoming article by Walliss.

References

Amnesty International. (2020, June 30). *Forced pregnancy: A commentary on the crime in international criminal law*. https://www.amnesty.org/en/documents/ior53/2711/2020/en/

Animal Welfare Act. (2006). Retrieved April 9, 2024, from https://www.legislation.gov.uk/ukpga/2006/45/contents

Banwell, S. (2023). *The war against nonhuman animals: A non-speciesist understanding of gendered reproductive violence*. Palgrave Macmillan.

Beirne, P. (1999). For nonspeciesist criminology: Animal abuse as an object of study. *Criminology*, *37*(1), 117–148.

Boon, K. (2001). Rape and forced pregnancy under the ICC statute: Human dignity, autonomy, and consent. *Columbia Human Rights Law Review*, *32*, 658–659.

Bourke, J. (2020). *Loving animals: On bestiality, zoophilia and post-human love*. Reaktion Books.

Brown, K. (2016). *A feminist analysis of human and animal oppression: Intersectionality among species*. Western Washington University. https://cedar.wwu.edu/cgi/viewcontent.cgi?article=1234&context=scholwk

Clay, N., Garnett, T., & Lorimer, J. (2020). Dairy intensification: Drivers, impacts and alternatives. *Ambio*, *49*, 35–48.

Compassion in World Dairy Farming (n.d). *About dairy cows*. Retrieved from https://www.ciwf.com/farmed-animals/cows/dairy-cows/.

Cusack, C.M. (2013). Feminism and husbandry: Drawing the fine line between mine and bovine. *Journal for Critical Animal Studies*, *11*(1), 24–45.

Dogs Trust. (2018). *Puppy Smuggling: When will this cruel trade end?* https://www.dogstrust.org.uk/downloads/2018%20Puppy%20smuggling%20report.pdf

Dunayer, J. (2013). The rights of sentient beings: Moving beyond old and new speciesism. In R. Corbey, & A. Lanjouw (Eds.), *The politics of species: Reshaping our relationships with other animals* (pp. 27–39). Cambridge University Press.

Flynn, M., & Hall, M. (2017). The case for a victimology of nonhuman animals. *Contemporary Justice Review*, *20*(3), 299–318.

Framework of Analysis for Atrocity Crimes. (2014). https://www.un.org/en/genocideprevention/documents/about-us/Doc.3_Framework%20of%20Analysis%20for%20Atrocity%20Crimes_EN.pdf

Gillespie, K. (2014). Sexualized violence and the gendered commodification of the animal body in Pacific Northwest US dairy production. *Gender, Place & Culture*, *21*(10), 1321–1337.

Gillespie, K. (2018). *Cow with ear tag #1389*. University of Chicago Press.

Grey, R. (2017). The ICC's first 'forced pregnancy' case in historical perspective. *Journal of International Criminal Justice*, *15*(5), 905–930.

Halbmayr, B. (2010). Sexualized violence against women during Nazi "racial" persecution. In S. Hedgepeth, & S. Saidel (Eds.), *Sexual violence against Jewish women during the Holocaust* (pp. 29–44). Brandeis University Press.

International Criminal Court. (1998). *Elements of crime*. https://www.icc-cpi.int/sites/default/files/Publications/Elements-of-Crimes.pdf
Kelty-Huber, C. (2015). *Towards a framework for reproductive violence*. University of Southern Maine-USM Digital Commons. https://digitalcommons.usm.maine.edu/cgi/viewcontent.cgi?article=1059&context=etd
Maher, J., & Wyatt, T. (2019). Rural-urban dynamics in the UK illegal puppy trade: Trafficking and trade in "man's best friend." *International Journal of Rural Law and Policy*, *9*(2), 1–20.
Maher, J., & Wyatt, T. (2021). European illegal puppy trade and organized crime. *Trends in Organized Crime*, *24*, 506–525.
McMillan, F.D. (2017). Behavioral and psychological outcomes for dogs sold as puppies through pet stores and/or in commercial breeding establishments: Current knowledge and putative causes. *Journal of Veterinary Behavior*, *19*, 14–26.
McMillan, F.D., Serpell, J.A., Duffy, D.L., Masaoud, E., & Dohoo, I.R. (2013). Differences in behavioral characteristics between dogs obtained as puppies from pet stores and those obtained from noncommercial breeders. *Journal of the American Veterinary Medical Association*, *242*(10), 1359–1363.
Miller, E., & Silverman, J.G. (2010). Reproductive coercion and partner violence: Implications for clinical assessment of unintended pregnancy. *Expert Review of Obstetrics and Gynecology*, *5*(5), 511–515.
Nibert, D.A. (2013). *Animal oppression and human violence: Domesecration, capitalism and global conflict*. Columbia University Press.
Noske, B. (1997). *Beyond boundaries: Humans and animals*. Black Rose Books.
PETA UK. (n.d.). *Cow's milk: A cruel and unhealthy product*. https://www.PETA.org/issues/animals-used-for-food/animals-used-food-factsheets/cows-milk-cruel-unhealthy-product/
Regan, T. (1983). *A case for animal rights*. University of California Press.
Rome Statute of the International Criminal Court. (1998). https://legal.un.org/ilc/texts/instruments/english/conventions/7_4_1998.pdf
Roscini, M. (2017). Animals and the law of armed conflict. *Israel Yearbook on Human Rights*, *47*, 35–67.
Sexual Offences Act. (2003). Retrieved October 24, 2023, from https://www.legislation.gov.uk/ukpga/2003/42/contents
Shuchat, S. (2016, June 15). *The rape rack. Their turn: The social justice movement of our time.* https://theirturn.net/2016/06/15/2016061420160613the-rape-rack/
Singer, P. (1975). *Animal liberation: A new ethics for our treatment of animals*. New York: HarperCollins
UN Office on Genocide Prevention and the Responsibility to Protect. (n.d). *Crimes against humanity*. Retrieved August 9, 2024, from https://www.un.org/en/genocide-prevention/definition
Walliss, J. (2024). '"You don't have the right to steal my life": Exploring the harms of puppy farming on ex-breeding dogs.' *Trace: Journal for Human-Animal Studies, 10*. https://trace.journal.fi/article/view/131170
Wauthier, L.M., & Williams, J.M. (2018). Using the miniC-BARQ to investigate the effects of puppy farming on dog behaviour. *Applied Animal Behaviour Science*, *206*, 75–86.
Wise, S.M. (2000). *Rattling the case: Toward legal rights for animals*. Perseus Books.
Wyatt, T., Maher, J., & Biddle, P. (2017). Scoping research on the sourcing of pet dogs from illegal importation and puppy farms 2016–2017. https://www.gov.scot/binaries/content/documents/govscot/publications/research-and-analysis/2017/11/scoping-research-sourcing-pet-dogs-illegal-importation-puppy-farms-2016/documents/00527436-pdf/00527436-pdf/govscot%3Adocument/00527436.pdf
Yeates, J. & Bowles, D. (2017). Breeding and selling of companion animals. In J. Maher, H Pierpoint & P. Beirne (Eds.), *The Palgrave international handbook of animal abuse studies* (pp. 15–38). Palgrave Macmillan.

14

WHICH ANIMALS DID NOAH EAT? AN ANIMAL-CENTRIC FOCUS ON FOOD CRIME

Matthew Robinson

Have you ever heard of the term "food crime?" Until you picked up this book, it is unlikely that you ever have. Even in academic disciplines that study crime, such as criminology and criminal justice, there is very little focus on the subject matter (Robinson, 2017). Though it is shame, given the wide range of harms produced by the behaviors included in food crimes, the lack of focus is not surprising given that academic disciplines such as criminology and criminal justice are almost exclusively focused on street crimes, things such as murder, rape, robbery, assault, theft, motor vehicle theft, burglary, arson, and other crimes such as drug abuse and terrorism. Such acts are unquestionably serious, as they produce significant loss of life, loss of property, and also diminish quality of life (Robinson, 2020).

What if I told you that the behaviors of focus in this chapter cause harms that easily dwarf those caused by the street crimes academic criminologists tend to study? It is true: Food crimes cause more death, injury, and loss of property than all street crimes combined, and it is not even close.

In the chapter, I provide a definition of food crime and illustrate the types of behaviors that are included in it. Then, I delve into the topic state–corporate criminality, which gives us an important context in which to discuss food crimes. Next, I examine harms against animals that should be part of the food crime literature.

What Is "Food Crime"?

Food Crime Defined

A growing body of literature on crimes of and within the food industry has emerged (e.g., see Gray & Hinch, 2018). The term "Food Crime" was first used by Hazel Croall (2007, p. 206) to describe the "many crimes that are involved in

DOI: 10.4324/9781003441908-18

the production, distribution, and selling of basic foodstuffs." Later, Croall (2013, p. 167) defined food crime as the behaviors in the food industry that are "morally dubious" and also sometimes illegal, which involves "manipulation and exploitation of the planet." Gray and Hinch (2015, p. 97) add that food crimes involve "serious harms ... addressed beyond the traditional definitions of crime ..." that have "negative consequence for a variety of both human & non-human victims."

Later, these scholars suggested that food crime includes "illegal, criminal, harmful, unjust, unethical or immoral food-related" behaviors and omissions of behavior (such as poor regulation of harmful foods by the government) (Gray & Hinch, 2018, p. 12). So, it is important to note at the outset that food crime includes many legal behaviors that are "lawful but awful" (Passas, 2005). Some will question the logic of studying legal behaviors in the context of criminology or criminal justice, since the academic disciplines are traditionally focused on criminal acts (i.e., acts that violate the criminal law). Yet, a social harm perspective along the lines of Zeminology would essentially mandate examining these behaviors given the enormous harms they impose on society (Canning & Tombs, 2021; Dorling et al., 2008). The nature and extent of those harms is the subject of this chapter.

Examples of Food Crimes

As examples of food crimes, Croall (2013) discusses food poisoning caused by negligence or recklessness of food producers (also see Leighton, 2018), fraud, misleading advertisements, inequality of food availability, illegal trade and pricing in the food industry, labor exploitation, and animal cruelty (also see Beirne, 1999). Monopolization and unfair trade practices can also logically be seen as unethical and even illegal behaviors, as can wage deflation, bullying of workers, and control of prices by suppliers.

Smith et al. (2017: np) define food fraud as "the deliberate and intentional substitution, addition, tampering, or misrepresentation of food, food ingredients, or food packaging; or false or misleading statements made about a product for economic gain." These behaviors are "carried out intentionally to avoid detection by regulatory bodies or consumers." With regard to fraud, Croall examines foods that are watered down as well as companies that sell branded foods that are no better or different than generic foods. Croall also examines foods that are actually deadly based on additives or inadequate regulation for safety. Poor regulation also produces widespread food poisoning, which produces illness and occasional death.

Other unethical and potentially illegal behaviors by food companies include exploitation of labor, such as in the case of seasonal and migrant workers—described as the "new slavery" by some (Lawrence, 2008, p. 111)—and forcing employees to work in hazardous conditions. Deaths produced by hazardous conditions amount to about 54,000 a year just in the United States, far more than are murdered every year in the country (Reiman, 2020). Factory farming is not only horribly cruel to animals, but also is often harmful to employees as well as the environment, as in the case where animal waste and chemicals pollute waterways as well as the air (e.g., see Mugni et al., 2012).

Gray and Hinch (2015) also discuss different forms of potential deviant and criminal behavior in the food industry. These include low pay for farm workers, child labor and slavery in the production of food (e.g., see Coe & Coe, 2013; Schrage & Ewing, 2005), the dominant power held by agribusiness over traditional farmers (e.g., see Culp, 2005; Fairley, 1999), pesticide use on crops (e.g., see Del Prado-Lu, 2018; Goodman, 2011), patented seeds and unintentional violations of those patents, laws that tend to better protect agribusiness than farmers and consumers, and weak enforcement of existing regulations. Later, the same authors (Gray & Hinch, 2018) add a focus on acts such as forced labor and slavery in the food industry (e.g., see Hinch, 2018; Satre, 2005), the use of hazardous chemicals in farming, food poisoning through corporate negligence, animal slaughter, unethical food labels and trade policy (e.g., see Lawrence, 2004), unhealthy school lunches, the impact of food production on climate change (e.g., see White & Yeates, 2018), and more.

Other scholars have written about monopolization of numerous parts of the conventional food system, including bio-technology involved in genetically modified foods (GMOs) (Walters, 2006), potential harms associated with GMOs themselves (Boone, 2013; Morgan & Goh, 2004; Nottingham, 2003; Walters, 2008), environmental harms such as threats to the quality of air, soil, and water (White, 2014), buying up or seizing land for food and biofuel production and extracting valuable natural resources, referred to as "land grabbing" (White, 2012), food waste (Long & Lynch, 2018), and numerous other crimes such as bribery and corruption that stem from activities of organized crime syndicates involved in the food industry (White, 2014).

The Important Context of Food Crime

All of these behaviors are confined in the context of "cheap capitalism," which is "characterized by degraded business morality, low prices and/or unsafe goods or services, and low-waged labour to maximize profits" (Asomah & Cheng, 2018, p. 194). This makes sense given that practices such as "industrialization, corporatization, and neoliberalization have drastically reformed the modern practices of the food industry, as well as the regulations which govern them" (Gray & Hinch, 2015, p. 97). In this system, corporations are designed to be greedy, and the result is much crime and deviance (Robinson & Murphy, 2008). Stated simply, global food companies, the main drivers of what is often called the "conventional food system" (Robinson, 2024) care about profit above and beyond all other things, including most especially assuring the food products they produce and sell are healthy as well as not the source of harms.

Though many harms discussed in the chapter may be understood at the individual level, a more productive way to comprehend the issues examined within is to house them in the context of global capitalism that has taken over much of the world (Frieden, 2020; Robinson, 2014). So, for example, whereas poor diet and nutritional deficiencies will be found among certain segments of the population, the reasons this is so are owing to the nature of the economy and food economy itself. That is, the system is built to assure certain outcomes, even if the resulting harms are not intentionally conceptualized or planned. Major food companies are designed to produce and profit from unhealthy food.

One simple example makes the point: it is widely known and understood that certain foods are linked to higher rates of illness and death. Specifically, diets high in fats, saturated fats, salts and sugars, as well as low in vegetables and fruits, whole grains, nuts and seeds, and vitamins and minerals are associated with increased rates of heart disease, obesity, and other deleterious health outcomes (Micha et al., 2017). Major food companies are in the business of providing all sorts of foods to consumers, including both healthy and unhealthy foods. Yet, the great bulk of foods produced and consumed are in the unhealthy variety, such as in the case of ultra-processed foods (Elizabeth et al., 2020; Monteiro et al., 2019). So, the system is set up to assure unhealthy outcomes in people. That the negative health consequences in human populations are not intended is irrelevant; the companies remain culpable—morally if not legally responsible—for the outcomes.

One unhealthy outcome is widespread death. A study examining data from 2012 found that just over 700,000 people in the United States died that year from a "cardiometabolic disease" (CMS). This included more than 500,000 from heart disease (including almost 375,000 from coronary heart disease, nearly 130,000 from stroke, just under 70,000 from diabetes (Micha et al., 2017). CMS is defined as "a combination of metabolic dysfunctions mainly characterized by insulin resistance, impaired glucose tolerance, dyslipidemia [elevated plasma cholesterol and/or triglycerides], hypertension, and central adiposity [high accumulation of fat in the lower torso]" (Saljoughian, 2016). According to Saljoughian (2016), both the American Society of Endocrinology and World Health Organization recognize CMS as a disease.

Of the 700,000 CMS deaths, about 395,000 (about 56%) were attributed to the foods that people ate and didn't eat (Micha et al., 2017). Roughly 66,500 of the deaths identified in the study were due to high sodium intake. Another 58,000 deaths were caused by eating too much processed meat. Many other deaths were attributable to the absence of certain foods in the diet, including 59,000 deaths due to a lack of nuts and seeds in the diet, 55,000 deaths due to too little omega-3 fats, 53,000 deaths from too few vegetables, and 52,500 deaths from too few fruits. Finally, about 52,000 deaths were from too many sugar-sweetened beverages (SSBs) (Micha et al., 2017).

The 395,000 deaths is very close to that reported by the US Centers for Disease Control and Prevention with regard to poor lifestyle patterns among Americans: "estimates are that some 300,000 deaths each year in the U.S. likely are the results of physical inactivity and poor eating habits" (US Centers for Disease Control and Prevention, 2018). For perspective, only about 42,000 people died from illicit drugs in 2012 (National Institute of Drug Abuse, 2018). And just under 15,000 people were murdered in the United States in 2012 (Federal Bureau of Investigation, 2013).

Stated simply, this means that, in 2012, the foods we ate (and didn't eat) killed more than nine times more people than all illegal drugs combined and more than 26 times as many people as who were murdered. Further, SSBs alone are more deadly than all illegal drugs and murder combined! Incredibly, evidence shows that SSBs are essentially drugs (Robinson, 2022), yet, of course, they are legal and widely available. Extrapolate such data across so much of the world, at least in places where people eat much of the same foods produced by the same global food companies, and you get a strong sense that this is a social problem worthy of serious consideration.

State–Corporate Crime

State–corporate crime refers to acts of elite deviance (behaviors committed by the powerful, some illegal, some not), conducted by corporations for profit, with the assistance, approval, reward, and/or complicity of government(s) (Kramer, 1990; Kramer & Michalowski, 1991; Kramer et al., 2002; Ross, 2017; Whyte, 2014). One fundamental purpose of scholarship on state–corporate crime is to illustrate that such crimes would not be possible without the assistance of the state. As an example, a study by Friedrichs and Rothe (2014, p. 146) illustrated how large-scale fraud committed by banks and investment banks is committed with the assistance of "government-sponsored enterprises and international financial institutions." Other scholarship supports this conclusion, implicating, at a minimum, individual actors, corporate actors and enablers, and failing regulatory agencies (Hansen, 2009; Pontell, 2005). A recent analysis of the collapse of the US economy in 2008 examined reports by the US Senate Permanent Subcommittee on Investigations and the Financial Crisis Inquiry Commission, and clearly illustrated that both corporate and state organizations were responsible for the "financial crisis" (Robinson & Rogers, 2018).

As globalization continues and the influence of capital continues to spread across the globe, it is increasingly clear that state–corporate crime is not limited to a nation-state such as the United States, but is instead international in nature (Tombs, 2012). One example is the rising rates of obesity and diabetes in the United States and across the world. One major reason this is happening is because large-scale companies, with international reach—assisted by states and international organizations—produce, manufacture, market, and sell their foods to citizens of nearly every nation.

One could even argue that the behaviors of powerful entities that assure the continuation of obesity and diabetes across the globe constitute violent state–corporate crime (Kramer, 1994), for the outcomes are widespread suffering, illness, and death.

Although some might take issue with the term *crime*—because most of what is involved in the food industry is not explicitly illegal (Tappan, 1947)—it is nevertheless dangerous and often deviant in nature (e.g., when companies lie about the contents of their food products or adulterate them in ways that are dangerous to human life).

What About Animals?

Of course, the focus of this book is on animals, so a special focus on animals in this chapter is warranted. Americans, like citizens from many other countries around the world, eat a lot of animals without ever considering where they even come from. In essence, meat in the grocery store is just that—meat—it has been completely disassociated from the animal from which it came.

It is the conventional food system itself that is responsible for this reality. According to the non-profit organization, FoodPrint (2021):

> The modern, industrialized way in which we produce meat, eggs and other animal products has turned animals into units of production, subject to inhumane treatment, terrible living conditions and cruel deaths. In this system, to maximize efficiency and profits, companies and operators of *concentrated animal feeding operations* (*CAFOs*) prioritize rapid growth and large scale production over animal health and welfare.

CAFOs serve *industrial food animal production* (*IFAP*) (Johns Hopkins Center for a Livable Future, 2021). IFAP feeds human consumers. The average American eats more than 274 pounds of meat every year, not including fish and other forms of seafood (Christen, 2021). While consumption of red meat has fallen since 1970, consumption of pork has remained steady, and consumption of chicken has skyrocketed. With each meal comes a new death. So, it is us—we are the reason—for CAFOs, or factory farms.

Issues With Animals in CAFOs and IFAPs

First, the number of animals killed as a result of our food system is enormous. At the end of 2012, the US population of poultry, swine, cattle, and other food-producing animals numbered over 2 billion—outnumbering US citizens by seven to one. Over the course of a typical year, over 9 billion animals are slaughtered (Johns Hopkins Center for a Livable Future, 2021).

According to People for the Ethical Treatment of Animals (PETA, 2021), each year:

- 9 billion chickens are killed;
- 245 million turkeys are killed, including 46 million at Thanksgiving and 22 million at Christmas;

- 31 million ducks are killed;
- 30 million cows die in meat and milk production;
- Millions of pigs are killed;
- Tens of billions of fish and shellfish are killed.

Second, there is the issue of animal suffering. PETA notes: "When they've grown large enough to slaughter or their bodies have been worn out from producing milk or eggs, animals raised for food are crowded onto trucks and transported for miles through all weather extremes, typically without food or water. At the slaughterhouse, those who survived the transport will have their throats slit, often while they're still conscious. Many remain conscious when they're plunged into the scalding-hot water of the defeathering or hair-removal tanks or while their bodies are being skinned or hacked apart."

Even before their deaths, the birds, pigs, cows, and other animals we eat are "are confined in densely stocked indoor facilities. A single IFAP facility, for example, typically houses over 5,000 hogs, or over 100,000 laying hens. Beef and dairy cattle spend at least part of their lives in similarly large-scale indoor or outdoor feeding operations. IFAP facilities are designed to maximize the amount of meat, milk, or eggs from a particular breed of animal in the shortest amount of time and at the lowest cost" (Johns Hopkins Center for a Livable Future, 2021).

Animals are largely raised indoors because this "eliminates weather from the equation, giving producers greater control over temperature, lighting, and other factors that influence productivity. Specialized confinement facilities maximize the number of animals per area, and allow for automated feeding, watering, ventilation, and waste removal" (Johns Hopkins Center for a Livable Future, 2021). The majority of animals in IFAP "rarely if ever see the outdoors, let alone feel sunshine, experience day or night, or walk on grass. They are unable to perform many of their natural behaviors; pigs raised on concrete are unable to root in the earth with their powerful nose discs, chickens are unable to dust bathe, cattle are fed unnatural diets of grain that cause painful liver abscesses. In most egg production facilities, hens are confined to 'battery cages' so small they are unable to spread their wings. Female pigs (sows) used for breeding are confined to 'gestation crates' in which they are unable to turn around or groom themselves" (Johns Hopkins Center for a Livable Future, 2021).

According to FoodPrint (2021), dairy cows are milked for 10 to 12 months a year: "Average US milk production is about 17,000 pounds per cow annually, though herds with averages of up to 24,000 pounds per cow are not unusual." They spend their lives making milk for humans, and are killed years before their normal life span of 15–20 years (Humane League, 2022).

Beef cattle are typically slaughtered at 12 to 16 months of age, and grass-fed beef cattle are killed between 22 and 28 months (FoodPrint, 2021). Often, cows are fed grains they cannot digest, leading to gastrointestinal problems. Male dairy calves are not useful for milk and so they are sold for veal or beef production, and nearly "half of female calves are raised as replacement milk cows, as the older ones slow down; the rest are sold for veal or beef" (FoodPrint, 2021).

As for chickens, layers (egg birds) and broilers (meat birds) both live lives of misery. Hens begin laying eggs at 18 to 20 weeks of age and lay an egg almost every day of the week, in battery cages, so small they are unable to even spread their wings (FoodPrint, 2021). Forced molting (e.g., denying food and water) is used to achieve more egg production. Male chicks are fed into industrial grinders. Chickens tend to be slaughtered after a year even though their life span is 6 to 12 years (FoodPrint, 2021). The same is true for pigs.

Then, add on debeaking chickens, and tail removal on cows and pigs (without anesthesia) to prevent animals from hurting other animals, and animals literally living in their own feces and urine. According to FoodPrint: "Animal behaviors, like pigs rooting in the dirt or chickens taking dust baths, are stifled when animals live in cages or in houses with metal floors; in some cases, such as with veal calves and nursing pigs, animals are unable to turn around. Animals also have to live in or on top of their own excrement, breathing in the toxic fumes, and in perpetual stress from the crowded conditions" (FoodPrint, 2021).

Third, there is the issue of waste. According to Caroline Christen, animals that we farm for food "only convert 2 to 13 percent of the calories they eat into edible body parts. Poultry wastes 77 percent of feed calories, pork 91 percent, lamb and mutton 94 percent, and beef 98 percent." Specifically, "only half of the body of a slaughtered cow, pig, lamb, chicken, or turkey ends up being eaten" (Christen, 2021). On top of that, Americans waste around a third of the food that we buy. This costs the average household $1,866 per year, or $240 billion for the whole population (Christen, 2021).

Professor David Cassuto (2013) claims that "farms have become factories, and the animals raised in those factories are just commodities. That's how they're viewed by the industry, and that's how they are treated by the industry. They are a vehicle through which to make money." He continues, claiming that "a factory farm does not preserve the stability, integrity, or beauty of the biotic community. It does precisely the opposite. It is, therefore, wrong. It is wrong on every ethical level."

The good news is a 2015 Gallup poll found that 94 percent of Americans believe "animals should have some protection from harm and exploitation, including 32 percent who believe they should have the same rights as humans" (Rifkin, 2015). Yet, the conventional food system continues to cause animals to suffer unimaginably, prior to killing them so that we can eat.

Discussion

We have a food system built on the backs of innocent animals. The suffering and untimely deaths of tens of billions of animals a year just to feed people in the United States is hardly justifiable given the availability of alternative nutrition available through non-meat products. Yet, we have system of CAFOs to serve IFAP, run by large corporations and empowered by the state to continue raising and slaughtering animals so that we can mindlessly purchase meat from our grocery stores while literally giving it no thought at all.

A suggestion of this chapter is that the behaviors discussed above—the suffering, slaughtering, and waste of animals that comprise the conventional food system—ought to be the focus of additional food crime research. Criminology and criminal justice professors ought to devote their time and energy to chronicle the realities of the lives of animals that suffer so needlessly so we can merely eat. As of now, only a small portion of even the food crime literature focuses on the welfare and suffering of animals. It is the hope of this author that this paper brings additional needed attention to the issue of animals as food crime.

References

Asomah, J., & Cheng, H. (2018). Food crime in the context of cheap capitalism. In A. Gray, & R. Hinch (Eds.), *A handbook of food crime: Immoral and illegal practices in the food industry and what to do about them*. Bristol, UK: Policy Press.

Beirne, P. (1999). For a nonspeciest criminology: Animal abuse as an object of study. *Criminology*, *37*(1), 117–147.

Boone, R. (2013). GMO wheat lawsuit: Idaho farmers sue Monsanto. *Huffington Post*, May. Retrieved from: https://www.huffingtonpost.com.

Canning, V., & Tombs, S. (2021). *From social harm to zemiology: A critical introduction*. New York: Routledge.

Cassuto, D. (2013). Environment, ethics and the factory farm. Retrieved from: https://www.academia.edu/110239769/Environment_Ethics_and_the_Factory_Farm.

Christen, C. (2021). Meat consumption in the U.S. is growing at an alarming rate. *Sentient Media*, March 17. Retrieved from: https://sentientmedia.org/meat-consumption-in-the-us/.

Coe, S., & Coe, M. (2013). *The true history of chocolate*. London, UK: Thames & Hudson.

Croall, H. (2007). Food crime. In P. Beirne, & N. South (Eds.), *Issues in green criminology: Confronting harms against environments, human and other animals*. Portland, OR: Willan.

Croall, H. (2013). Food crime: A green criminology perspective. In N. South, & A. Brisman (Eds.), *Routledge international handbook of green criminology* (pp. 167–183). New York: Routledge.

Culp, C. (2005). Monsanto assault on US farmers detailed in new report. *Center for Food Safety*, 13, January. Retrieved from: https://www.centerforfoodsafety.org.

Del Prado-Lu, J. (2018). Impact of hazardous substances and pesticides on farmers and farming communities. In A. Gray, & R. Hinch (Eds.), *A handbook of food crime: Immoral and illegal practices in the food industry and what to do about them*. Bristol, UK: Policy Press.

Dorling, D., Gordon, D., Hillyard, P., Pantazis, C., Pemberton, S., & Tombs, S. (2008). *Criminal obsessions: Why harm matters more than crime*. London, England: Center for Crime and Justice Studies.

Elizabeth, L., Machado, P., Zinocker, M., Baker, P., & Lawrence, M. (2020). Ultra-processed foods and health outcomes: A narrative review. *Nutrients*, *12*, 1–33.

Fairley, P. (1999). Farmers sue seed farms. *Chemical Week*, *35*, 13.

Federal Bureau of Investigation (2013). Crime in the United States 2013: Violent crime. Retrieved from https://ucr.fbi.gov/crime-in-the-u.s/2013/crime-in-the-u.s.-2013/violent-

crime/murder-topic-page/murdermain_final#:~:text=In%202013%2C%20the%20estimated%20number,from%20the%20number%20in%202004.
FoodPrint (2021). Farm animal welfare. Retrieved from: https://foodprint.org/issues/farm-animal-welfare/.
Frieden, J. (2020). *Global capitalism*. New York: W.W. Norton.
Friedrichs, D.O., & Rothe, D.L. (2014). State-corporate crime and major financial institutions: Interrogating an absence. *State Crime*, *3*(2), 146–162.
Goodman, B. (2011). Pesticide exposure in womb linked to lower IQ. Retrieved from: https://www.webmd.com.
Gray, A., & Hinch, R. (2015). Agribusiness, governments and food crime: A critical perspective. In R. Sollund (Ed.), *Green harms and crimes: Critical criminology in a changing world*. New York: St. Martin's Press.
Gray, A., & Hinch, Gray R. (2018). *A handbook of food crime: Immoral and illegal practices in the food industry and what to do about them*. Bristol, UK: Policy Press.
Hansen, L. (2009). Corporate financial crime: Social diagnosis and treatment. *Journal of Financial Crime*, *16*(1), 28–40.
Hinch, R. (2018). Chocolate, slavery, forced labour, child labour and the state. In A. Gray, & R. Hinch (Eds.), *A handbook of food crime: Immoral and illegal practices in the food industry and what to do about them*. Bristol, UK: Policy Press.
Humane League (2022). Dairy cows: How long do dairy cattle live? Do they suffer? Retrieved from: https://thehumaneleague.org/article/dairy-cows.
Johns Hopkins Center for a Livable Future (2021). Industrial food animal production. Retrieved from: https://www.foodsystemprimer.org/food-production/industrial-food-animal-production/.
Kramer, R. (1990). *From white-collar to state-corporate crime*. Paper presented to the Louisville: North Central Sociological Association, March 22.
Kramer, R. (1994). State violence and violent crime. *Peace Review*, *6*(2), 171–175.
Kramer, R., & Michalowski, R. (1991). *State-corporate crime*. Paper presented to the American Society of Criminology, Baltimore, November 7–12, 1990.
Kramer, R., Michalowski, R., & Kauzlarich, D. (2002). The origins and development of the concept and theory of state-corporate crime. *Crime & Delinquency*, *48*(2), 263–282.
Lawrence, F. (2004). *Not on the label: What really goes into the food on your plate*. London, UK: Penguin.
Lawrence, F. (2008). *Eat your heart out: Why the food business is bad for the planet and your health*. London: Penguin Books.
Leighton, P. (2018). Mass Salmonella poisoning by the Peanut Corporation of America: Lessons in state-corporate food crime. In A. Gray, & R. Hinch (Eds.), *A handbook of food crime: Immoral and illegal practices in the food industry and what to do about them*. Bristol, UK: Policy Press.
Long, M., & Lynch, M. (2018). Food waste (non)regulation. In A. Gray, & R. Hinch (Eds.), *A handbook of food crime: Immoral and illegal practices in the food industry and what to do about them*. Bristol, UK: Policy Press.
Micha, R., Penalvo, J., Cudhea, F., Imamura, F., Rehm, C., & Mozaffarian, D. (2017). Association between dietary factors and mortality from heart disease, stroke, and type 2 diabetes in the United States. *JAMA*, *317*(9), 912–924. Retrieved from: https://jamanetwork.com/journals/jama/fullarticle/2608221.
Monteiro, C.A., Cannon, G., Levy, R.B., Moubarac, J.C., Louzada, M.L., Rauber, F., Khandpur, N., Cediel, G., Neri, D., Martinez-Steele, E., Baraldi, L.G., & Jaime, P.C. (2019). Ultra-processed foods: What they are and how to identify them. *Public Health Nutrition*, *22*(5), 936–941. doi: 10.1017/S1368980018003762.
Morgan, D., & Goh, G. (2004). Genetically modified food labelling and the WTO agreements. *Review of European Community and International Environmental Law*, *13*(3), 306–319.

Mugni, H., Demetrio, P., Paracampo, A., Pardi, M., Bulus, G., & Bonetto, C. (2012). Toxicity persistence in runoff water and soil in experimental soybean plots following chlorpyrifos application. *Bulletin of Environmental Contamination and Toxicology*, *89*(1), 208–212.

National Institute on Drug Abuse (2018). Trends & Statistics Drug Overdose Death Rates. Retrieved from https://www.drugabuse.gov/related-topics/trends-statistics/overdose-death-rates

Nottingham, S. (2003). *Eat your genes: How genetically modified food is entering your diet*. New York: Zed Books.

Passas, N. (2005). Lawful but awful: "Legal corporate crimes. *The Journal of Socio-Economics*, *34*(6), 771–786.

People for the Ethical Treatment of Animals (2021). Animals used for food. Retrieved from: https://www.peta.org/issues/animals-used-for-food/.

Pontell, H. (2005). White-collar crime or just risky business? The role of fraud in major financial debacles. *Crime, Law and Social Change*, *42*(4-5), 309–324.

Reiman, J. (2020). *The rich get richer and the poor get prison: Thinking critically about class and crime*. New York: Routledge.

Rifkin, R. (2015). In U.S., more say animals should have same rights as people. *Gallup*, May 18. Retrieved from: https://news.gallup.com/poll/183275/say-animals-rights-people.aspx.

Robinson, M. (2017, November 15–18). *Food crime: An introduction to deviance in the food industries*. [Conference Presentation]. Paper presented to the American Society of Criminology, ASC, 2017 Convention, Philadelphia, PA, United States.

Robinson, M. (2020). *Criminal injustice*. Durham, NC: Carolina Academic Press.

Robinson, M. (2022). Eating ourselves to death: How food is a drug and what food abuse costs. *Drug Science, Policy and Law*, *8*, 1–21

Robinson, M. (2024). *Food crime: An introduction to deviance in the food industry*. New York: Routledge.

Robinson, W. (2014). *Global capitalism and the crisis of humanity*. Cambridge, MA: Cambridge University Press.

Robinson, M., & Murphy, D. (2008). *Greed is good: Maximization and elite deviance in America*. Lanham, MD: Rowman & Littlefield.

Robinson, M., & Rogers, J. (2018). Applying contextual anomie and strain theory to recent acts of corporate deviance. *Journal of Theoretical & Philosophical Criminology*, *10*, 71–92.

Ross, J. (2017). Protecting democracy: A parsimonious, dynamic and heuristic model of controlling crimes by the powerful. *Criminal Justice Studies*, *30*(3), 289–306.

Saljoughian, M. (2016). Cardioetabolic syndrome: A global health issue. US Pharmacist, February 16. Retrieved from: https://www.uspharmacist.com/article/cardiometabolic syndrome-a-global-health-issue.

Satre, L. (2005). *Chocolate on trial: Slavery and the ethics of business*. Athens, OH: University Press.

Schrage, E., & Ewing, A. (2005). The cocoa industry and child labour. *The Journal of Corporate Citizenship*, *18*, 19–112.

Smith, R., Manning, L., & McElwee, G. (2017). Critiquing the inter-disciplinary literature on food fraud. *International Journal of Rural Criminology*, *3*(2), 250–270.

Tappan, P. (1947). Who is the criminal? *American Sociological Review*, *12*(1), 96–102.

Tombs, S. (2012). State-corporate symbiosis in the production of crime and harm. *State Crime*, *1*(2), 170–195.

US Centers for Disease Control and Prevention (2018). Physical inactivity. Retrieved from: https://www.cdc.gov/healthcommunication/toolstemplates/entertainmented/tips/PhysicalInactivity.html.

Walters, R. (2006). Crime, bio-agriculture and the exploitation of hunger. *British Journal of Criminology*, *46*(1), 26–45.

Walters, R. (2008). Criminology and genenetically modified food. *The British Journal of Criminology*, *44*(2), 151–167.
White, R. (2014). Environmental instability and fortress mentality. *International Affairs*, *90*(4), 835–851.
White, R. (2012). Land theft as rural eco-crime. *International Journal of Rural Criminology*, *1*(2), 203.
White, R., & Yeates, J. (2018). Farming and climate change. In A. Gray, & R. Hinch (Eds.), *A handbook of food crime: Immoral and illegal practices in the food industry and what to do about them*. Bristol, UK: Policy Press.
Whyte, D. (2014). Regimes of permission and state-corporate crime. *State Crime*, *3*(2), 237–246.

PART V

The Ideological Architecture of Profitable "A-IC-Related-Violence"

15

INSIDE THE SPANISH ZOOLOGICAL PARK INDUSTRY

Worker Insights on Human–Animal Relationships and Shared Vulnerabilities in Spain

Olatz Aranceta-Reboredo and Júlia Castellano

There are more than 60 active zoos, aquariums, and animal theme parks in Spain. Most are not associated with organisations that establish and can conduct inspections on standardised levels of animal welfare and health, as well as educational and conservational duties: only 18 are part of AIZA (Iberian Association of Zoos and Aquariums).[1] This chapter is about the interconnections of zookeepers and nonhuman animals (NHAs) within the Spanish zoo industry. We consider as "zookeepers" all the personnel that directly work with the NHAs in captivity, and "zoological parks" as spaces of the zoo industry where "wild" animals are kept in captivity and exhibition. Industries within the Animal-Industrial Complex (A-IC) (Noske, 1989; Twine, 2012) share varying degrees of human-nonhuman relationships of harm, violence, care, and vulnerability. However, a distinguishing feature of the zoo industry is its exhibition of captive NHAs in the presence of human visitors.

Zoological parks exhibit what we could consider a pastoral, "benevolent" stewardship-based narrative "where care and violence exist in tension but also entangle" (Holmberg, 2022, p. 863). In this study, we understand harm and care relationally because care work "frequently involves exploitation and domination through the nurturing of some lives and relations at the expense of others," and it will often require "constant, gruelling work from those who provide it" (Holmberg, 2022, p. 862).

Vulnerability is shared by most of the population in capitalistic societies. Closely connected with the idea of dependence, it often means relying on others for support, being exposed and open to both suffering and "having one's engagement with the world frustrated" (Thierman, 2012, p. 26). While dependency has been used to justify marginalising and exploiting animals, a disability studies perspective of interdependence allows us to recognise the vulnerability of all beings outside of a victimising or mutual advantage paradigm (Taylor, 2014). This perspective is

DOI: 10.4324/9781003441908-20

necessary for an industry that renders NHAs totally dependent on zookeepers and, simultaneously, makes both vulnerable to exploitation, dependent and influenced by the public, their mutual bonds, the material conditions of the job, and their labour.

Much scientific inquiry into zoos has focused on visitors' perceptions, beliefs, and attitudes (i.e., Merino et al., 2010), yet the impact of visitors on NHAs and zookeepers remains understudied. On the one hand, NHA behaviour and physiology are altered in response to visitor presence (Davey, 2007). On the other hand, even if zookeepers are mostly passively exposed to the public (e.g., performing daily routines like cleaning and feeding), they also often directly interact with visitors, which involves emotional labour (Mesmer-Magnus et al., 2012). However, further research is necessary to comprehend these dynamics from NHAs' and zookeepers' perspectives altogether and the intersectionality of vulnerabilities in zoological parks.

A survey conducted on 311 zoo and aquarium professionals explored stressors, satisfaction, and feelings about their work and workplaces (Brando et al., 2023). Lack of opportunities for recognition, feelings of frustration, stress, lack of safety, overwork, and feeling worn out were common challenges; however, the main stressor was the inability to stop thinking about work while at home, unable to continue with the care of the NHAs they bond with.

Studies have shown that the bonds that zookeepers form with NHAs under their care can have operational and affective benefits for both (Hosey & Melfi, 2012). However, caregiving work can also cause empathic distress, a "strong aversive and self-oriented response to the suffering of others, accompanied by the desire to withdraw from a situation" (Dowling, 2018, p. 749). Typically described in the context of medical personnel, it has now become a fundamental aspect of jobs in the A-IC (Schlanser et al., 2021; Thompson-Hughes, 2019).

The material conditions of the job also influence empathic distress. Zookeepers encounter challenges in their work environment, training, staffing, and resource availability, which leads them to experience disengagement from enjoyable activities due to the cumulative effects of mental and physical exhaustion (Bemister-Bourret & Tawfik, 2023). These factors directly impact animal welfare, which they strive to enhance despite facing challenges such as financial limitations and limited support from management (Bacon et al., 2021). One reason for this could be the highly vocational nature of the profession.

Working with and taking care of other animals is a lifetime goal for many zookeepers. A person's love and dedication to achieving excellent animal care and providing well-being have led many to refer to working in zoos as their "calling" (Bunderson & Thompson, 2009). However, the conjunction of workers' devotion and the harsh conditions, such as the physical demands and precariousness within the sector, generates a characteristic vulnerability.

Vocational passion does not replace the necessity of performing paid work, neither does it exempt workers from being subject to the various oppressions and exploitations inherent in the capitalist system. Hence, it may be necessary to analyse the shared experience of zookeepers and NHAs with labour at zoological parks.

NHA work has been theorised and reflected on in many different contexts.[2] For the zoological parks, the framework of performative, symbolic and hidden labour of NHAs employed in the tourism industry (Rickly & Kline, 2021) helps shift the ascribed role as ambassadors for NHAs in public and captive environments. Embodying the task of representing their counterparts in nature, they are gazed upon, endure social, emotional, and care work, perform in shows, and sometimes interact with visitors in petting zoos and other activities. NHAs can be considered "educational ambassadors" who work for and with others (Speiran, 2021), but this would require us to acknowledge their agency and realities, as well as the interconnections in labour with the human workers of these spaces. Moreover, a point often overlooked is that zookeepers are also considered the embodiment of the zoo industry by public opinion, especially under a system where work is considered a core aspect of one's identity.

Methodology

We conducted 17 interviews with people directly working with "wild" NHAs in captivity and exhibitions across Spain. Participants were recruited using purposive sampling through a combination of social media outreach and the snowball method.[3] The average age of the zookeepers was 34 years old. 65% of the participants were women and 35% were men. The spaces interviewees have worked at include a wide range of zoological parks: from zoos, aquariums, and animal theme parks, to "wild" animal shelters, rescue centres, and foundations—where most NHAs are seized from smugglers and circuses, or are unreleasable. All interviewees but one had cleaning, feeding, enclosure maintenance, and caretaking chores. Simultaneously, they worked on other tasks: shows, training, enrichment, public interactions, department/area management, and being a veterinary assistant. Unless the distinction is necessary, we refer to them as zookeepers.

The individual interviews were semi-structured and designed to explore themes such as vulnerability, harm, emotional bonds, and caretakers' challenges in their interactions with NHAs. These interviews were conducted from November 2023 to February 2024. Except for one face-to-face interview, all were done via Zoom, recorded and immediately transcribed using Whisper.ai. Later, all transcripts were meticulously coded using Atlas.ti software. We employed inductive and deductive coding techniques to uncover nuanced patterns and themes. Throughout the chapter, we have included quotes from participants with an anonymised code for each, with the first letter assigned individually followed by information on their work (i.e., after the low bar, "z" stands for zoological park, "a" for aquarium, followed by "c" for caretaker, "e" for educator, and "i" for researcher).

We conducted our literature review with a grounded theory approach (Glaser & Strauss, 1967), which allowed us to incorporate literature on new themes and issues that emerged in the interviews.

Limitations

The sample size, while diverse, cannot be understood as representative of the experience of the entire population of zookeepers in Spain. Moreover, the focus on shared vulnerabilities of the chapter, and the limited extension, means that some of the results on the positive and rewarding aspects have been potentially shadowed by negative experiences.

Balancing Vulnerabilities, Compassion, and Vocation

> "Are you sure that what you're doing is ethical? Are you sure?" *(x3)* I used to say "You like animals, how can you enjoy seeing them there?" *(I_ze)*

In this section, we present the findings of our study organised into distinct themes that emerged from the interviews with zookeepers. These themes are structured as follows: the industry, insights on their work, interconnected experiences, and social (mis)perception.

The Industry

Working in zoological parks meant something different ethically, morally, and politically for many respondents. There were three ethical evaluations on feeling conflicted about working with NHAs in captivity. From most common to least: feeling some discomfort at times but accepting it for other reasons (conservation, educational potential); being critical but staying in the industry to make sure that NHAs receive the best possible care; and, sometimes, feeling "incoherent" in values.

Interestingly, one of the points raised was that regardless of the captivity model, there was no such big difference between zookeepers' work ethic and ideological alignment. According to one of the interviewees: "I used to be almost against zoos […], and now I see that the workers I work with have the same ideology as the workers I met in a sanctuary or a rescue centre; […] they are working to improve the welfare of the animal" (T_ac).

Hierarchical Processes

The most important challenges were the environment and the difficulties in requesting improvements. The most common environmental challenges were the lack of proper installations, badly behaved public, and noise. An interviewee reported that:

> I have seen animals that, every time a show starts, hide in their interior rooms to avoid the noise, and it is repetitive because the shows if they are done three or four times a day every day it is all the time, the same music, the same intensity.
>
> *(T_ac)*

Another source of reported frustration are the expectations of some visitors (i.e., touching NHAs, trying to feed them, getting closer than allowed, requesting to buy them). Instances of harm included dropping materials that risk NHAs swallowing things and getting injured; touching (i.e., petting or grabbing fishes so strongly they get crushed); feeding them foods against the rules; and screaming and purposefully triggering responses, sometimes violent, from NHAs at the zoo.

Regarding the second one, interviewees explained that most of the harm to NHAs happens because of bad decisions, prioritisation of the public and profit over the zookeeper's voice and the bureaucratic chain of command. An interviewee explained that caretakers' jobs are undervalued by the public and at the workplace: "In the work hierarchy, we are at the bottom of the ladder, the lowest of all, the one who is paid the least, the one who has the least. [...] this work is not visibilised" (IU_zc). According to some respondents, the owners, sometimes disregarding the advice of zookeepers, make decisions that jeopardise animal welfare. For instance, overcrowding enclosures to accommodate more individuals is a practice that results in harm to NHAs and increases stress and mental health problems for caretakers and veterinarians. Unable to establish or negotiate the timetable of exposure to the public, zookeepers lose the ability to do any medical or enrichment training during those hours and cannot provide proper medical care during that time (e.g., ointments).

Lack of (Transparent) Protocols or Over-Bureaucracy

Some interviewees expressed a pressing need for clearer guidelines and protocols. From daily decisions regarding enrichment to handling emergencies, they must navigate complex situations without standards; this lack of guidance results in inconsistencies and potential harm.[4] Several participants identified the need for a comprehensive registry documenting what works for each animal, particularly when relocating them to different installations.

Zookeepers have confided on how the problematic nature of protocols and bureaucratic obstacles within zoological facilities might culminate in malpractices. For instance, some interviewees considered that cost-based decisions are often made without considering welfare, which created frustrating situations—especially in cases where the requests involved affordable products with a significant positive impact (e.g., shampoo, balls). This is closely intertwined with staffing shortages, malpractices, and insufficient protocol implementation (e.g., veterinarians approving NHAs going to exhibition without checking on them).

Moreover, some interviewees reported that:

> many caretakers that worked in these places [with administrative limitations and over-bureaucracy] see the reality and say "I cannot handle this, I became a caretaker for animals to be better, not to fight with the administration"; [...] if you see you cannot help the animals and you're taking all that suffering with you, you leave.
>
> *(V_pe)*

On the other hand, several respondents also reported the rewarding aspects of being able to work with a supportive team of zookeepers who keep working and innovating to take care of NHAs so that they are as comfortable as possible and "don't show any stress or strange behaviour; [...] the most gratifying thing is to see them well" (T_ac) as a result of their work.

A Clash of Interests

Some zookeepers reported a clear prioritisation by the zoological parks for visible, marketing and public-oriented changes, especially with certain charismatic species and new enclosures, which can be frustrating and "wears out" (N_ac) when "they are just fixing the visual aspect" (A_cz). Other interviewees explained that many enclosures are built on concrete in cities or mostly artificial environments, so issues like the lack of appropriate ground consistency mean that, while unnoticeable to the visitors, zookeepers had extra medical check-ups to carry out during their shift even if they propose ways of addressing the source of harm.

Interviewees also reflected on the complex ethical situation of needing enough resources to provide adequate care but not wanting NHAs to be exploited. One interviewee worked in a private management facility where most NHAs were rescued or seized and did not receive government funding. They found working for TV to earn money and invest in their care more ethical. Other participants highlighted the importance of flexibility with the participation of NHAs in shows. The ability to accept that the NHAs could sometimes not participate (e.g., age, not wanting to, highly sexual periods) reduced frustrations for caretakers. It offered alternative ways of teaching visitors about the individuality and preferences of the NHAs. A recurring idea for improvements was the wish to avoid having to depend on visitors to invest, public funding, and the need for stricter laws and regulations to inspect and push for a change within Spain.

Insights on Their Work

Most interviewees shared a similar workload and types of tasks and reported many responsibilities and not enough time, resources or colleagues to complete them. The repeated displacement of essential tasks (e.g., enrichment, cleaning of drinking areas) for cleaning visitor-experience-focused areas was a source of distress. Others described the physical toll of working "exposed to the natural elements," intense and long working days, and the expectation to do unpaid extra hours. This, they explained, was often assumed because of their vocational commitment to other animals.

Lack of Staff and Professionalisation

Lack of continuity in staffing, the interviewees' reported, results in inconsistent care and unfamiliarity. Moreover, according to the respondents, in the Spanish zoological industry "favours" are common (i.e., assuming positions they are not qualified for), and there are instances where zookeepers work with species they have no knowledge about. As a result, they limit their work to cleaning and feeding and give NHAs little to no enrichment. People not experts or familiar with the species might be asked to prepare enrichment projects instead of hiring qualified people and dividing tasks. Furthermore, this staffing instability adversely affects zookeepers' mental health and well-being. Some constantly move between centres in search of better opportunities yet struggle to find satisfactory employment due to concerns over salary, working conditions, and facilities. When asked about what changes would improve the situation of both zookeepers and NHAs, most respondents answered: hiring more people. Specifically, and ideally, more trained people, and more veterinarians with training or experience in the sections they manage.

Interviewees noted a significant disparity between Spain and other European countries in specialised and regulated education availability. They stressed the importance of non-private official studies to enhance professionalisation in the field. Many also report that the existing educational curriculum primarily focuses on practical tasks (e.g., enclosure and kitchen maintenance), neglecting crucial animal care and behavioural aspects. Interviewees expressed concern about the inadequate preparation, which, together with the lack of protocols, poses a significant challenge: "there's not so much about how you deal with the animals, what signals you get from an animal that it's right or wrong, or that there are things you have to change" (N_ac). Most participants agreed on the priority of hiring more people and getting their work properly paid as the most significant change that would improve the lives of zookeepers and NHAs in captivity.

Sacrificial Labour and Shared Precarity

Several interviewees explained that this job comes with several personal sacrifices: economic, emotional, physical and time. Many are exposed to the elements for hours doing physically demanding tasks; the job's rewarding aspect compensates until their vocation is used against them. Referring to the sector, interviewees explained that "In Spain, it is clear that there is a great deal of precariousness and the animals are the first to suffer from this" (N_pc), and that "In Spain it happens more than in other places, they take advantage of the passion" (R_ac). Another one said that:

> First in precarity, it is them [NHAs][…]. There is precarity at an employment level because we work many hours doing a very physical job […] in charge of so many animals and installations, which means that you cannot attend.
>
> *(I_ac)*

Precariousness is considered one, if not the most, challenging aspect of their work by almost all the participants. Some zookeepers also considered that the industry takes advantage of their vocation because, otherwise, they would not accept such minimum wage, exhaustion and bad work-life balance. According to several zookeepers: "it is a tool for manipulation 100%, […] this compassion or this love for animals is what I think they have us hooked with because, within this precariousness, we let it happen" (I_ac); "the main problem is that being a calling, […] I'm not going to go home. If those two animals are going to be left without food, what happens? They don't pay you overtime, because they know you're going to do it" (IU_zc); "either you are totally fascinated by animals […] or you can't tolerate it" (IE_zc); "precariousness is an issue in all the centres I have been in. And I think they take advantage of […] the passion to work with animals" (N_pc).

Some interviewees reported that their vocation is "abused." They reported having to choose between sacrificing the wellbeing of NHAs—who would not get fed, their installations properly cleaned (e.g., water change), or any enrichment—and leaving work late, working extra unpaid hours, working sick, losing life-work balance and sacrificing their well-being: "I had to choose between the animals and me […] but if I don't go to work there's nobody there, so I had no other choice and I chose to do it for the animals" (IU_zc). Harshly, they report that:

> bosses don't care about destroying your vocation, because more people will be coming after you who are just starting […]. I've had colleagues who have left this work because they've lost it; imagine how fucked it is that you are stripped from a vocation that you've had since you were a child.
>
> *(IU_zc)*

According to some interviewees, this situation favours a competitiveness that creates toxic working environments: "[in] a job that requires so much empathy with animals to do it well, the fact that there is no empathy between each other is terrible" (M_pe). In the interviews, we learnt of people who, because of the harsh working conditions, experience depression and/or anxiety, sleeping problems and burnout—sometimes calling in sick because they are overwhelmed and overworked.

Furthermore, they reported their situation being more precarious than in many sectors because of not having a trade union, collective, or a specific labour agreement. In this shared precariousness, and by prioritising NHAs, their mental health was reported to decline drastically; when trying to deal with it, they realised that the individuals under their care suffered the consequences because

> when you have nothing better to give because they are taking it all away from you, that's when the worst moments come; […] making you choose between yourself and the animal.
>
> *(IU_zc)*

Limits and Boundaries at the Workplace

Most interviewees reported that there were no bad practices in their current workplace, and emphasised the frustration generated by the existence of centres that engage in misconduct, tarnishing the reputation of those who strive to do things right. A shared vulnerability in this regard is the ability to speak up against these bad practices; depending on prevailing economic conditions, levels of support, and individual confidence, the boundaries of what the interviewed zookeepers were willing to accept significantly. These challenges ranged from having to coerce animals into performances to participating in practices that clashed with their ethics (e.g., breeding animals for private persons).

When addressing what they considered malpractices, some interviewees recalled being unable to leave the job or do something about it due to fear of retaliation. Others shared their experiences of speaking up and facing the consequences of challenging certain practices; the response ranged from being ignored or ridiculed by colleagues or knowing that colleagues have been fired for speaking up to being able to stand up for a change in collectivity and forcing the directives to change their decisions (e.g., the number of shows on a day). Some interviewees reported they were thankful for never facing a situation or doing "anything outside of my principles" as they are very uncompromising, not caring that they are "at the bottom of the ladder" (IE_zc).

Interconnected Experiences

Interviewees described the connections with NHAs as rooted in trust, recognition, and even friendship—which they regarded as one of the most rewarding aspects of their profession. However, they had divergent perspectives regarding the nature of these relationships. While some emphasised the importance of maintaining a professional distance to prevent excessive emotional attachment, others embraced a more emotionally invested approach, viewing genuine connections with the animals and the ability to help NHAs as rewarding.

Moving Animals, Moving Connections

Moving NHAs from a place with their social group, mate, or caretakers to another place is an issue that appeared in several interviews. Moving is, according to the interviewees, motivated by: individuals of the same species that do not get along and cannot share space, being "spare" in another zoo and the owners of the place wanting another "specimen," reproductive abilities, and the closing of inappropriate spaces. However, there are also cases of purchase and sale of NHAs between zoological parks, for profit, that go unregistered.[5]

These situations might involve choosing who left the enclosure and who stay ed, a tough decision that, according to interviewees, made them think:

> I'm going to decide which one you keep and which you send to another centre. [...] What's their fault if you decide they leave, or that they can't have another facility in the same place and continue with the same caregivers.
>
> *(T_ac)*

The lack of a register of enrichment methods implied potential harms when moving installations. Interviewees described the pride of working to "naturalise" animal behaviours, breaking stereotypic behaviours, self-harm patterns and more; however, they have also mentioned cases of relocation or caretaker change that result in the NHAs exhibiting symptoms of psychological distress, all while being unable to follow-up on their situation. Since many effective enrichment tactics come from mouth-to-mouth knowledge, avoidable harms from this lack of registry provoke frustration: NHAs may experience prolonged periods with only physiological needs met and may become distressed when confronted with past traumas for which there is no record. However, if the working conditions were to improve, the interviewees considered the prioritisation of enrichment activities one of the changes with the most direct impact and improvement on the quality of life of NHAs.

On Life and Death

Many respondents regarded euthanasia as one of the most challenging aspects of their work. While some interviewees stated its importance in cases where an individual's quality of life is severely compromised, ethical concerns arose for the zookeepers when other factors drove the pressure to euthanise. The examples of ethical challenges they provided include financial considerations, space limitations, or a perceived lack of utility for the animal (e.g., cannot participate in shows). Such practices have been criticised by some respondents who advocated for the implementation of preventive measures such as birth control or sterilisation.

On the other hand, the death of NHAs is not always preceded by a conscious decision. Respondents report that the deaths of animals with which they had formed a bond were very painful, especially when the causes were avoidable.

> I have encountered very traumatic deaths that have affected me a lot. [...] It took me more than a year to overcome it [...] many times I still find myself, when I remember them, I start crying.
>
> *(L_ac)*

Speciesism further complicates ethical considerations. According to the interviewees, in some cases, injured individuals may be denied euthanasia due to their species being of interest. In contrast, certain species may be euthanised based on perceived lesser importance. This discriminatory practice results in unequal treatment, disproportionately affecting aquatic non-mammal animals.

One zookeeper said they had gone against direct orders by hiding and not sacrificing (aquatic) individuals to allow them to recover successfully. Moreover, in an example of intersecting discriminations, the company wanting the expositions to "look good" entailed routine ableist violence by sacrificing disabled NHAs, affecting individuals who happen to have neurological disorders, injuries or disabilities that "have no value for the public" (I_ac). On the other hand, the inability of some species to get used to captivity, accidents, poor quality installations and so on result in the death of many individuals. In the interviews, zookeepers described it as more common in aquariums than in zoos or other spaces. According to them, aquariums in Spain capture NHAs from nature in collaboration with fishermen; for example, even if the mortality rate for some species like octopi is enormous, park directors insist on abducting them from nature for the public. One of the most extreme cases of ethically conflicting situations an interviewee shared was:

> when the aquarium places an order [to fishermen] for squid, lobster and prawns, even shrimps, of course, [...] many animals don't adapt and die, and some workers took the ones that died home and made a *paella*; [...] it happened a lot with octopuses too.
>
> *(I_ac)*

Trying to reduce the number of NHAs imported from nature and, in the process, reduce the harm they experience, some zookeepers research how to breed species in captivity and offer it as a more cost-efficient practice to the zoological park.

Social (Mis)Perception

Most respondents talked about two distinct polar-opposite groups regarding social perception: those who appreciate and admire their work and those who criticise it. Zookeepers reported being accused of being pro-animal prisons and contributing to the exploitation by accepting to work in these spaces. This is a double crossroads situation; they are considered the face of zoological parks by the companies and the public, while their concerns are often not properly considered by the management.

They argued for reframing the debate on zoological parks and the transition into a more animal, conservation and education-centred model. According to the interviewees, the criticism should be not just focused on "extremism" but looking at what "cannot be improved yet" and not forgetting what "is being done well" to promote animal welfare. Interviewees explained their frustration because "many things need to change in Spain [...]. and we carers also criticise it, you know? It's not that we are supporting it just because we work there. We are the first ones concerned about changing those kinds of things" (N_pc). Nevertheless, zookeepers are still considered either "shit cleaners" (derogatory) or accused of all the harm that NHAs face in captivity.

Some zookeepers also highlighted the hypocrisy of criticism that overlooked the reasons for some individuals being in captivity—such as illegal trade for the "exotic pet" industry or anthropogenic ongoing harm (i.e., consumption of palm oil)—while ignoring their work of rehabilitation and reintroduction (I_zc).

Discussion

The challenges the interviewed zookeepers face in the Spanish zoological industry, such as workload, staffing issues, and lack of professionalisation, together with the vocational aspect of their profession—that might lead them to endure and sacrifice their well-being for the NHAs they care for— result in what we call "weaponised compassion." As we have seen, some zookeepers consider that their compassion is taken advantage of because, otherwise, they would not compromise and accept this minimum wage, exhaustion and bad work-life balance. This connection between self-sacrifice in care-based work and potential precarity and exploitation has been identified in other industries (Baines, 2016). In this shared vulnerability, the interconnectedness of human workers and NHAs in captivity stands out. The lack of union representation and adequate labour agreements further compounds their struggles.

Despite their dedication to animal welfare, the interviewed zookeepers sometimes encounter profit prioritisation conflicting with their efforts. Key challenges include environmental factors like excessive noise, public behaviour, and the struggle to advocate for improvements within hierarchical structures. Aquatic non-mammal animals are the most harmed individuals, replicating speciesist biases that affect their care in the zoo industry (Conley, 2008) and all over the A-IC (Mather, 2019). They struggle with extensive responsibilities, inadequate resources, and prioritising cleaning over essential tasks like enrichment (Hoy et al., 2010). Despite these challenges, zookeepers form strong bonds with the animals under their care, enhancing veterinary care, enrichment and reducing stress when possible (Birke et al., 2019). The interviewed zookeepers desire more resources, animal-centred installations, stricter regulations, more protocols and knowledge sharing between zoological parks and less dependence on visitor funding to ensure adequate animal care without exploitation.

Furthermore, the social perception of zookeepers adds another layer of complexity. According to the interviewees, zookeepers in Spain face a stark social divide: admiration or criticism. Despite their dedication to animal welfare, respondents explain that they often feel misunderstood and labelled as complicit in animal exploitation despite their daily and precarious dedication to NHAs, which might be exacerbated because they are used as the face and ambassadors for embodying the "values" of zoological parks.

Conclusion

In conclusion, this chapter has explored the intricacies of human–animal entanglements within the Spanish zoo industry and addressed a multitude of challenges, vulnerabilities, and precarity shared by zookeepers and NHAs in captivity. Through fifteen interviews, we gained valuable insights into the daily experiences, ethical dilemmas, and limitations inherent in the zoo industry. The profound dedication of zookeepers to the welfare of NHAs under their care leads them to endure precarity and sacrifice their well-being, creating a situation of weaponised compassion that harms both of them.

Amidst all the challenges, they forge deep bonds with the NHAs they care for, striving to enhance welfare and reduce stress. However, zookeepers face a stark social divide, grappling with both admiration and criticism while feeling misunderstood and misrepresented in their labour. The prevailing social perception often overlooks their dedication to animal welfare and labels them as complicit in animal exploitation.

The current model of zoos in Spain is untenable: it harms NHAs and human workers, renders them vulnerable to further exploitation and, sometimes, even violence. However, as long as harmful practices persist in our society—such as habitat destruction, "wild" animal traffic, or circuses—there will still be a need for spaces where non-releasable "wild" animals can receive care and rehabilitation. This chapter highlights the need to reconsider the intersectionalities of harm in zoos. Therefore, when advocating for an ethical transition in zoo models it is crucial to have a nuanced and non-polarised debate on NHA captivity, and involve collaborative efforts with zookeepers to ensure progress towards an animal-centric approach that challenges the model of most zoos. By acknowledging areas needing improvement and critically supporting positive aspects of their labour, we can work towards a more ethical and sustainable future for zookeepers and the NHAs they care for in the Spanish zoological industry while it exists.

Acknowledgements

We would like to express our special thanks to the volunteers for their time and trust. For their comments and suggestions, we are grateful to: Prof. Núria Almiron, Dr. Eze Páez, and Dr. Leire Morrás Aranoa.

Notes

1 See: "¿Cuántos zoos hay en España?" (2022); "Anexo:Zoológicos y acuarios de España" (2023).
2 See Hribal (2014).
3 Before initiating contact with potential participants, we obtained approval from the Pompeu Fabra University ethics committee, which involved reviewing our interview script, data management, and informed consent form. Informed consent was obtained from all participants prior to their involvement in the study.

4 The interviewed zookeepers that highlighted most of the positive aspects of the evolving, improving, and complex enrichment (e.g., social, physical and cognitive) and medical training were the ones working with marine mammals, large reptiles and rescued big cats.
5 One interviewee reported that they had heard their superiors discussing the need to rush the purchase of certain NHAs from the USA before the state-level Law for the Protection of Animal Rights and Animal Welfare (7/2023) was made effective and these practices became illegal (I_ac).

References

Anexo: Zoológicos y acuarios de España. (December 27, 2023). In *Wikipedia*. https://es.wikipedia.org/w/index.php?title=Anexo:Zool%C3%B3gicos_y_acuarios_de_Espa%C3%B1a&action=history.

Bacon, H., Vigors, B., Shaw, D.J., Waran, N., Dwyer, C.M., & Bell, C. (2021). The modern zoo: Demographics and perceptions of two international groups of zoo staff. *Journal of Zoological and Botanical Gardens*, *2*(4), 636–649. https://doi.org/10.3390/jzbg2040046.

Baines, D. (2016). Moral projects and compromise resistance: Resisting uncaring in nonprofit care work. *Studies in Political Economy*, *97*(2), 124–142. https://doi.org/10.1080/07078552.2016.1208793.

Bemister-Bourret, K., & Tawfik, M.H. (2023). Record, recall, reflect: A qualitative examination of compassion fatigue in Toronto zoo staff. *Journal of Zoological and Botanical Gardens*, *4*(2), 413–426. https://doi.org/10.3390/jzbg4020031.

Birke, L., Hosey, G., & Melfi, V. (2019). "You can't really hug a tiger": Zookeepers and their bonds with animals. *Anthrozoös*, *32*(5), 597–612. https://doi.org/10.1080/08927936.2019.1645504.

Brando, S., Dold, C., Goulart, V.D., & Robeck, T.R. (2023). Factors influencing the development of human–animal relationships at SeaWorld entertainment parks. *Aquatic Mammals*, *49*, 294–307. https://doi.org/10.1578/AM.49.3.2023.294.

Bunderson, J.S., & Thompson, J.A. (2009). The call of the wild: Zookeepers, callings, and the double-edged sword of deeply meaningful work. *Administrative Science Quarterly*, *54*(1), 32–57. https://doi.org/10.2189/asqu.2009.54.1.32.

¿Cuántos zoos hay en España? (April 4, 2022). In *Vida Silvestre Ibérica*. https://vidasilvestreiberica.org/cuantos-zoos-hay-en-espana/.

Conley, M.K. (2008). Caring for dolphins, otters, and octopuses: Speciesism in the regulation of zoos and aquariums. *Animal L*, *15*, 237–264.

Davey, G. (2007). Visitors' effects on the welfare of animals in the zoo: A review. *Journal of Applied Animal Welfare Science*, *10*(2), 169–183. https://doi.org/10.1080/10888700701313595.

Dowling, T. (2018). Compassion does not fatigue! *Canadian Veterinary Journal*, *59*(7), 749–750.

Glaser, B., & Strauss, A. (1967). *The discovery of grounded theory: Strategies for qualitative research*. Sociology Press.

Holmberg, M. (2022). Constructing captive ecology at the aquarium: Hierarchy, care, violence, and the limits of control. *Environment and Planning E: Nature and Space*, *5*(2), 861–880. https://doi.org/10.1177/25148486211014508.

Hosey, G., & Melfi, V. (2012). Human-animal bonds between zoo professionals and the animals in their care. *Zoo Biology*, *31*(1), 13–26. https://doi.org/10.1002/zoo.20359.

Hoy, J.M., Murray, P.J., & Tribe, A. (2010). Thirty years later: Enrichment practices for captive mammals. *Zoo Biology*, *29*(3), 303–316. https://doi.org/10.1002/zoo.20254.

Hribal, J. (2014). *Los animales son parte de la clase trabajadora*. Ochodoscuatro ediciones.

Mather, J.A. (2019). Ethics and care: For animals, not just mammals. *Animals*, *9*(12), 1018. https://doi.org/10.3390/ani9121018.

Merino, L., Lilienfeld, S.O., Malamud, R., Nobis, N., & Broglio, R. (2010). Do zoos and aquariums promote attitude change in visitors? A critical evaluation of the American zoo and aquarium study. *Society & Animals*, *18*(2), 126–138.

Mesmer-Magnus, J., DeChurch, L.A., & Wax, A. (2012). Moving emotional labor beyond surface and deep acting. *Organizational Psychology Review*, *2*, 53–56.

Noske, B. (1989). "The animal industrial complex." In *Humans and other animals: Beyond the boundaries of anthropology*. Pluto Press.

Rickly, J.M., & Kline, C.. (2021). "Introduction: Working for the (hu)man in the tourism industry". In *Exploring non-human work in tourism. From beasts of burden to animal ambassadors*. Eds. Jilliam M. Rickly and Carol Kline. De Gruyter Studies in Tourism; 5. De Gruyter Oldenbourg.

Schlanser, T.V., Rabinowitz, P.M., & Thompson-Iritani, S. (2021). Compassion fatigue and satisfaction in US army laboratory animal medicine personnel. *Journal of the American Association for Laboratory Animal Science*, *60*(4), 422–430.

Speiran, S.I.. (2021) Monkey see, monkey do: The work of primates in Costa Rican sanctuaries. In J. M. Rickly & C. Kline (Eds.), *Exploring non-human work in tourism. From beasts of burden to animal ambassadors*. De Gruyter Studies in Tourism; 5. De Gruyter Oldenbourg.

Taylor, S. (2014). Interdependent animals: A feminist disability ethic-of-care. In C. Adams & L. Gruen (Eds.), *Ecofeminism: Feminist intersections with other animals and the earth*, 109–126. Bloomsbury Academic.

Thierman, S. (2012). *Vulnerability, care, power, and virtue: Thinking other animals anew*. Doctoral thesis, University of Toronto. TSpace. https://tspace.library.utoronto.ca/bitstream/1807/34944/1/Thierman_Stephen_F_201211_PhD_thesis.pdf.

Thompson-Hughes, J. (2019). Burnout and compassion fatigue within veterinary nursing: A literature review. *Veterinary Nursing Journal*, *34*, 266–268.

Twine, R. (2012). Revealing the 'animal-industrial complex' – A concept & method for critical animal studies? *Journal for Critical Animal Studies*, *10*(1), 12–39.

16

"IF I BROKE DOWN THE WALL OF FLESH"

Blurring the Human/Animal Distinction in the Slaughterhouse through Ivano Ferrari's Poetry

Chiara Stefanoni

Since the advent of the industrial slaughterhouse in the 19th century, most of the animal advocacy movement has concentrated on the visibility/concealment dynamic inherent in non-human animal exploitation and its moral and political implications. This focus is exemplified by the well-known slogan "If slaughterhouses had glass walls, we would all be vegetarian". However, how can one truly see something that, by its very nature, persists as "hidden in plain sight" (Pachirat, 2011)? Are undercover investigations, such as photojournalism, documentaries or essays, sufficient? Or do they risk sensationalizing and didactically showcasing violence, leading to desensitization and double exploitation, overshadowing non-human animals' individuality and agency (Aaltola, 2014; Corman, 2017)?

Within Western culture, vision is traditionally considered the epistemic sense *par excellence*, the means of accessing reality and morality (Jonas, 1954; Tunstall, 2011). In this context, particularly insidious is the purported transparency of technologically mediated images (Sontag, 2003). The broader issue of representation also looms large: are non-human animals represented in their own right or do they stand symbolically for something else?

The vision of violence against non-human animals within the institutions of the animal-industrial complex, epitomized by the slaughterhouse where the most extreme violence, i.e., putting to death, occurs, does not necessarily ensure a challenge to the status quo, instigate moral shock,[1] or dismantle the human/animal distinction.[2]

Faced with these problems, this chapter proposes to enter the slaughterhouse through the medium of poetry – particularly the works of Italian poet Ivano Ferrari (1948–2022), known for his collection *Slaughterhouse* (2019).[3] The collection was penned during a period when, driven by economic necessity for the birth of his daughter and in the absence of an alternative (Budriesi, 2019, p. 63), Ferrari worked at the municipal slaughterhouse in Mantua from 1973 to 1978.

DOI: 10.4324/9781003441908-21

He started as a porter, tasked with transporting deceased cattle to the cold stores. Over the years, he transitioned through two other roles: first to a cleaner and subsequently to a stamper. The poems stemmed from his need to express his experiences poetically. Certainly, poetry can manifest itself through the language of vision and imagination, yet the poetic word, even when grounded in realism, transcends mere visual description. It is a nuanced medium capable of evoking without saturating or showcasing and, precisely because of this, it unveils horror more effectively. As Ferrari himself articulates when questioned about the decision to poetize from the slaughterhouse: "My motivation is the search for detachment. In poetry, words are heavy, dense. The weight of the word frees me from the image. It was a path of liberation for me, the word was an act of liberation" (Budriesi, 2019, p. 63).

Antispeciesist Resonance: "Betrayer of the (Human) Species"

Ferrari became a poet while, and as a result of, working at the slaughterhouse. Contrary to its seemingly a-poetic nature, the slaughterhouse serves as the site of Ferrari's poetic formation, initiating a process of research and reflection. In his own words: "After the slaughterhouse, I became a poet. I had never written before. It made me reflect, search. […] I was working on emptying the excrement. When the slaughter was over, I was free, and I read Baudelaire, Rimbaud and Lautréamont, right there in contact with the refuse and, above all, I read them all in the slaughterhouse. That was my education" (p. 66). Ferrari not only writes about the slaughterhouse but from and in the slaughterhouse, he is an intimate witness to it. He jots down notes in "bloodstained notebooks" (p. 65) in his carcass stamper's office adjacent to the killing floor.

Born in this place, Ferrari's poetry is both brutal, vulgar, scatological, crude, raw, anatomical, "expressionist" (Vago, 2018, p. 39) and absolutely lyrical and incredibly delicate. His poetry is sensorial, never sensational: it has a strong visual impact with outstanding chromatic contrasts, dominated by red[4] and white; it is impregnated with smell, vapors, odorous trails, "miasmatic path" (Ferrari, 2019, p. 41), stenches of all sorts; it resonates with "noise of the trucks/the screams of the drivers" (p. 43), of "love songs" sung by those who "fill the intestines with water" (p. 71), of calves' "stammering objections" (p. 23), of a bull's whispers and "lacerating silence" (p. 53), as those of a young heifer, "[…] the only one who doesn't fill with sound/the space of death" (p. 35). Ferrari's poetry chews in the mouth tastes of "guilt/vomited with wine" (p. 31), of "warm blood" (p. 81); it is markedly haptic: hands touching warm innards, foals brushing against each other, hair tousled by the wind, licking tongues, among others.

In sum, "the slaughterhouse is a hell of senses" (Babina, 2022) and Ferrari's poetic word adhere to the starker aspects of this reality. As commented by Cosci (2021), "Ivano Ferrari's style and lexicon are functional to the reality he represents and to the author's personal experience, they are not mere tautology, nor smug exhibitionism: they are life and its backhandedness and travail".

It is unsurprising, then, that Ferrari's unique and dissonant voice, marked by poetry that is never consolatory, uplifting, or reassuring, has remained marginal in the panorama of contemporary Italian poetry. On the other hand, how could it be otherwise when poetry is crafted about and derived from the slaughterhouse, when poetry makes sensible – not just visible – the violence inflicted upon human and non-human bodies, along with boundless suffering and death?

Ferrari did not identify himself as an animal advocate. When directly questioned on this matter, he ironically remarked: "I was a communist. I couldn't have chosen another loser species!" (Budriesi, 2019, p. 66). He preferred to define himself as a "betrayer of the (human) species, the one who communicates to his own kind what was happening" (p. 67). From this standpoint of witness, Ferrari refrains from delivering lessons; his poetry is never didactic nor educational, and animal advocacy-oriented goals are not the primary objectives or motivations of his writing. Nevertheless, Ferrari not only expressed profound compassion for all the "beasts" he encountered during his years at the slaughterhouse (p. 65) but explicitly regarded his poems as "useful for reflection, to not be indifferent". He further stated: "I am not a vegetarian […], but I believe that someone while cutting the Florentine steak will think about all this" (p. 67).

Besides the clear indications of Ferrari's own words and reflective statements, it is evident that the poems themselves carry an extraordinary strength, interest and source of inspiration when read from an antispeciesist point of view. Matteo Gilebbi (2016, 2018, 2019), the English translator of the collection *Slaughterhouse*, has provided a meticulous and insightful ecocritical reading. Additionally, credit is due to the efforts of Laura Budriesi (2019) for her interview with Ferrari, published in the Italian critical antispeciesist journal *Liberazioni*.[5] Gilebbi, in particular, interprets Ferrari through key concepts of Critical Animal Studies, "accompanying the reader along some of the many paths that Ferrari's poetry was able to map" (2016, p. 109). The first path traces the "carno-sado-pornocentric" dimension of and in the slaughterhouse exposed in all its violence and normalized obscenity in several of Ferrari's compositions by adopting Carol Adams' analysis of the "the absent referent" and intersection between sexism and speciesism (Gilebbi, 2018, pp. 48–51).

The second path analyzes a possible "constructive animalization" (2016, p. 109), a way of becoming animal in the dramatic interspecific contact zone of the slaughterhouse which is potentially able to reduce the distance between humans and non-humans and open spaces for empathy and solidarity (2018, pp. 51–56).

In what follows, I will propose a third path that shares a family area with Gilebbi's second path of constructive animalization, but deepens, broadens and consolidates its theoretical and practical implications. This will be accomplished through an elaboration on Calarco's (2011, 2015) indistinction approach to the animal question.[6]

A Poet of Indistinction

The philosophical reflection in the field of Critical Animal Studies – deeply influenced by a constellation of themes and authors in the continental tradition, poststructuralism, posthumanism and feminisms – maintains as a fundamental theoretical acquisition the framing of the animal question in terms of the human/animal distinction, defined in various ways, such as dualism (Plumwood, 1993) or anthropological machine (Agamben, 2002/2004). The common underlying idea, radicalized through the fully antispeciesist twisting of the concept of the anthropological machine, is that the human/animal distinction – with its ontological, ethical and political dimensions – is not a moral prejudice (a mistake of rationality and reasoning) nor a matter of "nature", rather it is the product of material and symbolic apparatuses, rooted in historically contingent, specific, structural social organizations.[7] Reframed in these terms, the question of the animal assumes an unprecedented depth and propagation, stepping outside the boundaries of animal ethics and hitting the foundations of the humanities, social sciences, biology and ethology. Moreover, this perspective enables the systematization of approaches to the animal issue and various forms of antispeciesism. Depending on the movement they advocate for transcending the human/animal distinction, three main approaches have been identified (Calarco, 2011, 2015): identity, difference, and indistinction.

The indistinction approach challenges the dualism of human/animal through a complete subversion of this very distinction(s). Once the possibility of identifying two distinct entities is eliminated, even the equal/different polarity ceases to function. This process involves an erasure of the abyss, a blurring of differences that moves toward human-animal continuity, or more precisely, indiscernibility. The concept of indiscernibility is closely tied to the notions of contamination and hybridization, all of which signify the crossing of normative boundaries, the interruption of saying and of being said, "human" and "animal", in favor of the pre-identitarian realm of relations and encounters. In the literature, indiscernibility lacks a precisely defined status – it is neither purely biological, nor solely ontological, nor completely psychological or sociological. Instead, it can be situated at the intersection of these disciplines. Often articulated in spatial terms, it refers to a space or "*zone of indistinction*" (Calarco, 2015, p. 58) between human and non-human animals that can be inhabited, giving rise to new practices and modes of thought.

Consistent with the critique of the logocentrism inherent in the identity approach, the indistinction approach remains receptive to novel languages that facilitate exploration and comprehension of this realm, including literature and various art forms. Consequently, an analysis of Ferrari's poetry from an indistinction-oriented perspective is justified not solely at the level of content (as I delve into), but also methodologically.

Shocking Reductions in the Slaughterhouse

To inhabit this zone of indistinction means, first of all, to be stripped of the (supposed) characteristics of the human, to be deprived of our status as a person, with our own privileged, subjective, coherent place in the world to the point of experiencing our "fleshy embodiment", the fact that "we, too, are flesh – that we, too, are meat" (Calarco, 2015, p. 58). Just like non-human animals. Recognizing such deep, intimate identity is the outcome of an extremely violent "shocking reduction", such as that experienced by Australian philosopher Val Plumwood who became prey and "mere meat" for a crocodile (2000), or the reduction of human bodies depicted by Francis Bacon into melting, deformed figures beyond their recognizable and organized human shape (Deleuze, 1981/2003).

A shocking reduction of this kind would be preferable if it did not occur within a slaughterhouse. In this context, it is crucial that the distinction between humans and non-human animals remains clear-cut, and that *only* the latter are considered mere meat. Everything within the slaughterhouse, from the spatial arrangements to the objectifying scientific language employed, is designed to (attempt to) ensure this distinction and strengthen the banks of the supposed abyss between human and animal. However, within the very confines of the slaughterhouse, this distinction is besieged from all sides. Thus, despite all precautions, there may be instances of experiencing such a shocking reduction and entering the corporeal, fleshy zone of indistinction, as was the case for Ferrari. The most paradigmatic poem in this regard is the following:

> If I broke down the wall of flesh
> and hanging from the hook I smiled
> what would he say who is paid to dismember
> the stamper of tongues
> what label would they put on me
> how many organs would they discard
> and would the vet think *panta rei*?
>
> (Ferrari, 2019, p. 82)

The lyrical self breaks down "the wall of flesh", he surpasses that barrier, perhaps, that prevents us from recognizing that we too are meat and takes the place of the non-human animal along the disassembly line. He is now hanging on the hook; the killing has already occurred. The smile alludes to the grimace resulting from the involuntary widening of the mouth after the killing. And the hook starts to move, Ferrari's carcass parades in front of the workers, each designated for a stage of the meat production process: the butcher, the stamper, the quality controller, the gutting operator, the veterinarian. "How will they behave?" Ferrari wonders. "Will they notice an ontological difference between my carcass and the others?", "Will they recognize me as human?" Or, as the tone seems to suggest to me, will they approach this new carcass to be transformed into meat with mere technical curiosity: which label is suitable for this one, how many organs need to be discarded?

Ferrari is acutely aware of such a lack of discernibility:

As hard as clotted blood
and as soft as calf's marrow
similar am I if not equal.

(2019, p. 94)

The last verse is particularly significant in interpreting the entire poem. The clotted blood and the calf's marrow are not merely employed as the second terms of comparison for "hard" and "soft". Instead, they become the elements of an equation: it is the same matter, the same hard blood, the same soft bone marrow flowing in the veins and the spinal channel that ground our being equal and indistinguishable. As Gilebbi (2018) epitomizes in commenting on this poem: "We are all this single animal" (p. 55).[8]

The lack of discernibility becomes an even more pronounced and profound impossibility in the dramatic moments preceding death. Death manifests itself in the same way in the case of non-human animals and in the case of Ferrari's sick wife. The phenomenology of death thus takes on a specific poetic phraseology centered on eyes: in the "Death Wife", "The eyes plummet/or rather are lost at the top" (p. 107) and in the "Imperfect Beasts", "The beast collapses/opening wide her white eyes" (p. 92). Then, after death, the only thing distinguishing two perfectly identical meats is the social use made of them:

At a few hundred meters
past the fresh shape of the lawn
and after houses with dull eyes
the cemetery for humans is located
where there is meat that does not feed.

(p. 62)

Ferrari is therefore immersed in this zone of indistinction, recognizing it and fully accepting its painful truth. This is not the case for other workers in the slaughterhouse, such as the butcher, who, by cleanly cutting off two fingers and seeing his own blood flow, experiences a momentary and revocable shocking reduction – "the blood of one like the other" – and the consequent disruption of the human/animal distinction.

Two fingers severed
almost a metaphor
the blood of one like the other
the cussing, the iodine.

An instant of stillness, like an afterthought
then the mandate of the pitchfork
that searches for justice in the sphincters of a heifer.
(p. 40)

But it is just a fleeting instant, the "mandate of the pitchfork" must be fulfilled. In Ferrari's poetry, work tools are often personified. The pitchfork gives orders, the knives "mate/in the monotonous color of blood/tired of the games of hemorrhagic gods" (p. 40), they are "methodical and obtuse" (p. 56), they can be damaged by "uttered regrets" (p. 38). It is as if the knives – as a metonymy for the slaughterhouse – take over the workers. At first glance, I suggest this impression is due to the fact that the human behaviors and qualities attributed to the objects in this personification (in particular, giving orders, not allowing speeches) are typical attributes of the role of the boss. In the second instance, however, by focusing on the "methodic and obtuse" attributes, on the sense of monotony, of tiredness, of repetition, the personification fades and gives way to the machinic, ever-same dimension of the capitalist labor process: organized according to the principles of Fordism along the disassembly chain in which the workers are assimilated into the automatic processes of the machine. In this system, there is no time for "an afterthought". Ferrari's poems illuminate the abattoir as a quintessentially capitalist institution: the centrality of labor time in the production process.[9] The imperative of respecting labor time is expressed in many lines: "With dozens of hits/we have crushed the last head/one respects the schedule not the method" (p. 88), or "(sadness does not prevent us/from starting the slaughter at seven thirty sharp)" (p. 40). Moreover, the class dimension is explicitly invoked:

On Labor Day
we are more than fifty
executioners, rippers
who slaughter and who collect blood
meat trimmers, skinners, movers
those who slaughter door-to-door
tanners, packers and gravediggers,
the working class.
(p. 95)

Vulnerability

Once we "broke down the wall of flesh" and enter the zone of indistinction, what happens?

We experience the glowing order of separation
the onset of the present in the flesh

here the knowledge of pain occurs
the unknown dimension of amazement.

(p. 122)

Which new thoughts and practices may emerge?

First, in the experiencing of the "onset of the present in the flash", occurs "the knowledge of pain". It is the awareness and feeling of vulnerability, which is no longer "mine" or "yours", of the "beasts" (calves, cows, pigs, bulls, horses, lambs, but also cats, a dog) or of the "humans" (Ferrari himself, Ferrari's wife, the workers) but is shared by all beings. The "shocking reduction" makes one realize that the human being is also mere flesh, bodily materiality and flesh to be eaten, slaughterable meat. It is the dimension of suffering, of pain, of vulnerability, of *pathos* (Derrida, 2006/2008, p. 26) that crosses this zone. All the more so if it is experienced in the slaughterhouse. What to do with this knowledge of pain? In line with main directions in the indistinction approach (e.g., Acampora, 2006; Donovan & Adams, 2007; Gruen, 2015), Ferrari seems to tell us that this experience situates us, both on an experiential and theoretical level, on a shared and common ground which can give rise to an ethic capable of crossing the limit of the human, both in its scope and in its method. It is the ground of moral sentiments which, once the body dimension is brought to the fore, also becomes the ground of "corporeal sentiments": compassion, sympathy, empathy, pity, care, attentiveness.

Ferrari's entire anthology of poems testifies to these attitudes, being itself a product of them and crossed by a sense of guilt. The following is but one example:

Where will he hide the tears?
If the question hangs on the smashed skull of a colt
I fade while laboring verses
while suffering animals and things.

(p. 24)

It is a compassion, a pity that is not flat, one-dimensional, tidy or heroic. The lived experience of shared vulnerability seems to enable Ferrari to dig into the complexity and shadows of these bodily/moral feelings, unlike some strains of antispeciesism, such as the feminist tradition of care, which may favor a unilateral positive and bright depiction. Compassion can be ambivalent; it is not a pure sentiment. "Identification with the victims can go together with an obscure complicity with their torturers: in other words, perverse forms of sympathy also exist" (Starobinski, 1982/1985, p. 249). Such as the one shown by slaughterhouse workers when they get on the trucks to drag the non-human animals out toward death and "put blankets on their backs/with" an oxymoronic "torturing tenderness" (Ferrari, 2019, p. 82).

Nevertheless, compassion establishes the possibility of a refusal. Compassion is not merely the sharing of suffering with the victim; it is a declaration that the persecutor's act is unacceptable, and it can lay the foundation for the uncompromising repudiation of violence (Starobinski, 1982/1985, p. 249).

Potentiality

The zone of indistinction, however, is not only crossed by the vulnerability and suffering of fleshy living embodiment. This same embodiment, indeed, simultaneously opens the dimension of potentiality, *potentia* (Deleuze, 1968/1992). Or, in other words, the recognition that animal life (human and non-human alike) is "more than meat" to be consumed (Plumwood, 2000). Potentiality lies precisely in "what is more" (Calarco, 2015, pp. 62–63). Reflecting on such potentiality gives a new meaning to the notion of agency, no longer modulated on anthropocentric (and logocentric) capabilities and attributes, but rather focused on the unforeseen potentials that animals open up with their language: play, inoperativity,[10] wandering, contact, love.[11]

I claim that potentiality is grasped by Ferrari (2019) as the "unknown dimension of amazement" occurring – together with the "knowledge of pain" – in the "present of the flesh" (p. 122). Ferrari's poetry captures and sings this vitalist dimension on all levels: life in all its manifestations, delicate and obscene (Cosci, 2021). Even the flesh, dead or close to death, pulsates. It rebels.

> The dead flesh lives again
> in its great misery
> with the wind that brings back scents
> to an extended order.
> The dead flesh is embroidered
> by those sinuous beings
> that others call larvae.
>
> (Ferrari, 2019, p. 26).

I argue that Ferrari's deep sensitivity precisely to the dimension of potentiality is what enables an effective *zoopoétique* reading of his work, as exemplified by Vago's analysis (2018). "Zoopoétique", as Vago defines it, examines those literary works – whether in prose or poetry – that are capable of producing non-anthropocentric "other" affects and perceptions through the possibilities inherent in the creative language itself (p. 36).[12] In certain poems by Ferrari, an authentic "animal word" (Vago, 2018, p. 37) surfaces, free from the confines of anthropological iconography or literary tropes. A genuine displacement of perspective occurs, with non-human animals emerging as bearers of a personal point of view.

> Who spread this
> farfetched rumor
> that animals are prejudiced
> against me?
>
> (Ferrari, 2019, p. 92)

The "animal word" challenges their reduction to mere meat, to commodity, a fate imposed upon them within the confines of the slaughterhouse. Non-human animals not only speak, judge, see – "she sees me" (p. 34), Ferrari notes, referring to a young heifer –, they also escape. In this case, the potentiality challenging the reduction becomes more palpable, more evident and disruptive. In a word, it becomes resistance.

> An escaped black bull
> wanders on the overpass
> scaring the traffic,
> we chase him
> branding knives
> stun rods and beers
> he runs stops returns
> cops with machine guns arrive,
> now he lies on a thin veil of grass
> and whispers something to the flies.
>
> (p. 36).

And as such it is recognized by Ferrari, who states (Budriesi, 2019): "Escape was resistance. There was hardly ever an episode like the one with the bull" (p. 67).

The fleeing bull, as "animal out of place", is not only reclaiming his own individuality and identity against his property status, of being a commodity in the industry of meat production (Gilebbi, 2018, p. 49), he is also embodying a political stance: the interruption and the sabotaging of the anthropocentric and capitalist dominant paradigm, the order of species allocating respective roles, through the transgression of borders by defying the conceptual and material boundaries that keep him captive (Colling, 2021).[13] The black bull wanders, runs, stops, returns, zigzags through the traffic, the workers of the slaughterhouse chase him with the tools of the trade, transformed for the occasion into rudimentary hunting weapons, but only the intervention of the police – with far more appropriate weapons – reestablishes the status quo, the public order, which is the hierarchical order of species. In this order, power annihilates the potentiality of life escaping for freedom. As often happens in such cases, the disproportion of forces on the field is too unbalanced, and resistance is soon broken. The bull finds no multispecies solidarity (Colling, 2021, pp. 115–130). Nevertheless, Ferrari, up to the very end, bears witness to a stubborn potentiality: the bull in agony on a thin layer of grass (the flowerbed grass of an urban roundabout, of a meadow near the highway crossing the countryside?), the last breath is not passivity; it is a whisper directed to the flies already buzzing over the dying animal. Once again, it is the animal word, a dialogue, perhaps, on the borderline of life and death, a secret from which the human is excluded.

Notes

1 For an in-depth study of the strategic and ethical limitations and potential of violent images and moral shock in the context of the animal advocacy movement see Fernandez (2021).

2 For an analysis challenging the centrality of sight and vision in animal studies and the animal advocacy movement from a historical-materialist point of view see Stefanoni, *infra*.

3 The poems collected in the first Italian edition of *Macello* were published in 2004, almost 30 years after they were written. Sixty of the eighty-six poems in *Macello* had already appeared in 1995 in the collection *Nuovi poeti italiani 4*, edited by Mauro Bersani and published by Einaudi. After this first partial publication, the author rediscovered twenty-six other poems also composed while he was working in the slaughterhouse and which he did not remember having written. These poems are featured in Ferrari's second collection, *La morte moglie*, published in 2013, specifically in the section entitled *Le bestie imperfette*. This collection also includes more recent poems written after his wife's death. The English edition *Slaughterhouse*, brought to fruition through the efforts of Matteo Gilebbi, responsible for editing, introduction, and translation, consolidates both collections into a single volume.

4 See the following verses: "Of red dawns, here,/we don't feel the need/the monotony of the color/thickens the already heavy air,/of scarlet sunsets/let's not talk about those" (Ferrari, 2019, p. 77).

5 See Losi and Bertuzzi (2020, pp. 81–89).

6 In the Italian context, an akin analysis has been developed by Massimo Filippi who converges with and expands upon Calarco's ideas. Filippi labels the third approach "antispeciesism of the improper" (2012) and "antispeciesism of the common" (2017).

7 For a critique of an anthropocentric residue in Agamben see Calarco (2008), pp. 96–102; for non-anthropocentric proposals: speciesist machine (Filippi, 2017), sacrificial *dispositif* (Stefanoni, 2020), anthropological form (Aloe & Stefanoni, 2021).

8 While my interpretation of this poem is consonant with Gilebbi, my analysis of "If I broke down the wall of flesh" is less optimistic than Gilebbi's one. See 2018, p. 55.

9 According to Marx (1867/1982, pp. 417–454) there are two possibilities for increasing surplus value: either by extending the working day (absolute surplus value) or, with the working day remaining constant, by reducing the necessary labor time (relative surplus value). The latter is possible by increasing the productivity of the means of production.

10 Inoperativity is a key concept in Agamben's philosophy. In *The Use of Bodies*, the last volume of the *Homo Sacer* project, great attention is devoted to this concept which seems here characterized in exclusively human terms. Agamben writes: "the properly human life is the one that, by rendering inoperative the specific works and functions of the living being, causes them to idle [It., *girare a vuoto*], so to speak, and in this way opens them into possibility. Contemplation and inoperativity are in this sense the metaphysical operators of anthropogenesis, which, in liberating living human beings from every biological and social destiny and every predetermined task, render them available for that peculiar absence of work that we are accustomed to calling "politics" and "art"? (2014/2015 pp. 277–278). Despite this definition, it is interesting to refer to a related Agambenian concept, that of profanation, an operation that is not exclusively human but is also given "in nature", as Agamben writes to introduce the following example: "The cat who plays with a ball of yarn as if it were a mouse [...] knowingly uses the characteristic behaviors of predatory activity [...] in vain. These behaviors are not effaced, but, thanks to the substitution of the yarn for the mouse [...], deactivated and thus opened up to a new, possible use. But what sort of use? For the cat, what is the possible use for the ball of yarn? It consists in freeing a behavior from its genetic inscription within a given sphere (predatory activity, hunting). The freed behavior still reproduces and mimics the forms of the activity from which it has been emancipated, but, in emptying them of

their sense and of any obligatory relationship to an end, it opens them and makes them available for a new use. The game with the yarn liberates the mouse from being prey and the predatory activity from being necessarily directed toward the capture and death of the mouse. [...] The creation of a new use is possible only by deactivating an old use, rendering it *inoperative* [emphasis added]" (2005/2007, pp. 85–84).

11 For an antispeciesist reading of these notions, see, for example, Filippi (2020, pp. 235–237).

12 *Zoopoétique* is a field devoted to the inquiry of life, animality and literature that originated in France as part of the research project "Animots" (2010–2014). This field is attentive to the political implications of zoopoétique analysis in the direction of a criticism of the animal-industrial complex. See Simon and Taïbi (2015).

13 On the notion of animal resistance see also Hribal (2010), Reggio (2022) and Wadiwel (2016).

References

Aaltola, E. (2014). Animal suffering: Representations and the act of looking. *Anthrozoös*, *27*(1), 19–31. https://doi.org/10.2752/175303714X13837396326297.

Acampora, R.R. (2006). *Corporal compassion: Animal ethics and philosophy of body*. University of Pittsburgh Press.

Agamben, G. (2004). *The open: Man and animal* (K. Attell, Trans.). Stanford University Press (Original work published 2002).

Agamben, G. (2007). *Profanations* (J. Fort, Trans.). Zone Books (Original work published 2005).

Agamben, G. (2015). *The use of bodies* (A. Kotsko, Trans.) Stanford University Press (Original work published 2014).

Aloe, F., & Stefanoni, C. (2021). Anatomia della nazione. Dalla formula trinitaria alle forme della produzione di popolazione. *CONSECUTIO RERUM*, *10*, 357–82. https://www.consecutio.org/wp-content/uploads/2021/07/aloe.pdf.

Babina, P. (2022, Feb. 2). *Macello – Un'azione poetica*. Il primo amore. https://www.ilprimoamore.com/macello-unazione-poetica/.

Budriesi, L. (2019). "Quel toro che scappa sono io". Intervista a Ivano Ferrari. *Liberazioni. Rivista di Critica Antispecista*, *38*, 63–68. https://www.liberazioni.eu/wp-content/uploads/2019/10/Budriesi-1-lib38.pdf.

Calarco, M. (2008). *Zoographies: the question of the animal from Heidegger to Derrida*. Columbia University Press.

Calarco, M. (2011). Identity, difference, indistinction. *In CR: The New Centennial Review*, *11*(2), 41–60. https://www.jstor.org/stable/41949742.

Calarco, M. (2015). *Thinking through animals: Identity, difference, indistinction*. Stanford University Press.

Colling, S. (2021). *Animal resistance in the global capitalist era*. Michigan State University Press.

Corman, L. (2017). Ideological monkey wrenching: Nonhuman animal politics beyond suffering. In D.A. Nibert (Ed.), *Animal oppression and capitalism. Volume 2: The oppressive and destructive role of capitalism* (pp. 252–269). Praeger.

Cosci, P. (2021, June 28). *Due lettere e un saggio per Ivano Ferrari*. Il primo amore. https://www.ilprimoamore.com/due-lettere-e-un-saggio-per-ivano-ferrari/.

Deleuze, G. (1992). What can a body do? In: *Expressionism in philosophy: Spinoza* (M. Joughin, Trans.) (pp. 217–235). New Zone (Original work published 1968).

Deleuze, G. (2003). *Francis Bacon: The logic of sensation* (D.W. Smith, Trans.). Continuum (Original work published 1981).

Derrida, J. (2008). *The animal that therefore I am* (D. Wills, Trans.). Fordham University Press (Original work published 2006).

Donovan, J., & Adams, C.J. (Eds.). (2007). *The feminist care tradition in animal ethics: A reader*. Columbia University Press.
Fernandez, L. (2021). *Imágenes que liberan. Shock moral y comunicación visual estratégica en el activismo por la liberación animal*. [Doctoral dissertation, Universitat Pompeu Fabra]. e-Repositori UPF://hdl.handle.net/10803/6730343.
Ferrari, I. (2019). *Slaughterhouse (Macello)* (M. Gilebbi, Trans.). Legas.
Filippi, M. (2012). Antispecismi. In M. Andreozzi (Ed.), *Etiche dell'ambiente. Voci e prospettive* (pp. 199–213). Led Edizioni.
Filippi, M. (2017). *Questioni di specie*. elèuthera.
Filippi, M. (2020). "*Il faut bien tuer*", or the calculation of the abattoir (C. Salzani & E. Monacelli, Trans.). In F. Cimatti & C. Salzani (Eds.), *Animality in contemporary Italian philosophy* (pp. 223–243). Palgrave MacMillan.
Gilebbi, M. (2018). Witnessing the slaughter: Human and nonhuman animals in Ivano Ferrari's poetry. In S. Iovino, E. Cesaretti, & E. Past (Eds.), *Italy and the environmental humanities: Landscapes, natures, ecologies* (pp. 47–56). University of Virginia Press.
Gilebbi, M. (2019). Becoming human in the slaughterhouse. In I. Ferrari, & M. Gilebbi (Eds.), *Slaughtehouse* (Trans.) (pp. 10–19). Legas.
Gilebbi, M. (2016). Testimoni dei macelli. Esseri umani e animali nella poesia di Ivano Ferrari. *Ecozon@: European Journal of Literature, Culture and Environment*, *7*(1), 94–111. https://doi.org/10.37536/ECOZONA.2016.7.1.981.
Gruen, L. (2015). *Entangled empathy: An alternative ethic for our relationships with animals*. Lantern books.
Hribal, J. (2010). *Fear of the animal planet: The hidden history of animal resistance*. AK Press.
Jonas, H. (1954). The nobility of sight. *Philosophy and Phenomenological Research*, *14*(4), 507–519. https://doi.org/10.2307/2103230.
Losi, G., & Bertuzzi, N. (2020). What is Italian antispeciesism? An overview of recent tendencies in animal advocacy. In F. Cimatti, & C. Salzani (Eds.), *Animality in contemporary Italian philosophy* (pp. 71–93). Palgrave MacMillan.
Marx, K. (1982). *Capital: A critique of political economy. Vol. One* (B. Fowkes, Trans.). Penguin (Original work published 1867).
Pachirat, T. (2011). *Every twelve seconds: Industrialized slaughter and the politics of sight*. Yale University Press.
Plumwood, V. (1993). *Feminism and the mastery of nature*. Routledge.
Plumwood, V. (2000, July-August). *Being prey*. Utne Reader. https://www.utne.com/arts/being-prey/.
Reggio, M. (2022). Do nonhuman animals resist? Critical geographies, decolonial theories, and the case for veganism as multispecies solidarity. In S. Springer et. Al (Eds.), *Vegan geographies. Spaces beyond violence, ethics beyond speciesism* (pp. 239–257). Lantern.
Simon, A., & Taïbi, N. (2015). Qu'est-ce que la zoopoétique? *Sens-Dessous*, *16*(2), 115–124. https://doi.org/10.3917/sdes.016.0115.
Sontag, S. (2003). *Regarding the pain of others*. Picador.
Starobinski, J. (1985). *Montaigne in motion* (A. Goldhammer, Trans.). University of Chicago Press (Original work published 1982).
Stefanoni, C. (2020). On the origins of the anthropological machine: Sacrificial dispositif and equality. *Animal Studies Journal*, *9*(2), 284–310. https://dx.doi.org/10.14453/asj/v9.i2.12.
Tunstall, K.E. (2011). *Blindness and enlightenment: An essay: With a new translation of Diderot's "Letter on the blind" and La Mothe Le Vayer's "Of a man born blind"*. Bloomsbury Publishing USA.
Vago, D. (2018). Tori in fuga e parole cercate. Una lettura 'zoopoétique' della poesia di Ivano Ferrari. *SEMICERCHIO*, (58-59), 36–40. https://semicerchio.bytenet.it/files/36_vago.pdf.
Wadiwel, D.J. (2016). Do fish resist? *Cultural Studies Review*, *22*(1), 196–242. https://doi.org/10.5130/csr.v22i1.4363.

17

EMBODYING NON-SPECIESISM THROUGH ALTERED STATES OF CONSCIOUSNESS

Cindy Brooks Dollar

As this book emphasizes, the life of all beings – human and non-human – are intertwined. Our ability to recognize this fundamental truth unites us through a collective existence. Such an argument implies a non-dual perspective, which has vast implications for how we perceive, organize, and engage in our social relations.

Speciesism, through its postulation that humans are favored as the most powerful earthly beings (Sollund, 2008), exemplifies a dualist perspective. Speciesism contributes to unequal treatment across species and is linked to humans' (ab)use of non-human animals (Fitzgerald & Taylor, 2014; Weitzenfeld & Joy, 2015). It arguably explains impudent treatment of other non-human entities as well. For instance, "nature" or the environment may be characterized as a mere resource available for human manipulation and consumption.

When speciesist thought is systemic, as it is in the United States, violent practices against non-human species become institutionalized, upheld by conventional power structures that ensure the priority of humans (Adams & Donovan, 1995). Such arrangements justify and legitimize the exploitation and oppression of other-than-human beings. Indeed, speciesist practices allow humans to harm non-human entities with relative impunity, effectively encouraging actions that destroy non-human beings for economic profit and socio-political power. As this book seeks to demonstrate, the consistent, patterned violence toward non-human animals largely occurs *because of* the "othered" status attributed to non-human entities. This chapter, however, seeks to accentuate the processes involved in questioning and challenging speciesism.

By presenting information collected through in-depth interviews with persons who reported entering an altered state of consciousness following psychedelic substance ingestion, my hope is that this chapter will provoke conversations about why, when, how, and where speciesism may be resisted. Most interviewees

DOI: 10.4324/9781003441908-22

unexpectedly voiced that their psychedelic-induced state encouraged them to interrogate the presence and purpose of speciesist principles and practices. Through various psychedelic encounters, they sensed an inextricable relationship to other humans and non-human entities, which compelled them to question their previous held assumptions of and beliefs about human superiority. This newfound non-speciesist awareness was met with the questioning of other common cultural norms. In this way, the data denote non-speciesism as a transformative life event, which has immense socio-political, and even metaphysical, implications.

Non-Speciesism as the Personal and Political

Individual experience is inseparable from socio-historical context. The feminist movement of the mid-20th century coined the phrase "the personal is political" to call attention to the ways in which patterned socio-political arrangements, especially unfair systems of power, situate personal experience (e.g., Hanisch, 1970). Such an argument is relevant here in that the interpretations presented herein rely on individuals' in-depth interviews. While I concede the limitations of narrative-driven data, including issues of scope, memory (re)making, and potential social desirability bias, they do not preclude an advancement of knowledge about social patterns and sociological insights.

The Data

Most psychedelic drug use in the West takes place in social settings (SAMHSA, 2020) even though contemporary media and political attention focus on consumption in regulated, clinical settings (e.g., LaMotte, 2022; Moniuszko, 2022). Arguing that the disjuncture between actual use and publicized use distorts information about psychedelic use(rs), I initiated a project in 2019 to gather information from persons who ingested psychedelic substances outside of a medicalized setting. Throughout 2020, I conducted 37 in-depth interviews.

Participants age ranged from 18 to 71. All identified their gender using discrete categories, 19 identified as male or men and 18 as female or women. Though every participant was a current resident of the United States, the sample includes persons from six different nationalities. Most interviewees identified as White, but five identified as Latinx, two identified as Black or African American, and two identified as "mixed" race. Sexual identities varied, being described as heterosexual, bisexual, gay, lesbian, pansexual, and asexual. Interviewees held different religious affiliations – Jewish, Christian, Buddhist, Taoist, Hellenistic, and pagan. Some indicated no religious affiliation (e.g., atheist or agnostic), and in many of these cases further commented that they were "spiritual but not religious." Most interview participants had at least some college education and were currently employed. Occupations included relatively low-skilled service industry jobs, skilled trade work, and highly specialized professions. Those who identified as unemployed explained their job status as either retired or out of work due to COVID-related layoffs (see Dollar 2021, 2023 for additional sample descriptions).

The project was not intended to investigate (non)speciesism, and the interview instrument contained no questions meant to elicit such storylines. That said, ingesting psychedelics is known to bring about feelings of harmony. This is apparent, for example, when the founder of LSD, Alfred Hoffman (1979), recounted his first guided experience with LSD. Writing of the event, he noted "the shimmering beauty of unity." Still, I was surprised by the consistent non-speciesist rhetoric across the interviews. It was one of the most prevalent themes evidenced in the data! Well over half of the interview participants brought up issues related to (non) speciesism as they open-endedly recalled their psychedelic histories. Such a finding reflects Mario Small's (2010) notion of "unanticipated gains." Although Small coined this phrase to explain the interpersonal benefits of organizationally situated social networks, the concept could be applied to scholars who beneficially encounter surprise discoveries in research.

In the following pages, I share excerpts from the interviews to illustrate how respondents discussed their confrontations with problematic speciesism and their subsequent considerations of non-speciesism. The narratives reveal that under the influence of psychedelic compounds, they perceived an expansive and undeniable connection to other entities – human and non-human alike. The participants reported that, in this way, psychedelics acted as a teacher that showed them how their life, as humans, were inseparable from environmental life and the life of other animals. Statements further specified speciesism as a hegemonic expectation, thus proposing non-speciesism as a rejection of dominant cultural values. As such, their psychedelic experiences were framed as transformative. The altered state of consciousness ushered in by psychedelic substances altered their perceptions, their understandings of culture, and often their routine day-to-day behaviors.

Emerging Concerns of Speciesism

Thirty of the thirty-seven persons interviewed spoke about how psychedelic-induced altered states of consciousness challenged previously unquestioned speciesist thoughts and behaviors. Their emerging concerns were discussed in various ways. In the following, I group them into two distinct categories: reconsidering humans' relationship with and treatment of the environment, and reconsidering humans' relationship with and treatment of other animals.

Reconsidering Humans' Relationship With and Treatment of the Environment

In our modern world, the environment is often taken for granted. Nature remains largely invisible unless it is deemed a "resource" available for human use as a potential commodity. Ecological consciousness has been linked to psychedelic use since the early days of psychedelic science (e.g., Masters & Houston, 1966). The participants in this study voiced this link as well. Sixteen interviewees mentioned their psychedelic experience as changing their understanding and treatment of the environment. Importantly, the changes did not seem to be correlated with the frequency or type of substance ingestion.

The interview participants all indicated that their reconsideration of humans' relationship with and treatment of the environment occurred because they *felt* a sense of connection to specific environmental objects while under psychedelics' effects. It was experiential; changes were not seemingly provoked by cognitive understanding or rationalized analysis. One person, who reportedly used psychedelics simply "to have fun," commented that their psychedelic state of consciousness,

> definitely changed how I view the world. Like that tree that fell at my house yesterday [gestures to fallen tree in yard], I feel that more. I feel like I can appreciate the environment more. I don't know why.

Another participant, who had used psychedelics only once in her life, also indicated difficulty in making sense of her psychedelic experience. She relayed the following:

> I know this sounds crazy, but I actually experienced going to the coral reef...I felt very free... I don't really know if I liked that or not [laughs] ...I've thought about it and wondered why I would I ever imagine going there. Environmentally, they're saying that the coral reef is dying. This is where I'm from, ya know? And I feel like it was symbolism. It's where we're going as people, and this is what we're going to leave behind. It was on a deeper level for me. Unfortunately, we're killing our own planet. If the coral reef doesn't exist, there goes the whole entire life of fish and that type of stuff.

In this excerpt, the interviewee specified that she "actually experienced" going to a place that she associates with home. She also mentioned the freeness she felt while in that location. Following this experience, she noted that this area is "dying" because "we," as humans, are "killing our own planet." Although she later described this event as a moment of "significant learning," she also said that even after having thought about it, she was unable to understand the meaning. In fact, she explained "it wasn't good" because "it produced a lot of anxiety." When I asked her further questions about her single psychedelic experience, she returned to her visit to the underwater ecosystem, saying that the experience nagged at her. It was this unresolved nagging that made her adamant she would never use psychedelics again. Her visions of the coral reef perplexed her. "I just don't know what to do with this information," she told me.

Showing a bit less uncertainty is another participant who had accumulated well over 100 psychedelic experiences. In describing one of his last psychedelic experiences, he told me, "My consciousness changed with a tree." When I asked him if he could tell me, he continued:

> So, there's a big scale and a little scale, but I'm different even throughout every little conversation throughout the day. So, for me, there were a couple of variables that play into this whole thing, meaning the perspective of the environment – the world. One of the variables for me– Part of me knows I'm a little different, but I wanted to be normal if that makes sense. So, politically I had grown up around a lot of conservative men and stuff like that, and most of them were successful, and that's what I wanted to be, successful. So, I would put on this façade like [laughs] I'm a hard worker and Mr. Conservative, but really that's not me at all. I mean, I am a hard worker, but there was a huge façade there that I painted around myself to feel normal, like if I could fake this then I'm good, right? But I'm over that now, dude, I understand myself these days. When I was that tree, I feel like I was able to see many different political perspectives at the same time. I feel like I'm able to better put myself into the shoes of others without actually walking in their life path if that makes sense.

Toward the end of the interview, in response to the prompt "Can you tell me a bit more about yourself?" which I asked all interviewees, this participant once again referenced his tree-consciousness encounter. When describing his current life situation, he told me "When I was that tree, I got something. I don't know exactly what it was, but I got something." Like the participant who visited the coral reef, this participant viewed his psychedelic-induced experience as a significant transformative event even though he was unable to communicate with linguistic precision. Unlike the coral reef visitor, however, this participant pinpointed the tree-consciousness encounter as altering his life path. He told me that it became clear that we, as humans, have responsibility to do what we can to protect the environment, and that he gained "the confidence to know it was right." Although he confessed that prior to this psychedelic encounter he had never considered environmental issues as political issues, the tree-consciousness incident led him to enroll in a university to study environmental engineering. He reported no knowledge of anyone with this degree, training, or career, he was not of traditional college age, and did not have college-educated parents. Despite these circumstances and his living in financial and residential instability, he described having faith in his plans and aspirations because of his psychedelic-induced interaction. At the time of the interview, he was in his third year of college and was employed on a project that tested water quality in rural areas of the southeastern United States.

Other examples of speciesist confrontations were evident in the data. As indicated in the following excerpts, the participants' recollections, like those mentioned earlier, imply their newfound awareness stemming from a *feeling*, not from cognitive reasoning. The perceived distance between themselves and the environment was diminished, and in this way, humans and the natural environment became indistinguishable, something previously unthinkable to them.

> It makes you feel in tune with the whole environment. You can see the sun from the inside because you're a part of it.
>
> The other thing I felt, and this is a hallmark of my experience with mushrooms, is the world seems more magical, for lack of a better word. It seems like the environment is interacting with me in a way that is unusual. Like we're one thing working together.
>
> It put me in touch with the environment. I didn't always realize it, but I've always had a connection to something bigger than me.
>
> I don't remember any specific thought, but I guess you could say I was appreciating life in a different way, especially the beauty of it. I wasn't on the outside of anything. That's when I felt happiest and noticed things I don't normally notice. I was appreciating things that I normally took for granted. I feel a deeper connection with the environment, so I definitely think that it can broaden you – make you more in tune with nature.

Across these expressions, the participants identify as inseparable or "at one with" the environment. Although this recognition came from a visceral experience, participants sought to make sense out of their feelings, arguably to make them useful. One person remarked, "I had to figure out what to do with them." It was here where cognition, reasoning, and rationalization surfaced, which made the participants concerned about their previously held assumptions and enactments of dominance over the environment. One interviewee acknowledged this when he said,

> So before, I would have been more collapsed and responded much more adversarially, while also being detached from it. Now, there's definitely more harmony in how I respond and relate to the environment.

Another participant who reported using psychedelics "for therapeutic reasons" shared,

> I think it made me appreciate things that I never looked at or thought about before. It made me want to appreciate and be more respectful of the environment. It's just a love of the Earth. (laughs). Breathing clean air, eating clean food, cleaning up after yourself, picking up other people's trash if you see it out, being kind to others.

This latter excerpt demonstrates how environmental responsibility, self-care, and interpersonal consideration may intersect. Such rhetorical alignment was common across narratives. One interview participant, who had over 100 ingestions of various psychedelics, closed the interview by saying,

> I think it helps us respect everything – your environment, the world, people, caring about what is happening in your community, and your state, and country, and also other countries.

These storylines reflect the expansiveness inherent in non-speciesism, thus suggesting how one manifestation of speciesism, the animal-industrial complex, implicates both environmental destruction and the systematic exploitation of animals.

Reconsidering Humans' Relationships With and Treatment of Other Animals

Psychedelic consumption not only modified users' perceptions and treatment of the environment, it also altered participants' understanding of other animals. As suggested earlier, psychedelic consciousness was said to reveal the intricate entanglement between one's own life and the life of other animals, both human and non-human.

Non-Human Animals

Participants regularly voiced apprehensions about factory food production and animal meat consumption as a detrimental practice. For instance, the youngest interview participant told me, "Psychedelics helped me see things differently. It really makes you focus on the bad things and makes you want to fix them." When I asked if she could give me an example of what she meant, she immediately replied, "Well, the grossness of meat-eating…I would think about meat and how gross it is. Like, we have to kill an animal and eat the meat, and that's really gross." Later in the interview, she mentioned that this realization led her to search for information about the meatpacking industry, which resulted in her and her friends searching for "alternative ways of eating" and ultimately adopting a vegan diet.

Even among interviewees who reportedly consumed meat and other animal parts, they expressed thinking differently about doing so following their psychedelic explorations.

> Before [ingesting psychedelics], I thought vegetarians and vegans were weird. And that was just boring to me – how could you live without meat? That just wasn't the way things were. It wasn't quite right. But exploring from that point [after psychedelics], I explored big pharma and agriculture stuff. I had been taking Prozac. And that's kinda what opened me up to being [pause], uhm – it helped me open my mind and be more receptive to alternatives. I would say

> after mushrooms, after my first psychedelic experience, that was an introduction to that way of thinking. For me, it was more like, okay, there's something up with what I've taken as true before. That's how I took my experience.

Other participants spoke about changes in dietary preferences in other ways. For instance, approximately four minutes into his interview, one remembered going to a fast-food restaurant while taking LSD. Saying psychedelics changed his perception of food, he recalled the food "looked like plastic." He framed the event as a significant turning point in his life. The interviewee had been chronically ill since age 14 and fast food was easily accessible and convenient, but he "just couldn't eat it…because it was not real food." Despite its adverse effects, eating fast food is extremely common in the United States. About half of U.S. residents eat fast food each day (Fryar et al., 2018), so to forego doing so, especially through one's teen and young adult years, is unusual.

Interestingly, two young males, each with well over 100 psychedelic ingestions, reportedly confronted their self-identified eating disorders following psychedelic revelations. One indicated that he was "severely obese" prior to trying psychedelics and attributed his compulsive eating habits as resolved because psychedelics "showed him some deep stuff" about himself and his preference for unhealthy, but culturally popular, foods. He purportedly changed his "relationship with food" and lost nearly 200 pounds. The other participant spoke about his extreme diet restrictions prior to his experimentation with psychedelics. That, coupled with his excessive workout routine was tempered, he said, because of the confidence and likeability he gained in himself following multiple high dose psychedelic ingestions.

Although food production and consumption were commonly cited as frameworks by which interviewees reconsidered their relationship and treatment of non-human animals, other storylines also occurred. One interviewee described seeing a reflection of herself while on LSD. The reflection was that of a non-human animal, which raised questions about how closely related she was to other species. She concluded from this and subsequent experiences that "We are all connected…Everything is a part of everything else in the universe…so we have to get back to us planting and taking care of animals."

Human Animals

Participants also expressed visceral feelings of connection to their own species following psychedelic ingestion. Many indicated that their psychedelic consciousness urged them to reassess their relationships with other humans and human institutions. The consequences of such reevaluation occurred on multiple levels.

Some interview participants spoke specifically about how psychedelics helped them build healthier interpersonal relationships with others. One person promoted psychedelic consciousness as helping "get your mind right, so you can be a good friend, a good spouse, a good citizen." He continued by claiming that psychedelics gave him the ability to comprehend how others may be feeling and perceiving his actions, which he said was useful to resolve interpersonal conflict.

> It pulls me back to the truth of humility, like if I'm having a disagreement with someone, my first order of business is now to check with myself and see what I could do differently…The drugs really spark those "oh yeah" moments because you see the other's point of view more clearly.

Providing more specific details about how psychedelics provided insights into interpersonal relationships was one of the younger respondents. She reported that her LSD experiences allowed her to "better gauge" how she treated her mother. "I just remember seeing myself as being crazy and mean to my mom. Like, why did I have to be that way? I had thought I was upset, but really, I was just sad. I could see that." The participant admitted that she still struggled to treat her mother courteously, but through tears she told me that her mom was her closest friend and most stable supporter. Although the participant was hesitant to fully credit psychedelics with this realization, given her complicated history with addiction and her months of talk therapy, she spoke about psychedelics' unique ability to help her "see many things differently."

Psychedelic encounters also seemed to induce criticism of the intent and procedures of human-made institutions. Some participants framed their psychedelic experiences as heightening concerns about how humans were controlled, especially through religious means, legal policies, and medical industries. Illustrating macro-social scrutiny, one participant recalled becoming angry when he was confronted with the realities of the gender hierarchy during one of his psychedelic encounters. He told me, "I saw how males had used history to claim dominance over females." This vision appalled him. He elaborated,

> And that's what I saw. I guess for the first time really. The logic was to prevent women from rising up and having equal rights, so men said, "Let's rewrite history to make women the problem." That it blew my mind. I was so upset about it. It seemed so hypocritical to me, and at that moment, I decided I'm done. I actually left my religion because of it.

Another participant, who defined psychedelics as a "gamechanger" for his health, spoke at length about how criminalizing psychedelics meant naturally occurring substances, like psilocybin mushrooms, became illegal, while industrially synthesized substances, made for economic profit, became overused. He closed the interview by stating,

> You know, there's a certain level of access that gets taken away from people when it's medicalized. I probably paid $200 or $300 on mushrooms and $200 on ayahuasca. That's a good investment in my mental health. I've had insurance

> through a few jobs, but we don't always have it, so the return on this investment is not so bad, you know? I'm thankful for mushrooms, in particular. To be able to use them on my own terms. We have to keep that in mind.

Underscoring the sentiment of safe, legal access to "natural medicine" as opposed to relying on manufactured pharmaceutical intervention, another participant shared that after decades of being on numerous prescribed medications for chronic health issues, psychedelics "lifted the fog," showing him that his prescription regimen was hampering his "ability to live life [while] making lots of money for other people."

Speaking explicitly about concerns with the military and police forces' harmful practices, of which he was a part, one participant found that psychedelics delegitimized his job tasks. After describing his intent as a military and police officer was to "keep people safe," he continued,

> But they [psychedelics] show you things. Things that you've forgotten, that you didn't even know still bothered you…It was like, "why are you always chasing this being a soldier and being a cop and all this other stuff?" It's to impress your father, and he's never gonna be impressed anyway, so just leave it alone. Do something that's going to make you happy.

After going into some detail about the relationship he had with his father and grandfather, both of whom were career military men, the participant suggested that his psychedelic experiences brought an "awakening" that helped him "say goodbye to some tragic memories," which included being assaulted at school multiple times. During the interview, he expressed tinges of shame in some of his past work activities, although he also conveyed tension between that shame and his continued allegiance to be "a protector."

Concluding Remarks

The interview data suggest that psychedelic-induced altered states of consciousness raise questions about (non)speciesism. Interviewees unexpectedly but consistently spoke about how their psychedelic experience manifested concerns with previously held assumptions regarding the superiority of humans. While the content of these discussions varied, they were often voiced as concerns about their past participation in polluting or destroying the environment, eating industrialized food, and negative treatment of other (non)human animals. Participants also implied that their non-ordinary consciousness states provoked apprehensions about violence, including but not limited to legalized forms of dominance and aggression. This was evident when interviewees spoke about the injuries associated with a for-profit [capitalist] economy, and the patriarchal, racist, ethnocentric and xenophobic projects contained within it. These sentiments have been raised in earlier studies (see Ayres, 2020; Courtwright, 2001; Dávalos et al., 2011; Fitzgerald, 2010, 2015; McSweeney, 2015; Nierenberg, 2005; Paley, 2014; Ritzer, 1995; Schlosser, 2001; Sinclair, 1946); however, the participants in this study specified that it was their psychedelic consumption that introduced them to their newfound awareness.

The narratives indicate *visceral experiences of empathy* as motivation to move toward more inclusive and compassionate life practices, including non-speciesism.[1] Perhaps visceral or emotional empathy is required to embody non-speciesism. Participants commonly mentioned that non-speciesist ideas arose from *feeling* a deep connection to the world around them. The feelings in turn prompted participants to problematize assumptions about separation, hierarchy, and competition – all of which are present in speciesist worldviews. With the rise of questioned assumptions, participants became conscious of their conformity to systemic social ideologies and practices. What is more, the narratives suggest that conformity by participating in social life without doubting the validity of existing institutionalized arrangements (a la Schütz, 1962; Schütz and Luckmann, 1973).

As a result of their newfound psychedelic confrontations, interviewees describe a transformative process wherein the presence, purpose, and consequences of speciesism were perceived, interrogated, and perhaps resisted. In this way, non-speciesism should be perceived as a process – one that involves initially doubting what was previously held as commonplace, if not natural. In some cases, the interviewees speak of rising doubt about abstract ideals like the need for dominance and negative competition, and in other cases, interviewees are more specific about how their doubt altered their consumption patterns and treatment of other animals. In any event, when the transformative process was initiated, opportunities for change became possible. While I am not aware of research that explicitly investigates empathy's influence on non-speciesism, Blair (2018) concludes that highly empathetic people are less likely to display aggression and cause harm. This finding coupled with the findings I present here and prior psychedelic research alluding to a negative association between empathy and aggression could be suggestive for future research. For instance, it may be possible that the negative association between empathy and aggression and aggression and harm is mediated by non-speciesist understandings.

Of course, even if psychedelic consumption is one entryway toward non-speciesism, it is certainly not the only way. Seeking, experimenting with, and entering altered states of consciousness has occurred for centuries. Across a variety of methods, practitioners often claim significant psycho-social, existential, *and* spiritual consequences. Among those who tout the benefits of non-ordinary consciousness states, stories about its ability to (re)ignite our capacity to heal is prominent, and most of these stories speak of the necessary enactment of integration – with oneself and others.

Integration may be described as a process of combining, unifying, consolidating, or desegregating. Non-speciesism, then, is arguably an integrative act. By rejecting the assumption of human superiority, it questions the naturalness and need to separate and unevenly value species. Through its interrogation of convention, non-speciesism can equalize, or at least, align human interests with the interests of other species.

Given the pervasiveness of non-speciesism as conventional Western conduct, it is worth asking if speciesism is an entirely socio-cultural phenomenon. In considering this question, I turn to the work of Erich Fromm. Fromm (1964) coined the term biophilia to explain the fundamental desire for humans to develop harmonious relationships. Combining the ancient Greek words bio (life) and philia (love), he sought to capture human's innate love of life by arguing that we are drawn to *all* things that are alive and/or seemingly necessary to life (Fromm, 1994). Fromm (1964; 1956) rationalized that this attraction explains humans' tendency to fear death and their search to overcome loneliness. Importantly, Fromm (1973) saw the potential for realizing biophilia *only when social conditions did not thwart its presence*. He proposed that our natural inclination toward biophilia was suppressed via institutionalized social arrangements and cultural practices. In fact, Fromm (1973) reasoned that necrophilia, or an attraction to death, reigned supreme in our contemporary world, resulting in a desire to destroy, rather than foster, life. Throughout his writings, Fromm aptly establishes serious psychological, social, and spiritual implications of having conflicting innate and social desires, values, and behaviors. So again, to the extent that speciesism advocates hierarchy and dominance and legitimates disconnectedness, it functions to perpetuate separateness through "othering" processes. These phenomena have been linked to a plethora of harmful, destructive acts.

Even though the interview participants in my research expressed no previous awareness of the existence of speciesism prior to their psychedelic consumption, most told unprompted stories of how their psychedelic-induced state of consciousness confronted them with entanglement – specifically, a feeling of mutuality between themselves and other living things and a responsibility to rethink and/or rectify their future orientations. With this, feelings of concern about speciesism were launched. In short, their psychedelic consciousness problematized duality more generally, and speciesism specifically.

Note

1 This statement is meant to highlight differences between empathy and compassion, whereby visceral empathy entails an emotional response to a problem, and compassion empowers change that seeks to develop a solution to the problem (e.g., Jinpa, 2021).

References

Adams, C.J., & Donovan, J. (1995). *Animals and women: Feminist theoretical explorations*. Duke University Press.
Ayres, T. (2020). *The war on drugs and its invisible collateral damage*. Routledge.
Blair, R.J.R. (2018). Traits of empathy and anger: Implications for psychopathy and other disorders associated with aggression. *Philosophical Transactions of the Royal Society B*, *373*. Available at https://royalsocietypublishing.org/doi/epdf/10.1098/rstb.2017.0155.

Courtwright, D. (2001). *Forces of habit: Drugs and the making of the modern world*. Harvard University Press.
Dávalos, L. M., Bejarano, A. C., Hall, M. A., Correa, H.L. et al. (2011). Forests and drugs: Coca-driven deforestation in tropical biodiversity hotspots, *Environmental Science and Technology*, 45(5), 1219–1227.
Dollar, C.B. (2021). Recreation and realization: Reported motivations of use among persons who consume psychedelics in non-clinical settings, *Journal of Qualitative Criminal Justice and Criminology*, 10(4). Available at https://www.qualitativecriminology.com/pub/vkpkloso/release/1
Dollar, C.B. (2023). Heads, seekers, psychonauts, and one-timers: Patterns of psychedelic consumption. *Journal of Drug Issues*, 53(3), 402–421.
Fitzgerald, A. (2015). *Animal advocacy and environmentism: Understanding and bridging the divide*. Polity Press.
Fitzgerald, A.J. (2010). A social history of the slaughterhouse: From inception to contemporary implications. *Research in Human Ecology*, *17*, 58–69.
Fitzgerald, A.J., & Taylor, N. (2014). The cultural hegemony of meat and the animal industrial complex. In Nik Taylor & Richard Twine (Eds.). *The rise of critical animal studies*. Routledge.
Fromm, E. (1956). *The art of loving*. Harper & Row, Inc.
Fromm, E. (1964). *The heart of man: Its genius for good and evil*. Harper and Row.
Fromm, E. (1973). *The anatomy of human destructiveness*. Fawcett Crest.
Fromm, E. (1994). Some beliefs on man, in man, for man. In R. Funk (Ed.), *On being human* (pp. 99–104). Continuum International.
Fryar, C.D., Hughes, J.P., Herrick, K.A., & Ahluwalia, N. (2018). *Fast food consumption among adults in the United States, 2013–2016*. NCHS Data Brief 322.
Hanisch, C. (1970). *The personal is political*. https://www.carolhanisch.org/CHwritings/PIP.html.
Hoffman, A. (1979). *LSD: My problem child*. Translated by Jonathan Ott. Oxford Press.
Jinpa, T. (2021). *The Power of Awareness and Compassion to Transform our World*. https://www.compassioninstitute.com/blog/the-power-of-awareness-and-compassion-to-transform-our-world/.
LaMotte, S. (2022). How psilocybin, the psychedelic in mushrooms, may rewire the brain to ease depression, anxiety, and more. *CNN Report*. Available at https://www.cnn.com/2022/06/11/health/psilocybin-brain-changes-life-itself-wellness-scn/index.html.
Masters, R., & Houston, J. (1966). *The varieties of psychedelic experience: The classic guide to the effects of LSD on the human psyche*. Holt, Rinehart, and Winston.
McSweeney, K. (2015). The impact of drug policy on the environment. *Open Society Foundations*. Available at https://issuelab.org/resources/36013/36013.pdf.
Moniuszko, S.M. (2022). The next big trend in mental health treatments? Psychedelic therapy. *USA Today*.
Nierenberg, D. (2005). *Happier meals: Rethinking the global meat industry*. Worldwatch Institute.
Paley, D. (2014). *Drug war capitalism*. AK Press.
Ritzer, G. (1995). *The McDonaldization of society*. Sage.
SAMHSA. (2020). *Key substance use and mental health indicators in the United States: Results from the 2019 National Survey on Drug Use and Health*. PEP20-07-01-001, NSDUH Series H-55.
Schlosser, E. (2001). *Fast food nation: The dark side of the all-American meal*. Houghton Mifflin Company.
Schütz, A. (1962). *The problem of social reality: Collected papers volume 1*. The Hague: Martinus Nijhoff.
Schütz, A. & Luckmann, T. (1973). *The structures of the life-world, volume 1*. London: Northwestern University Press.

Sinclair, U. (1946 [1905]). *The jungle*. The Viking Press.
Small, M.L. (2010). *Unanticipated gains: Origins of network inequality in everyday life*. Oxford University Press.
Sollund, R. (2008). *Global harms: Ecological crime and speciesism*. Nova Science Publishers.
Weitzenfeld, A., & Joy, M. (2015). An overview of anthropocentrism, humanism, and speciesism in critical animal theory. In A.J. Nocella, et al. (eds.). *Defining critical animal studies: An intersectional social justice approach for liberation*. Peter Lang.

18

A TIME TO KILL

Cruelty and Compassion with Companion Animals and Urban Wildlife

Stephen L. Muzzatti and Kirsten L. Grieve

Humans' lethal violence against nonhuman animals is sometimes highly visible, with eyewitness accounts and/or other visual evidence supporting legal action under existing animal cruelty laws. These highly visible cases may result in arrest, prosecution, and sentencing. However, these examples are not considered institutionalised violence against nonhuman animals; that is, sanctioned violence, administered *justifiably* (i.e., legally and with the implicit support or consent of the populace). Despite an increasing public consciousness regarding the necropolitical relationships amongst the state, consumer capitalism, and nonhuman animals, the Animal-Industrial Complex's (A-IC's) foundational violence against food production animals, laboratory animals, certain wildlife *nuisances* or *pests* and other *bad animals* does not often stir discussion amongst the public. Unlike transgressive violence, that is, violence that is illegal, taboo or socially unacceptable (neglect, torture, bestiality, crush videos, dogfighting, etc.) and punishable under existing animal protection legislation, the institutionalised violence of the A-IC is permitted, socially sanctioned, and almost completely removed from the public gaze precisely because it is necessary for the continuation of the global capitalist project. Driven by profit margins and efficiency metrics the A-IC's perpetual violence against nonhuman animals takes many forms.

This chapter employs a discourse analysis of media accounts to address some of the issues surrounding the killing of nonhuman animals, specifically domestic companion animals (i.e., family pets) and wildlife *nuisances* or *pests* (e.g., coyotes, raccoons, rats, etc.) in urban settings. To ensure that our study focused on the most relevant sources we employed a purposeful sampling technique focusing on Canadian legacy media's premier news sources based in major cities, including, the *Toronto Star*, *National Post*, *Calgary Herald*, *Edmonton Journal*, *Vancouver Sun*, Global News, CTV News, and CBC News. We examined 86 articles published

DOI: 10.4324/9781003441908-23

between 2019 and 2023. This was supplemented by a number of older Canadian articles ($n = 13$) and a few articles from US sources ($n = 9$) to increase validity, for a total of 108. Drawing upon both highly visible instances of socially unacceptable forms of violence that result in the death of nonhuman animals, as well as on institutionalised (and often shrouded) killings committed by authorised agents (e.g., law enforcement, animal control or wildlife officers, etc.), the news items presented herein illustrate the use of lethal means on nonhuman animals running a gamut of cruelty to compassion. This chapter will utilise specific examples to highlight the spectrum of lethal violence and some of the responses elicited. Informed by the critical trajectories of green criminology and human–animal studies, this chapter unpacks the narrative scripting and competing discourses of harm and suffering within the context of interspecies cohabitation.

Arluke and Sanders (1996) seminal sociozoological scale suggests that humans rank nonhuman animals on how well they serve our interests and fit into the roles we assign them under late capitalism. Companion animals, zoo attractions and livestock are designated "good animals" because they are of use to humans, whilst "bad animals" either not knowing, or openly resisting, *their place,* pester, inconvenience or frighten us. The latter group includes a small number of domesticated species such as feral cats (Grieve & Muzzatti, 2023: Wald & Peterson, 2020), but is mostly comprised of certain types of invasive species (e.g., Burmese Pythons in Florida, but not House Sparrows in Ontario) and what the wildlife management and ecotourism industries call apex predators. However, most of us in Canada or the USA will not come into conflict with bears, cougars, or sharks. Urbanites frequently have conflicts with common indigenous wildlife such as coyotes, mice and racoons. It is these bad animals that are at greatest risk of marginalisation and destruction at the hands of humans.

Some of this human enmity has ancient roots, such as fear of being preyed upon, but much of it is more recent and is tied to agriculture and food production animals. In short, it is often about protecting the property of humans from nonhuman animals. Fear and hatred of certain kinds of nonhuman animals was reified by the State. Laws were established as early as 1424 in Scotland to control bird damage, and some of the earliest laws passed in North America were bounties on foxes and wolves (Nyhus, 2016).

Theriocide by Any Other Name

Socially acceptable methods of killing nonhuman animals are often framed to highlight the benefit this death will bring to humans, employing terminology such as *humane* or *necessary*. For example, a plan to cull deer in British Columbia (BC) was favourably reported as a necessary course of action in order to preserve the delicate ecosystem from being eradicated by overgrazing. Reassurances were made to the readers that the killings would be enacted by "professionals" utilising "globally accepted" methods of dispatch so that the deaths would be as "humane" as possible. After the killings, the article emphasised that the recovered meat was distributed to Indigenous communities, thus illustrating how the necessary death of these deer further benefited humans (Little, 2023). The article did not elucidate on what the "globally accepted" methods of killing were, although it did mention that the deer were shot, nor what is considered a humane method of killing. (Single or multiple shots? Shooting from a hide or helicopter? Herding the deer first into an enclosure?) Far from uncontested, the term humane has varying definitions and connotations dependent on the species, dispatching method, institution, jurisdiction, and a host of other variables.

Coexistence between wild species and urban residents is a delicate balance, and humans who engage in outdoor recreation (e.g., vegetable gardening, feeding birds, walking dogs in ravines, etc.) can encounter wildlife that disrupt these pursuits. These encounters often result in angry humans seeking violent revenge on the unwelcomed wildlife *invader*.

Two separate stories in Toronto described the actions and ultimate arrests of two men engaging in violence against raccoons that had, presumably, caused problems or damaged their property. One man beat raccoons in his backyard with a shovel (Hume, 2011), and another man trapped and attempted to drown a kit in a water filled garbage bin (Aguilar, 2017; Westoll, 2017). In the first instance, the homeowner was angered by the raccoons' damage to his vegetables, bludgeoning several kits to the point of severe injury and death whilst the mother raccoon hovered on the perimeter and attempted to rescue the injured and dying babies (Hume, 2011). Dong Nguyen later pleaded guilty to animal cruelty and was sentenced to 100 hours of community service at a humane society and made to pay restitution for the veterinary bills accrued for the treatment of the surviving kit (Consiglio, 2013). In the second story, it was reported that the man had been "disturbed" by the raccoons in his backyard and captured the baby raccoon in a homemade trap with the intent of killing it (Aguilar, 2017). Both stories painted a sympathetic portrayal of these *nuisance animals*, with quotes from neighbours, a police constable, and a spokesperson from the Toronto Wildlife Centre that condemned the respective men. When media reports include compassionate statements by officials attending a call where a nonhuman animal is in distress, or have eyewitnesses describe the sounds of suffering or crying, it guides the narrative to a more sympathetic portrayal for nonhuman animals.

Torontonians have a love/hate relationship with raccoons. Whilst there are advocates for racoon protection, and most people react with revulsion to instances of violence described above, there are those who would like to see them eliminated. A poll showed half of Torontonians want raccoons culled to control the population (Cecco, 2018). Discursive framing of potentially dangerous nonhuman animals roaming the city, damaging property, confronting humans, chasing companion animals and spreading disease connotes wildlife not as something to be understood and respected, but as a threat to be feared and controlled: lethally, if necessary.

There is little doubt that language used in media accounts influences public opinion. One story that warned residents to "stay away from the park and use caution in the neighbouring areas" because an "aggressive" coyote bit two people (CBC News, Nov. 21, 2021) suggests a dangerous predator. Though this kind of territorial behaviour is extremely rare in coyotes (see Poessel, Mock, & Breck, 2017), the sensational language is unduly cautionary at best, and fear-mongering at worst. It is not surprising that responses in instances such as these mirror those of people who overestimate the prevalence of violent street crime and their likelihood of victimisation; calls for harsher and more punitive responses from the criminal justice system. In cases of human and nonhuman animal conflict, people who are afraid for their companion animals or themselves are more likely to support lethal intervention to eliminate *dangerous predators* (Alagona, 2022).

Slightly more contentious was a story involving a coyote attack in Collingwood, Ontario, which killed a family dog. Ontario and some other Canadian provinces (Alberta and BC), have what is known as a "coexistence policy", whereby residents are encouraged to responsibly protect their property and companion animals themselves and share the land with the native species. Instead, residents complained increasingly about encounters with aggressive coyotes putting their companion animals at risk and swayed public opinion swayed toward trapping and killing the coyotes.

One resident even filed a class action lawsuit asserting that the city failed to protect residents' property and was in violation of the animal cruelty section of Canada's *Criminal Code*. Opinions proffered by interviewees in the story differed on whether or not trapping and killing is a reasonable method to deal with the coyotes. Some residents recognised the problem is largely a result of suburban encroachment on a native species' territory and by curbing human behaviour that entices these coyotes to approach (i.e., closely supervising companion animals outside, keeping them inside overnight, clearing their faeces from yards, and covering garbage) the problem might be mitigated. Others contend that such responsible stewardship is an unfair burden on the residents (Ireland, 2019).

In a separate article about a planned cull of coyotes in Vancouver, BC, a far more sympathetic tone is struck. After a series of complaints in which human were bitten by coyotes in Stanley Park, action was taken which resulted in the closure of the park in order to trap and euthanise them (Griffin, 2021a; Pawson, 2021). There were a number of reports that focused exclusively on the facts (the number of incidents, how many coyotes would likely be culled, the dates in which this would take place, etc.). One considerably more critical article, stated that, "…more people are killed by errant golf balls than attacked by coyotes in North America each year", and continued that "…humans have broken an unspoken contract forged between coyotes and humans" and, as a result of that, "….it's the coyotes paying the price" when describing how humans' deliberate (and illegal) feeding of the wild animals had habituated the coyotes to the point where they no longer feared and were far more comfortable approaching humans (McKeen, 2021). Such non-speciesist coverage, whilst reflecting established zoological knowledge, featured infrequently.

The method of death and its representation to the public through media accounts also sways opinion. For the Stanley Park coyote cull, reports stating the coyotes were "trapped" and "humanely euthanised" are far more palatable than reports stating modified leg-hold traps were used, and that the coyotes were killed by bolt guns, devices used to dispatch food production animals (Griffin, 2021b). The imagery conjured by "leg hold trap" and "bolt gun" is more gruesome than that by "humanely euthanised". The word *euthanasia* itself is utilised to describe a gamut of nonhuman animal death, from carbon monoxide poisoning and decapitation in some laboratory animals (Hickman, 2023), to bolt guns for agriculture and food production animals (Government of Ontario, n.d.), to injection for companion animals, and shooting in farm animals and other large animals (American Veterinary Medical Association [AVMA], 2016, 2020; Canadian Veterinary Medical Association [CVMA], 2021a). As a blanket term, it loses some power of its meaning (from the Greek, *eu* for "good" and *thanatos*, "death"), when applied broadly and frequently to all nonhuman animal death. These news stories do the vital ideological work of likening the killing of nonhuman animals to the human beneficence afforded beloved companion animals: a weighty decision to orchestrate peaceful methods of death by veterinarians in controlled environments.

Euthanasia in Various Forms

Euthanasia of companion animals is most often performed by veterinarians either in a clinic or hospital or in a private home by a mobile veterinarian. An examination is performed to confirm that euthanasia is warranted (based on a disease or illness presence, age-related decline, pain level, and suffering of the companion, likely continued decline, and a host of other considerations) and an appropriate level of anaesthesia is obtained, usually through injectable drugs, prior to the euthanasia agent being administered. It should be noted that, whilst some veterinarians may perform euthanasia without the use of pre-sedation or anaesthesia, it is strongly recommended by veterinary medical associations in Canada and the USA for euthanasia methods that may cause distress, e.g., intra-organ injections (AVMA, 2020; CVMA, 2021b). Whilst emotionally distressing to both the guardians and the veterinarian, the actual euthanasia itself, when carried out appropriately and without complications, is quite literally a process in which the companion gradually and gently *goes to sleep* and does not wake up. In other words, calm, peaceful and without physical distress. This construction of euthanasia serves the interests of the A-IC by obfuscating the true processes behind the majority of nonhuman animal deaths. However, this term is also used to describe the deaths of nonhuman animals in very different scenarios, in which the death itself may not be calm, peaceful and without distress. Neoliberalism has accelerated the processes by which certain nonhuman animals are increasingly vulnerable and made "killable", that is, killed remorselessly because it serves the interest of humans (Haraway, 2008). Companion animals enjoy certain protections, but other good animals (e.g. chickens, cows, pigs) do not. Driven by a tremendous infusion of finance capital (Feedback, 2024) and hidden by Cartesian veils the A-IC exposes ever increasing numbers of nonhuman animals to fatal violence which could hardly be further from a *good death*.

For example, as of 2021, the AVMA recommends maceration as the preferred method of euthanasia for male layer chicks that are malformed, fail to hatch, or are not "economically viable" (Wang et al., 2021). Maceration is a method of death using high speed, rotating blades or projections for rapid disruption of the brain to produce immediate death. This method, whilst approved by the AVMA and considered "humane", is undoubtedly a disturbing image for people: baby chicks on a conveyer belt heading toward a receptacle at the end with spinning, razor-sharp blades into which they will fall and be ground up alive. The CVMA allows for euthanasia techniques in cattle through gunshot or penetrating captive bolt. In the event death is not instantaneous via this first method and an adjunctive method is required, a procedure known as "pithing" may be utilised, in which a rigid but flexible metal rod or wire that is long enough to reach the brain and spinal column is inserted through the hole created by the gunshot or bolt. "The rod is thrust caudally through the brain, brainstem and finally spinal cord. It is then slid back and forth to cause maximum damage to the brain and upper spinal cord, a practice that is known as 'fiddling'" (CVMA, 2021a). Further, according to the AVMA (2020), appropriate methods of euthanasia for meat rabbits involve stunning (using an appropriate level of electricity to ensuring the period of unconsciousness lasts through the actual slaughter), shackling, and exsanguination (bleeding out). Decapitation and cervical dislocation by "properly trained individuals" are less common but acceptable methods of euthanasia as well (2016). It is safe to say that these examples of euthanasia are most likely not the typical methods with which the public is familiar. Images of cows being shot and then having their brain and spinal cord destroyed manually with a metal rod, or rabbits being shocked, shackled upside down, and having their throats slit is not necessarily the type of death with which the public associates the seemingly benevolent and liberally used term "euthanasia".

Public Outrage over Nonhuman Animal Deaths

Public opinion varies widely and emotions run high when discussing deaths of good animals (particularly companions such as dogs and cats) through methods that do not involve a carefully orchestrated euthanasia appointment with a veterinarian. Heightened emotions are less common when discussing the deaths of rarer (and perhaps less identifiable and relatable) companion animals such as rabbits, guinea pigs, and other small mammals. There appears to be little to no public outrage over media accounts of rodent culling, particularly as it pertains to methods of *vermin* control, typically in large urban environments. The following three events provide a telling contrast.

In 2023, Aleeta Raugust, pleaded guilty to nine counts of animal abuse after admitting to adopting and then torturing, nine cats, seven of whom she killed, and recording the events to watch for entertainment. She was sentenced to six and a half years in prison, the longest sentence for animal abuse in Canada to date. (Martin, 2023). In 2021, she contacted a veterinarian stating that she had a kitten with "injured legs at home" but refused to bring him in for treatment. According to an article, she later "…admitted to striking the Siamese kitten with a stick, shooting hairspray into its eyes and breaking its legs, covering the animal's mouth so her roommates wouldn't hear" (Kaufmann, 2023). The online comments on this article included statements such as, "There are people who should not be loose in society. She doesn't need a finite prison sentence, she needs permanent accommodation where she will not be free to interact with normal people".; "Just shoot the stupid btch [sic]"; and, "I didn't read past the headline. My decision is death for her. Straight into the lime pit". Clearly, there were people so enraged by Raugust's actions that they advocated killing her as retribution for the suffering and deaths she inflicted on these cats.

Another article described what was meant to be a transfer of small mammals (guinea pigs, rabbits, hamsters, rats, and mice) by the San Diego Humane Society to the Humane Society of Southern Arizona. According to the article, many of the surrendered animals were instead sold to a reptile breeding operation and were then frozen or fed alive to the reptiles (Bellware, 2023). The online comments for this story were decidedly less vehement and violent than those for the preceding one. One comment read, "Animals eat animals, that's just the way nature is, a lot of these pets were put up for adoption from people who stopped caring for them". Another read, "Reptiles need to eat!", and a third, "Uproar about the cycle of life. Each specie [sic] has a use in the world some to eat and some to be eaten".

Clearly, the public felt that having these small mammals (some of who were alive) fed to reptiles (whilst arguably less horrific than the torture described in the previous article, is an unpleasant and potentially prolonged death) is acceptable because "it's just nature's way". The online comments lead one to believe that there is decidedly less public sympathy for the fates of these domesticated species. Despite the fact that this event was illegally orchestrated by humans who sold socialised, former companion animals (surrendered under the auspices of adoption) to a company that would use them as reptile food, there were few sympathetic comments surrounding the deaths. Perhaps the fact that smaller mammal species are not common companions, and that they feature so prominently as prey in highly stylised, mediated presentations of "nature red in tooth and claw", creates an emotional detachment that elicits less sympathy to their plights even though their deaths occurred in a manner so obviously resulting from human greed and malfeasance.

Finally, an article about an increase in the number of rats in an Ottawa suburb drew mixed comments, ranging from support for the rats and penalising humans' actions that encourage their presence: "Rats are clean and smart animals. It's people that are not clean and attract these smart little guys. Get rid of your garbage and they'll leave you alone", to calls for poisoning: "Toronto also has a serious rat problem which the politicians are 'studying'. How about an aggressive program to kill them? Search out their nests and lay poison. Warn people to keep their kids and animals away and go for it. Rats are rodents. They carry disease. They need to be exterminated" (Molina, 2019). Methods of death for pest species are painful and cruel (poisoning can take hours or days before death occurs, spring-load and glue traps can result in broken limbs, starvation or suffocation or starvation) but are more casually discussed via online comments, and with considerably less vehemence than those elicited by stories of cruelty against more common companion animals.

Public Opinion on Wildlife Deaths and Sponsored Hunts

Sport and trophy hunting are not only legal in Canada, but the government also engages in wildlife control and population management techniques itself and also pays hunters to kill nonhuman animals considered a "nuisance" (Animal Justice, n.d.). For example, a controversial government initiative to kill wolves using sharpshooters in helicopters in an attempt to aid caribou recovery began in BC in 2015. The program was renewed in 2021–2022, despite follow up surveys showing that 59% of respondents were opposed to predator reduction tactics to aid in caribou recovery. Further, conservation groups argue the methods of dispatch are cruel, with some wolves requiring multiple shots when the first bullet fails to immediately kill (Morton, 2022).

Another story in Belleville, Ontario, describes a lawsuit filed by three advocacy groups against the province claiming that a contest sponsored by a local hunting outfitter since 2018 violates the province's laws against incentivised killing of animals (Morton, 2023). Prizes and cash are offered by Chesher's Outdoor Store to hunters who kill the heaviest coyotes (Kennedy, 2022). Whilst hunting coyotes and wolves with appropriate small-game licenses is legal in much of the province, the lawsuit alleges that, by offering cash and prizes, Chesher's violated Section 11 of the *Fish and Wildlife Conservation Act* that states a person shall not hunt or induce others to hunt for gain or the expectation of gain (Kennedy, 2022; Morton, 2023). The groups also stated concern about the geographic location of the sponsored hunt, because of the potential confusion between coyotes and Algonquin wolves, a threatened species in Ontario (Morton, 2023). The owner of the store claimed he received threats of, "…death, bodily harm, arson and other forms of property damage" after the public learned of the contest (Kennedy, 2022).

In the USA, wildlife killing contests still exist in most states. These contests are organised by a host of different groups (civic groups, social clubs, schools, etc.), and offer prizes and cash rewards. They are often sponsored by hunting outfitters or weapons and ammunition companies (The Humane Society of the United States [HSUS], n.d.). The "pest" species targeted in these contests can be killed in high numbers and can include bobcats, foxes, prairie dogs, rabbits, squirrels, and host of other species, depending on the state in which the contest is occurring. Contest organisers and participants argue that these kills help control predator (often large predator, such as coyote) populations, reduce attacks against livestock, and encourage stability in the levels of game population, despite an increasing amount of data showing these arguments are false (ibid.). In fact, contest kills do not target specific "problem" predators, nor do they typically occur in areas where livestock interactions are taking place, but rather in grasslands and woodlands where coyote conflicts are minimal. Furthermore, populations tend to increase after social disruptions such as these for a host of reasons. In addition, the methods used are considered by some to be cruel (luring coyotes with electronic devices that mimic other coyotes or pups in distress), and the indiscriminate killing of adults where young pups are left orphaned (ibid.). Once again, public perception is playing an increasing role in the banning of these events, with Arizona, California, Colorado, Maryland, Massachusetts, New Mexico, New York, Oregon, Vermont, and Washington banning the practice (Featherstone, 2024; Pettigrew, 2023). State wildlife and hunting commissions are coming out against the contests, believing they cast a negative light on *legitimate hunting* (Somvichian-Clausen, 2021). A national poll found that 80% of respondents opposed wildlife killing contests (Block, 2023). A similar poll of Illinois residents, a state in which these contests are still legal, found over 73% of respondents also supported a ban (Zimmerman, 2023). Clearly, many people deem the indiscriminate killing of wildlife for cash and prizes disreputable, yet a speciesist hierarchy of moral consideration persists, justifying harm if the corpses are used for clothing, meat, or serve more *noble* purposes such as population control or environmental protection.

Indeed, the killing of bad animals, whether through hunting contests, government sponsored culls or other forms of pest control continues, despite considerable evidence that they are expensive, ineffective and horribly cruel. Their focus on the symptoms of the problem, rather than on the anthropocentric cause, has done considerable harm, created problems anew and produced tremendous suffering (Alagona, 2022).

Concluding Thoughts: Genuine Ignorance or Fetishistic Disavowal?

There is a great deal of nonhuman animal death that occurs beyond media attention and the public gaze. Slaughterhouses, zoos, and a host of other environments are poorly regulated, protected by laws against independent investigation (e.g., "Ag-Gag" laws), and are kept hidden from public scrutiny. Sometimes the answer to the subheading – question above is obvious – such as why slaughterhouses don't have glass walls – other times, it is more complex.

In closing, consider the following two reports involving the deaths of many people's favourite North American apex predator, the grizzly bear. The articles tell dramatically different stories regarding humans' capacity for narcissism and duplicity when it comes to nonhuman animal death.

The first described the recent death of Samson, a grizzly bear found abandoned in Alaska as a six-month-old cub and transported to the Toronto Zoo where he spent the next twenty-four years. Due to age-related decline and pain from arthritic changes, he was euthanised and honoured with a special ceremony based on various Indigenous teachings, recognising the importance that the grizzly bear has to multiple Indigenous communities. He was recognised as a "gentle, intelligent bear who had a deep bond" with zoo staff (Stevenson, 2023). The second article describes how the mortality rate for young grizzlies in Elk Valley, BC, is higher than any other region in the province. Bears aged two to six in this area have the lowest survival rates in North America. The primary causes of death for bears include collisions with vehicles or conflict with people due to unsecured attractants such as livestock and garbage on private property (CBC News, October 15, 2023). In other words, death due to capitalist development, human encroachment, and behaviour patterns.

Samson and his numerous Elk Valley cousins are the same species, but experienced very different lives, and deaths, as a result of conflict with humans. Samson whose very presence at the zoo was, on several levels, attributable to specious ideology, was a *good animal* who was given a compassionate death and honoured in a respectful ceremony. In contrast, many grizzlies elsewhere are *bad animals*, whose deaths are also a result of human behaviour and intervention, but are painful, prolonged, and occur without ceremony and dignity.

Late capitalism's voracious appetite results in a tremendous amount of lethal violence against nonhuman animals, and this violence manifests both visibly and invisibly: in legal, socially sanctioned methods, but also transgressive violence. Whilst media framing and reporting of violence against nonhuman animals, and the way in which humans view nonhuman animal death and its *legitimacy* all vary widely, it is all largely viewed through the myopic lens of anthropocentrism tinged with Cartesian notions. Our views are determined, in part, by the kinds of relationships that we have with these creatures. These views then dictate policy, procedure, and, ultimately, how these nonhuman animals with whom we share the Anthropocene live and die.

References

Aguilar, B. (2017, June 10). Man charged after baby raccoon rescued from near-drowning in water-filled trap. *The Toronto Star*. https://www.thestar.com/news/gta/man-charged-after-baby-raccoon-rescued-from-near-drowning-in-water-filled-trap/article_6b1895ca-16ad-5ca3-b8d1-5ce7a30535ff.html.

Alagona, P.S. (2022). *The accidental ecosystem: People and wildlife in American cities*. University of California Press.

American Veterinary Medical Association. (2016). *AVMA guidelines for the humane slaughter of animals: 2016 edition*. https://www.avma.org/sites/default/files/resources/Humane-Slaughter-Guidelines.pdf.

American Veterinary Medical Association. (2020). *AVMA guidelines for the euthanasia of animals: 2020 edition*. https://www.avma.org/sites/default/files/2020-02/Guidelines-on-Euthanasia-2020.pdf.

Animal Justice. (n.d.). *Wild animals*. Retrieved January 10, 2024 from https://animaljustice.ca/issues/wild-animals.

Arluke, A., & Sanders, C. (1996). *Regarding animals*. Temple University Press.

Bellware, K. (2023, November 17). Hundreds of small animals meant for adoption may have been frozen, fed to reptiles. *The Toronto Sun*. https://torontosun.com/news/world/hundreds-of-small-animals-meant-for-adoption-may-have-been-frozen-fed-to-reptiles.

Block, K. (2023, April 14). *You can help stop one of the cruelest threats facing wildlife*. The Humane Society of the United States. Retrieved January 10, 2024 from https://www.humanesociety.org/blog/you-can-help-stop-one-cruelest-threats-facing-wildlife.

Canadian Veterinary Medical Association. (2021a). *Position statement on euthanasia*. https://www.canadianveterinarians.net/policy-and-outreach/position-statements/statements/euthanasia/.

Canadian Veterinary Medical Association. (2021b). *Guidance for veterinarians on euthanasia methods that do not include pentobarbital sodium*. https://www.canadianveterinarians.net/media/kobbdn3h/guidance-vets-euthanasia-methods-do-not-include-pentobarbital-sodium.pdf.

CBC News (2021, November 21) *Coyote 'still at large' after biting 2 people in North York park, city says*. https://www.cbc.ca/news/canada/toronto/coyote-bayview-village-park-

CBC News (2023, October 15) *This B.C. valley has become a death trap for young grizzly bears: report*. https://www.cbc.ca/news/canada/british-columbia/elk-valley-grizzly-mortality-rate-1.6992673.

Cecco, L. (2018, October 5). Raccoons v Toronto: how 'trash pandas' conquered the city. *The Guardian*. https://www.theguardian.com/world/2018/oct/05/canada-toronto-raccoons.

Consiglio, A. (2013, March 12). Man who attacked raccoons with a shovel pleads guilty. *The Toronto Star*. https://www.thestar.com/news/crime/man-who-attacked-raccoons-with-a-shovel-pleads-guilty/article_e32b08cb-36c4-550a-b87e-feabc4eebc21.html.

1.6257750.

Featherstone, S. (2024, January 4). *Gov. Hochul signs law banning wildlife killing contests in New York*. Syracuse.com. Retrieved January 10, 2024 from https://www.syracuse.com/outdoors/2024/01/gov-hochul-signs-law-banning-wildlife-killing-contests-in-new-york.html.

Feedback (2024). *Still Butchering the Planet: The big-name financiers bankrolling livestock corporations and climate change – 2024 update*. Retrieved March 17, 2024 from https://feedbackglobal.org/wp-content/uploads/2024/03/Feedback-2024-Still-Butchering-the-Planet-Report.pdf.

Government of Ontario. (n.d.). *On-farm euthanasia of cattle and calves*. Ontario.ca. https://www.ontario.ca/page/farm-euthanasia-cattle-and-calves.

Grieve, K.L., & Muzzatti, S.L. (2023). Trap, snip, repeat: Cat overpopulation and gendered labour in a feral sterilisation programme. *Zoophilologica*, *11*, 1–23. https://doi.org/10.31261/ZOOPHILOLOGICA.2023.S.07 (Accessible as of [07/09/2024]).

Griffin, K. (2021a, December 26). Stanley Park coyote cull recommended last summer as they were 'dangerous wildlife': B.C. government. *Vancouver Sun*. https://vancouversun.com/news/local-news/stanley-park-coyote-cull-recommended-last-summer-as-they-were-dangerous-wildlife-b-c-government.

Griffin, K. (2021b, September 8). 'Modified' leghold traps being used to cull coyotes in Stanley Park. *Vancouver Sun*. https://vancouversun.com/news/modified-leghold-traps-being-used-to-cull-coyotes-in-stanley-park.

Haraway, D.J. (2008). *When species meet*. University of Minnesota Press.

Hickman, D.L. (2023). Euthanasia of neonatal rats and mice using carbon monoxide. *Journal of the American Association for Laboratory Animal Science*, *62*(3), 274–278. doi: 10.30802/AALAS-JAALAS-22-000103.

Hume, J. (2011, June 1). Man arrested for allegedly attempting to kill raccoons with a shovel. *National Post*. https://nationalpost.com/posted-toronto/man-arrested-for-allegedly-attempting-to-kill-raccoons-with-a-shovel.

Ireland, N. (2019, April 22). *Coyote co-existence policies mean pets are never safe, group of homeowners say*. CBC News. https://www.cbc.ca/news/science/collingwood-lawsuit-against-living-with-coyotes-1.5105638.

Kaufmann, B. (2023, August 29). 'PSYCHOPATHIC': Serial cat killer's sentencing adjourned for review of psych report. *Toronto Sun*. https://torontosun.com/news/local-news/psychopathic-cat-killer-sentencing-delayed-psychiatric-report-review/wcm/7c5468e3-2f8e-4025-8704-34ad2981c671.

Kennedy, B. (2022, February 4). Cash-for-coyotes hunting contest deemed legal, but animal rights groups say Ontario ignoring its own laws. *The Toronto Star*. https://www.thestar.com/news/canada/cash-for-coyotes-hunting-contest-deemed-legal-but-animal-rights-groups-say-ontario-ignoring-its/article_af45f0d8-19bc-56ba-a57a-0b4ee0a9f244.html.

Little, S. (2023, December 19*). 84 deer killed on Sidney Island in controversial eradication program*. Global News. https://globalnews.ca/news/10181259/sidney-island-deer-kill-update/.

Martin, K. (2023, September 14). Psychopathic serial cat killer given stiffest animal abuse sentence on record. *Calgary Herald*. https://calgaryherald.com/news/crime/calgary-serial-cat-killer-sentencing.

McKeen, A. (2021, September 20). Vancouver's coyote relationship turns rocky. *The Toronto Star*. https://www.pressreader.com/canada/toronto-star/20210920/281702617852447.

Molina, K. (2019, July 16). *Rats plaguing Overbrook, tenants say*. CBC News. https://www.cbc.ca/news/canada/ottawa/acorn-petition-landlord-licensing-rat-problems-1.5213667.

Morton, M. (2023, December 21). *$10M spent on B.C. wolf cull, FOI documents reveal alongside details of shootings*. CBC News. https://www.cbc.ca/news/canada/british-columbia/10m-spent-on-b-c-wolf-cull-foi-documents-reveal-alongside-details-of-shootings-1.7065846.

Morton, M. (2022, March 6). *Coyote hunting contest prompts animal advocates to file lawsuit against province*. CityNews. https://toronto.citynews.ca/2022/03/06/ontario-lawsuit-coyote-hunting-contest/.

Nyhus, P.J. (2016). Human-wildlife conflict and coexistence. *Annual Review of Environment and Resources*. 41: 143–171. doi: 10.1146/annurev-environ-110615-085634.

Pawson, C. (2021, September 3). *B.C. to cull up to 35 coyotes from Vancouver's Stanley Park following attacks*. CBC News. https://www.cbc.ca/news/canada/british-columbia/stanley-park-coyote-cull-1.6164627.

Pettigrew, J. (2023, September 15). *Oregon Fish and Wildlife Commission bans 'abhorrent' killing contests*. KOIN. Retrieved January 10, 2024 from https://www.koin.com/news/animals/oregon-fish-and-wildlife-commission-bans-abhorrent-killing-contests/.

Poessel, S.A., Mock, E.C., & Breck, S.W. (2017). Coyote diet in an urban environment: Variation relative to pet conflicts, housing density and season. *Canadian Journal of Zoology*, *95*(4), 287–297. doi: 10.1139/cjz-2016-0029.

Somvichian-Clausen, A. (2021, March 23). *The controversy over wildlife killing contests*. thehill.com. Retrieved January 10, 2024 from https://thehill.com/changing-america/sustainability/environment/544533-the-controversy-over-wildlife-killing-contests/.

Stevenson, J. (2023, April 14). Toronto Zoo says goodbye to grizzly bear Samson with Indigenous ceremony. *Toronto Sun*. https://torontosun.com/news/local-news/toronto-zoo-says-goodbye-to-grizzly-bear-samson-with-indigenous-ceremony.

The Humane Society of the United States (HSUS). (n.d.). *Facts about wildlife killing contests*. Retrieved July 19, 2023 from https://www.humanesociety.org/wildlifekilling contests.

Wald, D.M., & Peterson, A.L. (2020). *Cats and conservationists: The debate over who owns the outdoors*. Purdue University Press.

Wang, X., Zhao, D., Milby, A.C., Archer, G.S., Peebles, E.D., Gurung, S., & Farnell, M.B. (2021). Evaluation of euthanasia methods on behavioral and physiological responses of newly hatched male layer chicks. *Animals*, *11*(6), 1802. doi: 10.3390/ani11061802.

Westoll, N. (2017, June 8). *Toronto man charged after attempted drowning of trapped baby raccoon: police*. Global News. https://globalnews.ca/news/3514449/raccoon-attempted-drowning-toronto-police/.

Zimmerman, B. (2023, March 15). *'An abomination and a disgrace': Humane Society calls for end to killing contests in IL*. WCIA. https://www.wcia.com/news/state-news/an-abomination-and-a-disgrace-humane-society-calls-for-end-to-killing-contests-in-il/.

19

THE LENNIE SMALL PARADOX

Loving Animals to Death

Michael D. Briscoe

In Steinbeck's (1937) novel *Of Mice and Men,* two men are traveling looking for work. One of these men, Lennie Small, is a huge and strong man who suffers from an unnamed developmental disability. Lennie has a good heart, but not understanding his own strength can sometimes cause harm to others – this is in fact why they had to leave their previous job. In particular, Lennie is obsessed with animals. His dream is to have a farm with his traveling companion and friend George Milton and to take care of the rabbits on that farm. Lennie loves animals and how they feel when he pets them, but this has led to him killing mice in the past, as their small bodies cannot handle his massive and uncontrolled strength. Eventually this killing of mice escalates to killing a puppy and then a woman. The novel ends with George deciding that Lennie has gone too far and killing him before the other men in town can.

Like the fictional character of Lennie, many humans unwittingly cause harm to nonhuman animals by loving them and wanting them close, not realizing the damage they are causing. Just as Lennie squeezed nonhuman animals to death, humans who try to squeeze animals into unnatural and often highly domesticated settings cause great harm to nonhuman animals. This love for nonhuman animals that brings harm is often exploited by the animal-industrial complex (A-IC). Three examples of this in the modern A-IC are zoos and aquariums, the declawing of cats, and the domestication of dogs with an emphasis on looks rather than lifestyle. People who frequent zoos and aquariums, declaw cats, and domesticate dogs would likely all describe themselves as nonhuman animal lovers. However, their behavior creates harm for nonhuman animals, and serves as an important reminder that good intentions are insufficient when it comes to improving nonhuman animal welfare.

DOI: 10.4324/9781003441908-24

Zoos and Aquariums

One major component of the A-IC are zoos and aquariums. According to the Association of Zoos & Aquariums, their 215 accredited zoos and aquariums contributed $24 billion dollars to the economy in 2018 (AZA, 2019). Around $3 billion of this was before and after spending at zoos and aquariums themselves (AZA, 2019). While clearly a part of the A-IC at such a scale, zoos and aquariums are complicated industries in terms of nonhuman animal outcomes. Critical animal scholars have drawn parallels between human prisons and nonhuman animal zoos (Morin, 2015). In fact, there were even instances where humans from marginalized groups were kept in zoos, such as Ota Benga, an African man imprisoned at the Bronx Zoo (Newkirk, 2015). Zoos themselves have recognized the connection between prisons and zoos and tried to distance themselves from it through renaming of "cages" as "habitats" (Dunayer, 2001), and in designing enclosures that look less cage-like – for example creating walls out of rock and wood rather than iron bars (Marris, 2021).

On the other hand, zoos and aquariums argue they provide critical conservation for vulnerable and endangered species (AZA, 2023a). The Association of Zoos and Aquariums' data suggests there are more than 183 million annual visitors to zoos in the United States as of 2023 (AZA, 2023b). The majority of these visitors are with children, and "93% agree their family enjoys seeing animals up close at zoos and aquariums" and "94% feel that zoos and aquariums teach children about how people can protect animals and the habitats they depend on" (AZA, 2023b).

This is where the Lennie Small Paradox comes into stark focus with zoos and aquariums. Visitors to zoos clearly enjoy nonhuman animals and are likely to care about their conservation to an extent. However, like Lennie Small, they want to be close to nonhuman animals, which results in harm to nonhuman animals in several ways. The first is in the size of enclosures. Visitors to zoos want to be able to see nonhuman animals, which means enclosures must be small enough to bring nonhuman animals into close proximity with visitors. One analysis of zoos in the United Kingdom found enclosures were on average 100 times smaller than a nonhuman animal's natural range (Casamitjana, 2003). Nonhuman animals with wide natural ranges fare worse psychologically and have higher infant mortality in zoos where they are confined to small enclosures (Clubb & Mason, 2003) – and with a wide natural range, even the largest enclosures can be constricting. The problems associated with small enclosures can be even more extreme for fish and whales held in captivity. Consider dolphins, a popular nonhuman animal often confined to aquariums. Dolphins can travel up to 100 kilometers in a single day and have a natural range of around 100 square kilometers (WAP, 2023). Even the largest aquariums in the world pale in comparison. Zoos and aquariums are trying to improve nonhuman animal welfare through other means, such as increasing the complexity of their enclosures and giving nonhuman animals more stimulating enrichment activities (de Azevedo et al., 2023). However, the core issue remains a challenge: visitors want to see and be close to nonhuman animals, and nonhuman animals do not want to see and be close to humans. The frequent result is small enclosures that hamper nonhuman animal welfare.

It is not just enclosure size where we see the Lennie Small Paradox in effect. The way these enclosures are designed also provides an example of the paradox. Studies confirm that for many nonhuman animals, human visitors at zoos cause them stress (Fernandez et al., 2009). Being observed up close by a predator (humans) in such a confined space is an unnatural and stressful situation for nonhuman animals to be in. In nature, direct eye contact can be a sign of aggression to nonhuman animals and elicit aggressive responses from those nonhuman animals (see for example Fuentes & Gamerl, 2005). For example, research shows having humans at the same level of primates and staring at them creates stress in them and increases their aggression (Chamove et al., 1988). Of course, staring is the primary activity done by humans at zoos. One change to enclosures suggested by researchers to address this fundamental problem is offsetting observation areas and enclosures so humans are not making direct eye contact with nonhuman animals. Chamove et al. (1988) suggest raising primate enclosures so only human heads and shoulders are visible from the enclosure, and humans therefore look smaller and less threatening. Still, this issue remains a challenge for nonhuman animal welfare.

Zoo and aquarium visitors are often motivated by a love of nonhuman animals. They most frequently attend with children and want to foster and encourage love of nonhuman animals in these children. Understandably, this love of nonhuman animals makes these visitors want to be close to nonhuman animals and to be able to see them clearly. Zoos and aquariums accommodate this, but this comes at the expense of nonhuman animal welfare as these creatures are forced into enclosures much smaller than their natural range and often forced to confront humans, predators, face-to-face in stress-inducing daily encounters.

Declawed Cats

Cat declawing is a second example of the Lennie Small Paradox. Humans who own cats and declaw them no doubt love these nonhuman animals, but misguidedly try to squeeze the cats into particular and unnatural lifestyles, resulting in harm to the cats. It may be that they love their lifestyle (including furniture) more than their cat. However, 48% of cat "owners" in the United States consider their cats as family members as much as a human member, and another 48% consider them family but on a slightly lower level than human members (Brown, 2023). These humans do care deeply about cats, but that love leads to harm as humans try to mold them (literally) into their lifestyles.

The "pet" industry is growing rapidly and comprises another one of the major parts of the A-IC, even larger than that of zoos and aquariums. Much of this growth is fueled by a love of nonhuman animals, and these nonhuman animal lovers are willing to spend for their pets. According to the American Pet Products

Association, the United States spent \$136.8 billion in 2022 on pet expenditures, up from \$90.5 billion in 2018 (APPA, 2023). About a quarter of this spending, \$35.9 billion, was in veterinary care (APPA, 2023). While most of this spending was likely on important care for nonhuman animals, another portion of this spending was on procedures to declaw cats.

Estimates on the number of cats that are declawed vary. One large survey from 2001 found 24.4% of cat-owning households declawed their cats (Patronek, 2001). Other estimates range as high as 43% of cats being declawed (The Paw Project, 2023). With an estimated 58 million cats as pets in the United States (see AVMA, 2023), that means between about 14 and 25 million cats have been declawed. Each procedure costs anywhere from several hundred to more than one thousand dollars, again illustrating the place of this practice in the A-IC as a multi-million, potentially even billion-dollar industry.

While 72.7% of veterinarians perform cat declawing surgeries (Ruch-Gallie et al., 2016), the practice has received increasing scrutiny for the harm it does to cats. In fact, of the 72.7% of vets who do declaw cats, 74.6% recommend non-surgical alternatives (Ruch-Gallie et al., 2016). Many people think of claws as similar to nails, and therefore think of declawing as a non-invasive procedure. However, the actual onychectomy (declawing) procedure includes the removal of the end part of a cat's paws – the distal phalanges. In humans the equivalent would be removing the tips of each finger and toe. This is done to prevent the claws from growing back. According to a review of studies done by the American Veterinary Association (AVMA, 2019), the harms to cats as a result of declawing include: severe pain as a direct result of surgery, indirect pain as a result of things like changed posture and gait, and a high rate of surgical complications (as high as 50%).

Besides pain, one major welfare impact of declawing is behavioral. Cat scratching is a natural behavior. This is one of the reasons cat owners resort to declawing – they fail to stop cats from scratching things like expensive furniture. However, the urge to scratch which is deeply embedded in cats does not go away after declawing surgery. "Between 59 and 78% of declawed cats will continue to exhibit scratching behavior" (AVMA, 2019, p. 4). Additionally, cat claws are used not only in scratching but also in other normal behaviors like climbing and balancing (AVMA, 2019). There is also some evidence of the development of new behavioral problems after declawing, though this evidence is mixed (AVMA, 2019; Patronek, 2001).

Given the welfare issues with cat declawing, why do cat owners, who presumably love cats, continue to do it? Rather than recognizing scratching as a normal and natural behavior done by cats, humans normalized the idea of declawing cats as an acceptable intervention (Grier & Peterson, 2005). The most common reason owners choose to declaw their cats is damage to furniture (Wilson et al., 2016). Other common reasons owners gave for electing declawing surgeries include wanting to avoid injury to humans or other pets (Yeon et al., 2001).

There are some alternatives to declawing to deal with cat scratching. Cats can be given designated scratching items and areas like scratching pads or posts, and training can help reduce the risk of scratching humans or other pets. However, even with these types of interventions, scratching is a natural cat behavior that they will likely continue to do. Rather than trying to squeeze cats into an unnatural lifestyle, cat owners may have to acknowledge owning a cat will require changes to their own lifestyle and expectations.

If the cat owners did not love cats, the alternative would be easy – they could simply abandon or surrender their cat to a shelter, as many do (or not get a cat to begin with). However, these owners want a cat badly enough in their life that they are willing to spend hundreds, and potentially thousands, of dollars to make the cat fit into their life. Unfortunately for the cats, this love and desire to squeeze them into human lives comes at the expense of their welfare.

Domesticated Dogs

Like cats, dogs have become popular pets in many households. In fact, more households own a dog than own a cat – 54% compared to 35% (HSUS, 2023a). For many people, dogs are more than just pets – they are considered part of the family. A survey from the American Veterinary Medical Association found 85% of dog owners considered them members of their family (Burns, 2018). Dogs are clearly loved in the United States, but as with cats, this love has come with a hefty price paid for by the dogs.

The desire to have dogs as pets brings harm to dogs in multiple forms. To begin, an estimated 3.3 million puppies sold each year in the United States come from puppy mills (HSUS, 2023b). Puppy mills are breeding facilities for dogs that often act as suppliers for chain stores selling puppies and other nonhuman animals as pets. Like industrial animal agriculture, these facilities often keep puppies in cruel conditions, including wired flooring that can injure dogs' paws (PETA, 2023), and stacked cages that allow waste from dogs on top to fall on those below (Towsey, 2010).

In 2020 Senator Dick Durbin introduced the Puppy Protection Act, which would require more humane conditions for commercial dog breeders, but as of 2023 the bill has not made any progress beyond being introduced (Puppy Protection Act of 2020), and puppy mills remain a large and harmful part of the A-IC. Like cats, a large amount (390,000) dogs are euthanized in shelters every year in the United States (ASPCA, 2023). So, while pet stores in malls attract genuine pet lovers with their cute puppies on display, these customers are often unwittingly participating in great harm to dogs.

Harm to dogs extends beyond where nonhuman animals are bred and euthanized. The genetic ways in which dogs have been bred can also create harm. Rollin (2018) argues:

> only a demented or psychopathic person would choose their child's traits in ways that would or even could harm the child in the future, or compromise their abilities and their quality of life. In fact, most of us shudder at the very thought

of such a world. Yet we proceed to demonstrate that it is precisely this very way that we choose the traits of the dogs we profess to love and cherish.

(p. 4)

The most infamous example of this is in pugs. Pugs are bred to have stocky chests and flat faces, giving them a distinct "ugly-cute" look. The most commonly known problems for pugs include trouble breathing (Bartels et al., 2015; Gough et al., 2018) and eye ulceration (O'Neill et al., 2022), but they also more frequently have an abnormal gait, or walk (Rohdin et al., 2018). The health problems for pugs are so severe that they led researchers to conclude pugs "can no longer be considered as a typical dog from a health perspective" (O'Neill et al., 2022, p. 2). People have become more aware of the health issues associated with pugs and similar breeds of dogs such as the French Bulldog, and some governments are considering banning these types of dogs from being bred because of the welfare problems they have (The Animal Reader, 2023). On the other hand, the French Bulldog just surpassed the Labrador retriever as the most popular dog breed in the United States in 2022, according to the American Kennel Club (Haid, 2023).

Other breeds of dog have issues with space and climate similar to nonhuman animals in zoos. For instance, consider the Siberian husky – another popular and iconic breed. Huskies gained popularity as racing sled dogs in arctic climates. They have been bred to withstand cold weather and to run for many miles every day. Estimates vary, but huskies can run at least several dozen miles every day – something extremely difficult for humans to accommodate. Veterinarians recommend huskies receive at least two hours of exercise every day (PDSA, 2023), again something very difficult to achieve for working people. As dogs adapted to the arctic cold, huskies also tend to overheat at relatively low temperatures, yet are still purchased by people living in warmer climates (the exact numbers on precisely how many live in climates too hot for them are not available). Other dogs may not have as high of needs when it comes to exercise or climate, but still have higher needs than most people are able to accommodate. In the United States, 33% of people living in apartments or condominiums have a dog (AVMA, 2022). While not impossible to provide dogs in these living situations adequate exercise, it certainly presents a greater challenge and requires more extensive planning.

Although humans increasingly see dogs as members of their own family, and the number of dog owners is rising, this love comes with a cost. First, in the inhumane breeding conditions on puppy mills and similar breeding facilities, and later in living situations that fail to adequately meet dogs' needs.

The Paradox and the A-IC

Love for animals translating to their harm is not an accidental outcome. Rather, it is a feature of the A-IC. For animals to be treated as commodities, whether in zoos or as pets, the A-IC must manufacture love for these nonhuman animals. In discussing the dynamics of the A-IC and farm animals, Noske (1997) argues "capitalism tends to eliminate anything in the animal which cannot be made productive" (p. 101). Although different than nonhuman animals produced to be eaten, these same dynamics are at play with pets. Nonhuman animals kept as pets must produce feelings of love and joy in their owners. When something gets in the way of those feelings, such as a cat destroying a piece of furniture, then the A-IC will find a way to remove these barriers to productivity – in this case through surgical removal of cats' claws.

While the A-IC manufactures love for certain nonhuman animals that it uses to fuel consumption, it simultaneously reinforces the idea that nonhuman animals are disposable. Bauman (2004) addresses similar dynamics in modern society, arguing that modern societies claim values of individual worth and respect. Yet, paradoxically, these same societies work to actively marginalize populations and maintain inequalities, thereby creating groups of humans who are both loved and disposable (Bauman, 2004). These same societal mechanics are at play within the A-IC and the Lennie Small Paradox.

One way in which the A-IC reinforces the disposability of nonhuman animals, even those that humans love, is through reinforcing the socially constructed gap between humans and nonhuman animals. This deeply entrenched dualistic view of human versus nonhuman is harmful to both nonhuman animals and humans alike (see Peterson, 2013). Breaking down this harmful dynamic cannot be done through deeper love or affection for animals. And, even if animals were given standing equal to humans, they would still be viewed as loved but disposable as Bauman (2004) points out humans are in modern society. Addressing the Lennie Small Paradox therefore requires deeper changes to the structure of our society and the A-IC.

Harms, Helps, and a Warning for the Future

Humans are fascinated with nonhuman animals and want to be near them. Although often motivated out of love for nonhuman animals, much like the fictional character from Steinbeck's *Of Mice and Men* – Lennie Small, humans end up hurting nonhuman animals by trying to be close to them both in zoos and aquariums, as well as nonhuman companion animals. To conclude, I explore the ethical and practical dilemmas this relationship creates for the A-IC and groups who want to dismantle it.

In this chapter I have outlined the ways in which human love for nonhuman animals can actually cause them harm. However, this does not necessarily mean the harm done in these situations is greater than what might otherwise happen to the nonhuman animals – that is an issue that will need to be debated and explored at greater length. For instance, is it better to keep a species in cramped and stressful conditions in a zoo rather than risk them going extinct in the wild? Is it better to declaw a cat in order for someone to keep it rather than not adopt a cat and have it be euthanized at a shelter like the other half a million cats each year (see ASPCA, 2023)? Or is it better for a dog to be adopted and live in an unnatural climate or small apartment rather than at a shelter?

The role zoos play in conservation is contested, but there is some research to suggest visiting zoos can positively affect a person's self-reported conservation behaviors (Miller et al., 2013) as well as their conservation intent (Hacker & Miller, 2016). Even renowned animal activist Carl Safina seems to see a place for zoos in the future, arguing zoos "need to better inspire visitors' active engagement for animals, welfare, and conservation" (Safina, 2018, p. 10). Similar arguments are made in favor of keeping nonhuman companion animals. Research shows living with pets shapes positive attitudes toward other nonhuman animals, including more humane attitudes toward non-pet animals and even toward other humans (Daly & Suggs, 2010; DeMello, 2021; Hawkins & Williams, 2017). Animal liberation advocates would also likely counter that zoos and petkeeping, although they promote positive feelings toward nonhuman animals, reinforce a hierarchical and speciesist view of nonhuman animals that views putting them in cages as acceptable. There is no question these sorts of human-animal relationships can develop from a place of love for nonhuman animals, and they can even have positive impacts on nonhuman animals. However, the main point here is that loving nonhuman animals alone is insufficient for ensuring their welfare, and can even create new forms of harm for them. The question for zoos and pet owners is whether their positive outcomes outweigh the harms done to nonhuman animals.

These questions of the costs and benefits of certain human behavior for nonhuman animals are relevant in our current world, and worthy of further consideration and debate, but they also fail to address the underlying issues causing species loss and why so many dogs and cats are sent to shelters where many are then euthanized. Or in other words, zoos claim to be necessary for conservation, but what is driving the critical need for species conservation? The A-IC is a driving force behind both, as breeding nonhuman animals for pets creates a supply that outstrips demand, leaving nonhuman animals unadopted in shelters, and as habitat and biodiversity are destroyed to make space for the expansion of animal agriculture (Austin, 2010; Kaimowitz et al., 2004). These issues, and the fundamental nature of our relationship with nonhuman animals, will need to be addressed to really challenge the harm of the A-IC.

The key point illustrated by the Lennie Small Paradox is loving nonhuman animals is not enough to ensure their well-being. Animal advocates optimistically point to shifts in public opinion toward nonhuman animals as evidence of reflexive modernization characterized by increasingly zoocentric views (Franklin & White, 2001). However, the Lennie Small Paradox warns increasing love and sentimentality toward nonhuman animals is not sufficient to address the problems facing them, and in fact can contribute to new harms toward nonhuman animals. Bauman's (2004) *Wasted Lives* also illustrates here that elevating nonhuman animals' status may not change the essential dynamics that lead to better treatment, as modern society claims to value individual human life while simultaneously viewing entire human populations as disposable. Increasing concern for nonhuman animals is likely a positive societal change, but it needs to be coupled with changes to social structures that enable the A-IC to operate and reinforce the disposability of nonhuman animals.

Real concern for nonhuman animal welfare will require a change in the way we think about our fundamental relationships with nonhuman animals. This will almost certainly mean sacrifice as well on the part of humans. Although many people enjoy seeing nonhuman animals up close at zoos and aquariums, the harm to nonhuman animals likely outweighs the good of that experience for them. The nonhuman animal issues outlined in this chapter need to be addressed, but they also raise difficult social questions. With zoos and wildlife sanctuaries, larger spaces for nonhuman animals tends to mean higher costs, which again would exclude people with lower incomes from interacting with nonhuman animals. Would enacting stricter welfare guidelines lead to nonhuman animals becoming even more of a commodity – in this case only enjoyed by those of higher class? The answer to some of these ethical dilemmas is unclear, but the Lennie Small Paradox acts as a warning to animal advocates that love for nonhuman animals will not be sufficient to address, and indeed may contribute to, the challenges posed by the A-IC.

References

APPA. (2023). *Pet industry market size, trends, & ownership statistics.* American Pet Products Association. https://www.americanpetproducts.org/news/News-Public-Relations/pet-industry-market-size-trends-ownership-statistics.

ASPCA. (2023). *Pet statistics*. American Society for the Prevention of Cruelty to Animals. https://www.aspca.org/helping-people-pets/shelter-intake-and-surrender/pet-statistics.

Austin, K. (2010). The "Hamburger connection" as ecologically unequal exchange: A cross-national investigation of beef exports and deforestation in less-developed countries. *Rural Sociology*, *75*(2), 270–299.

AVMA. (2019). *Literature review on the welfare implications of declawing of domestic cats.* American Veterinary Medical Association. https://www.avma.org/sites/default/files/resources/declawing_bgnd.pdf.

AVMA (2022). *2022 AVMA pet ownership and demographics sourcebook*. American Veterinary Medical Association.

AVMA. (2023). *U.S. pet ownership statistics.* American Veterinary Medical Association. https://www.avma.org/resources-tools/reports-statistics/us-pet-ownership-statistics.

AZA. (2019). *AZA zoos and aquariums contribute $24 billion to U.S. economy*. Association of Zoos & Aquariums. https://www.aza.org/aza-news-releases/posts/aza-zoos-and-aquariums-contribute-24-billion-to-us-economy-.

AZA. (2023a). *Conservation*. Association of Zoos & Aquariums. https://www.aza.org/conservation.

AZA. (2023b). *Visitor demographics*. Association of Zoos & Aquariums. https://www.aza.org/partnerships-visitor-demographics.

Bartels, A., Martin, V., Bidoli, E., Steigmeier-Raith, S., Brühschwein, A., Reese, S., Köstlin, R., & Erhard, M. (2015). Brachycephalic problems of pugs relevant to animal welfare. *Animal Welfare*, *24*(3), 327–333.

Bauman, Z. (2004). *Wasted lives*. Cambridge, UK: Polity Press.

Brown, A. (2023). About half of U.S. pet owners say their pets are as much a part of their family as a human member. Pew Research Center. https://www.pewresearch.org/short-reads/2023/07/07/about-half-us-of-pet-owners-say-their-pets-are-as-much-a-part-of-their-family-as-a-human-member/.

Burns, K. (2018). *Pet ownership stable, veterinary care viable*. American Veterinary Medical Association. https://www.avma.org/javma-news/2019-01-15/pet-ownership-stable-veterinary-care-variable

Casamitjana, J. (2003). Enclosure Size in Captive Wild Mammals: A Comparison Between UK Zoological Collections and the Wild. *Captive Animal Protection Society*. https://animal-protection-consult.com/enclosures.pdf.

Chamove, A.S., Hosey, G.R., & Schaetzel, P. (1988). Visitors excite primates in zoos. *Zoo Biology*, *7*(4), 359–369.

Clubb, R., & Mason, G. (2003). Captivity effects on wide-ranging carnivores. *Nature*, *425*(6957), 473–474.

Daly, B., & Suggs, S. (2010). Teachers' experiences with humane education and animals in the elementary classroom: Implications for empathy development. *Journal of Moral Education*, *39*(1), 101–112.

de Azevedo, C.S., Cipreste, C.F., Pizzutto, C.S., & Young, R.J. (2023). Review of the effects of enclosure complexity and design on the behaviour and physiology of zoo animals. *Animals*, *13*(8), 1277.

DeMello, M. (2021). *Animals and society: An introduction to human-animal studies* (2nd ed.). Columbia University Press.

Dunayer, J. (2001). *Animal equality: Language and liberation*. Ryce Publishing.

Fernandez, E.J., Tamborski, M.A., Pickens, S.R., & Timberlake, W. (2009). Animal–visitor interactions in the modern zoo: Conflicts and interventions. *Applied Animal Behaviour Science*, *120*(1-2), 1–8.

Franklin, A., & White, R. (2001). Animals and modernity: Changing human–animal relations, 1949–98. *Journal of Sociology*, *37*(3), 219–238.

Fuentes, A., & Gamerl, S. (2005). Disproportionate participation by age/sex classes in aggressive interactions between long-tailed macaques (*Macaca fascicularis*) and human tourists at Padangtegal monkey forest, Bali, Indonesia. *American Journal of Primatology: Official Journal of the American Society of Primatologists*, *66*(2), 197–204.

Gough, A., Thomas, A., & O'Neill, D. (2018). *Breed predispositions to disease in dogs and cats*. John Wiley & Sons.

Grier, K.C.K., & Peterson, N. (2005). Indoor cats, scratching, and the debate over declawing: When normal pet behavior becomes a problem.

Hacker, C.E., & Miller, L.J. (2016). Zoo visitor perceptions, attitudes, and conservation intent after viewing African elephants at the San Diego Zoo Safari Park. *Zoo Biology*, *35*(4), 355–361.

Haid, M. (2023). *The most popular dog breeds of 2022*. American Kennel Club. https://www.akc.org/expert-advice/dog-breeds/most-popular-dog-breeds-2022/.

Hawkins, R.D., Williams, J.M., & Scottish Society for the Prevention of Cruelty to Animals (Scottish SPCA). (2017). Childhood attachment to pets: Associations between pet attachment, attitudes to animals, compassion, and humane behaviour. *International Journal of Environmental Research and Public Health*, *14*(5), 490.

HSUS. 2023a. *Pets by the numbers*. Humane Society of the United States. https://humanepro.org/page/pets-by-the-numbers.

HSUS. 2023b. *Puppy mills: Facts and figures*. Humane Society of the United States. https://www.humanesociety.org/sites/default/files/docs/HSUS_puppy-mill-facts-figures.pdf.

Kaimowitz, D., Mertens, B., Wunder, S., & Pacheco, P. (2004). *Hamburger connection fuels Amazon destruction* (pp. 1–10). Center for International Forest Research.

Marris, E. (2021). Modern zoos are not worth the moral cost. *The New York Times*. https://www.nytimes.com/2021/06/11/opinion/zoos-animal-cruelty.html.

Miller, L.J., Zeigler-Hill, V., Mellen, J., Koeppel, J., Greer, T., & Kuczaj, S. (2013). Dolphin shows and interaction programs: Benefits for conservation education?. *Zoo Biology*, *32*(1), 45–53.

Morin, K.M. (2015). Wildspace: The cage, the supermax, and the zoo. In *Critical animal geographies* (pp. 73–92). Routledge.

Newkirk, P. (2015). When the Bronx Zoo exhibited a man in an iron cage. *CNN*. https://www.cnn.com/2015/06/03/opinions/newkirk-bronx-zoo-man-cage/index.html.

Noske, B. (1997). *Beyond boundaries: Humans and animals*. Black Rose Books.

O'Neill, D.G., Sahota, J., Brodbelt, D.C., Church, D.B., Packer, R.M., & Pegram, C. (2022). Health of Pug dogs in the UK: Disorder predispositions and protections. *Canine Medicine and Genetics*, *9*(1), 4.

Patronek, G.J. (2001). Assessment of claims of short-and long-term complications associated with onychectomy in cats. *Journal of the American Veterinary Medical Association*, *219*(7), 932–937.

PDSA. (2023). *Siberian husky*. People's dispensary for sick animals. https://www.pdsa.org.uk/pet-help-and-advice/looking-after-your-pet/puppies-dogs/large-dogs/siberian-husky.

PETA. (2023). *Puppy mills*. People for the Ethical Treatment of Animals. https://www.peta.org/issues/animal-companion-issues/pet-trade/puppy-mills/.

Peterson, A.L. (2013). *Being animal: Beasts & boundaries in nature ethics*. Columbia University Press.

Puppy Protection Act of 2020. S.4757, 116th Congress. (2020). https://www.congress.gov/bill/116th-congress/senate-bill/4757.

Rohdin, C., Jäderlund, K.H., Ljungvall, I., Lindblad-Toh, K., & Häggström, J. (2018). High prevalence of gait abnormalities in pugs. *Veterinary Record*, *182*(6), 167.

Rollin, B.E. (2018). "We always hurt the things we love"—Unnoticed abuse of companion animals. *Animals*, *8*(9), 157.

Ruch-Gallie, R., Hellyer, P.W., Schoenfeld-Tacher, R., & Kogan, L.R. (2016). Survey of practices and perceptions regarding feline onychectomy among private practitioners. *Journal of the American Veterinary Medical Association*, *249*(3), 291–298.

Safina, C. (2018). Where are zoos going—Or are they gone? *Journal of Applied Animal Welfare Science*, *21*(sup1), 4–11.

Steinbeck, J. (1937). *Of mice and men*. Covici Friede.

The Animal Reader. (2023). Netherlands to ban pugs 'no animal should suffer'. *The Animal Reader: Daily Animal News*. https://www.theanimalreader.com/2023/01/20/netherlands-to-ban-pugs-no-animal-should-suffer-from-their-appearance/#:~:text=The%20Netherlands%20wants%20to%20ban,suffer%20from%20various%20health%20conditions.

The Paw Project. (2023). *Frequently asked questions about feline declawing*. The Paw Project. https://pawproject.org/about-declawing/faqs/.

Towsey, M. (2010). Something stinks: The need for environmental regulation of puppy mills. *Villanova Environmental Law Journal*, *21*(1), 159.

WAP. (2023). *Dolphin facts*. World Animal Protection. https://www.worldanimalprotection.org.au/education/animal-facts/dolphins.

Wilson, C., Bain, M., DePorter, T., Beck, A., Grassi, V., & Landsberg, G. (2016). Owner observations regarding cat scratching behavior: An internet-based survey. *Journal of Feline Medicine and Surgery*, *18*(10), 791–797.

Yeon, S.C., Flanders, J.A., Scarlett, J.M., Ayers, S., & Houpt, K.A. (2001). Attitudes of owners regarding tendonectomy and onychectomy in cats. *Journal of the American Veterinary Medical Association*, *218*(1), 43–47.

INDEX